THE MEASUREMENT OF AIR FLOW

Other Titles of Interest

ANGUS
The Control of Indoor Climate

BATURIN
Fundamentals of Industrial Ventilation

BRADSHAW
An Introduction to Turbulence and its Measurement

CLARK
Flow Measurement by Square-Edged Orifice Plates Using Corner Tappings

CROOME & ROBERTS
Airconditioning and Ventilation of Buildings

DANESHYAR
One Dimensional Compressible Flow

DORMAN
Dust Control and Air Cleaning

ECK
Fans: Design and Operation of Centrifugal Axial Flow and Cross Flow Fans

KUT
Warm Air Heating

OSBORNE
Fans

SACHS
Wind Forces in Engineering

THE MEASUREMENT OF AIR FLOW – OWER AND PANKHURST – 5th EDITION (September 1977)

CORRIGENDA

P. 47, l. 4: *for* increases *read* decreases

P. 128, l. 19: *for* are fluctuating . . . will not usually *read* are not fluctuating . . . will usually

P. 139, l. 21: *for* Fig. 3.6 *read* Fig. 3.16

P. 149, equation (2): *for* $v_2 - v_1^2$ *read* $v_2^2 - v_1^2$

P. 186, Table heading: *for* p_γ/p_v *read* p_v/p_s

P. 217, l. 1: *for* $T/\rho v^2 D^3$ *read* $T/\rho V^2 D^3$

l. 4: *for* $Q/\rho v^2 D^3$ *read* $Q/\rho V^2 D^3$

P. 224, l. 3 from foot: *for* 1.65 *read* 1.75

P. 231, l. 3 from foot: *for* $I_0^2 rKv^{0.45}$ *read* $I_0^2 + Kv^{0.45}$

P. 249, l. 16: *for* place *read* plane

P. 252, l. 11: *for* limit *read* limb

P. 292, l. 2: *for* N and H *read* M and N

P. 302, l. 3 from foot: *for* Strom *read* Stroom

P. 324, l. 8: for Chapter III read Chapter VI

P. 325, l. 11 from foot: *for* 300−13.56 *read* 300÷13.56

P. 328, l. 3: *for* $\rho\gamma p v^2 l/d$ *read* $2\gamma\rho v^2 l/d$

P. 330, equation (3): *for* I_1 *read* I

P. 336, l. 1: *for* p *read* ρ

P. 337, l. 2 from foot: *delete* for

P. 347, l.16: *for* in gm/cm^{-3} is 10^3 times *read* in gm/cm^3 is 10^{-3} times
l. 17: *for* vl/ν *read* $p/\rho v^2$
l. 20 *for* $lb/mass)/ft^3$ *read* $lb(mass)/ft^3$

see overleaf

P. 348, l. 2: *for* k.m.s. *read* kg m sec^{-2}
l. 3: *for* g.c.s. *read* g cm sec^{-2}
for lb(mass)-ft-sec^2 *read* lb(mass) ft sec^{-2}
l. 4: *for* lb(force)ft^1 sec^2 *read* lb(force) ft^{-1} sec^2

P. 351: *for* ANDERSON *read* ANDERTON
for COLLES *read* COLLIS

P. 353, REDDING, T. H.: *for* 159 *read* 195

P. 354: *for* SPARKS *read* SPARKES
for VENTURI, G. G. *read* VENTURI, G. B.

P. 356, col. 1, l. 12: *for* 242 *read* 253
l. 30: *for* 1 *read* 2

P. 357, col. 1, l. 1: *for* 103 *read* 193
l. 18: *for* 237 *read* 235
l. 27: *for* 239 *read* 229
col. 2, l. 31: *for* 134 *read* 139
l. 14 from foot: *for* 247 *read* 297

P. 358, col. 1, l. 19 from foot: *for* 385 *read* 335
col. 2, l. 20 from foot: *for* 164 *read* 169

P. 359, col. 1, l. 4: *for* 84 *read* 89
col. 2, last line: *for* 134 *read* 139

P. 361, col. 1, l. 6: *for* 85 *read* 83
col. 2, l. 15: *insert* 162

P. 363, l. 5: *for* 119 *read* 199
l. 9: *for* 119 *read* 199

THE MEASUREMENT OF AIR FLOW

5th Edition (in SI Units)

BY

E. OWER, B.SC., A.C.G.I., C. ENG.

AND

R. C. PANKHURST,
PH.D., A.R.C.S., C. ENG., F.R.A.E.S.

PERGAMON PRESS

OXFORD · NEW YORK · TORONTO
SYDNEY · PARIS · FRANKFURT

U. K.	Pergamon Press Ltd., Headington Hill Hall, Oxford OX3 0BW, England
U. S. A.	Pergamon Press Inc., Maxwell House, Fairview Park, Elmsford, New York 10523, U.S.A.
CANADA	Pergamon of Canada Ltd., 75 The East Mall, Toronto, Ontario, Canada
AUSTRALIA	Pergamon Press (Aust.) Pty. Ltd., 19a Boundary Street, Rushcutters Bay, N.S.W. 2011, Australia
FRANCE	Pergamon Press SARL, 24 rue des Ecoles, 75240 Paris, Cedex 05, France
WEST GERMANY	Pergamon Press GmbH, 6242 Kronberg-Taunus, Pferdstrasse 1, West Germany

---

First published 1927 by Chapman & Hall
Second edition 1933 by Chapman & Hall
Third edition 1949 by Chapman & Hall
Fourth edition 1966 revised
Reprinted with corrections and amendments 1969
Fifth edition completely revised 1977

Library of Congress Cataloging in Publication Data

Ower, Ernest, 1894–
The measurement of air flow.

Includes bibliographies and indexes.
The measurement of air flow.
1. Air flow—Measurement. I. Pankhurst, R. C., joint author. II. Title.
TJ1025.08 1977 620.1'074 76–27372
ISBN 0–08–021282–4 (Hardcover)
ISBN 0–08–021281–6 (Flexicover)

Typeset by Cotswold Typesetting Ltd, and printed in Great Britain by Page Bros (Norwich) Ltd.

CONTENTS

PREFACE

As STATED in the prefaces to earlier editions, this book is intended to serve as a textbook both for students and for engineers and other practitioners concerned with the measurement of the speed of air in motion relative to solid boundaries or surfaces, and of the associated pressures.

For the first time, the entire text now appears in metric units. We have adopted throughout the units of the Système International (S.I.) and have used its abbreviations except that we have preferred to retain "sec" as the abbreviation for "second" instead of "s". Brief notes on the S.I. are provided in Appendix 1. Recognizing that instruments reading in British units are likely to remain in use for some time to come, we have included in this Appendix the numerical values of the conversion factors for the units that occur most commonly in the measurement of air flow.

The change to metric units necessitated the re-working of much of Chapter XIII (Examples from Practice), but the principal result of using the S.I. is general simplification. In particular, the S.I. removes the risk of confusion between units of mass and force, provided that no attempt is made to use the kilopond (kilogram-force) in place of the proper S.I. unit (the newton): this point is emphasized in Appendix 2.

Appendix 2 also discusses the numerical evaluation of non-dimensional quantities: the task of metrication served to drive home once again the great advantages of expressing results in this form.

We have taken the opportunity to replace several physical terms by their equivalents in modern terminology, notably "specific gravity" (nowadays called "relative density") and "specific heat" (now "specific heat capacity"); and, in view of the increasing use of pressure transducers in place of liquid-column manometers, we have tried to avoid referring to pressures as "heads" except where this term is specially appropriate or convenient.

We have also re-written much of Chapter VI (which deals with flow measurements by pitot-traverse methods) so as to conform with the complete overhaul of the relevant part of British Standard 1042 (Part 2A) issued in 1973; and reference is made throughout the book to a number of significant developments that have taken place in air-flow measurement techniques since the fourth edition was reprinted in 1969.

Extract from Preface to Fourth Edition

The title under which the book has established itself since the first edition appeared in 1927 has been retained; but we recognize that the text does not cover all matters that a modern interpretation of the title might be expected to include. As in past editions, our aim has been to concentrate on aspects, both practical and theoretical, primarily of interest to engineers. We have, therefore, not dealt with problems of measuring the flow of air at hypersonic speeds, when such high temperatures can be attained that dissociation and ionization effects change the physical character of the air; or at very low pressures, when the mean free path of the molecules becomes comparable with the dimensions of the measuring instruments. Both these states concern physicists or space scientists rather than engineers dealing with air-flow measurements in typical industrial situations.

E.O.
R.C.P.

January, 1977

ACKNOWLEDGEMENTS

TAKEN together, many of the measurement methods described in this book constitute an important part of the experimental resources of wind-tunnel technique, to which one of us (R.C.P.) devoted a Sabbatical term at Cambridge in 1970. He is glad of this opportunity to acknowledge his indebtedness to the Master and Fellows of Emmanuel College for the award of the Fellow-Commonership which enabled him to do so.

We are both deeply indebted to Dr. E. A. Spencer and Dr. F. C. Kinghorn of the National Engineering Laboratory and to Mr. R. W. F. Gould of the National Physical Laboratory for their valuable co-operation in answering our numerous queries, mainly concerning measurements with pitot–static tubes and orifices. We wish also to record our special appreciation to Mrs. Sheila Bradshaw of the Aeronautical Research Council Library for her help in supplying us with books, papers, and reports we wished to consult.

Our thanks are also due to the following for providing material from which a number of illustrations have been made:

Air Flow Developments Ltd., Fig. 10.2; Mr. A. M. Binnie, Fig. 5.16; C. F. Casella & Co. Ltd., Fig. 10.8; G.E.C. Elliott Process Instruments Ltd., Figs. 11.5 and 11.6; Willh. Lambrecht K.G., Göttingen, Fig. 11.10; Lowne Instruments Ltd., Fig. 8.1; Salford Electrical Instruments Ltd., Fig. 11.7; H. Tinsley & Co. Ltd., Fig. 9.4.

Figures 3.14, 4.3,† 4.4,† and 10.10 have been reproduced with permission from publications of H.M. Stationery Office, and Crown Copyright is reserved. Figs. 3.15, 5.10, and 5.12 have been reproduced by permission of the Royal Aeronautical Society; and Fig. 3.20 by permission of *Aircraft Engineering*.

† With slight modifications due to the change from British to S.I. units.

ABBREVIATIONS USED IN LISTS OF REFERENCES

Aero.	Aeronautical
AGARD	Advisory Group for Aerospace Research and Development
A.I.A.A.	Aircraft Industries Association of America
Aircr.	Aircraft
App.	Applied
A.S.M.E.	American Society of Mechanical Engineers
B.S.I.	British Standards Institution
Chem.	Chemical
Engr(s)	Engineer(s)
Engng	Engineering
f.	für
Forsch. IngWes	Forschung auf dem Gebiete des Ingenieurwesens
I.C.A.O.	International Civil Aviation Organization
Inst.	Institute or Institut
Instn.	Institution
Instrum.	Instruments
I.S.A.	Instrument Society of America
I.S.O.	International Organization for Standardization
J.	Journal
Jb.	Jahrbuch
Mag.	Magazine
Mech.	Mechanical
Mechs	Mechanics
Mitt.	Mitteilungen
N.A.C.A.	(American) National Advisory Committee for Aeronautics
NASA	National Aeronautics and Space Administration
NAVORD	Naval Ordnance Laboratory (U.S.A.)
N.E.L.	National Engineering Laboratory
N.P.L.	National Physical Laboratory
Phil.	Philosophical
Phys.	Physics
Proc.	Proceedings
R.A.E.	Royal Aircraft Establishment
R. & M.	Reports and Memoranda of the Advisory Committee for Aeronautics (1909–20), or the Aeronautical Research Committee (1920–44), or the Aeronautical Research Council (1945 onwards)
Rep.	Report
Rev.	Review
Roy.	Royal
Scient.	Scientific
Soc.	Society
Tech.	Technical
T.M.T.	Technische Mechanik und Thermodynamik
Trans.	Transactions
u.	und
V.D.I.	Verein(es) deutscher Ingenieure
Z.	Zeitschrift

CHAPTER I

INTRODUCTION

ALTHOUGH this book deals primarily with the measurement of air flow, the methods to be described will in general apply also to the flow of other gases with little, if any, modification, except as regards the numerical values of the various physical properties occurring in the equations. Further, much of the theory will apply to the flow of liquids as well as gases, although, for practical reasons, the methods of measurement may not always be applicable. Two-phase flows are excluded.

When physical measurements have to be made, a choice of methods is often available; and the experimenter must then decide which of these is best suited to his particular purpose. His choice will be guided by considerations of simplicity, directness, and the degree of accuracy he requires; he should always avoid a complicated method when a simpler one will equally serve his ends. Let us therefore consider what means are available for the problem with which we are here concerned, namely the measurement of the speeds of streams of gases, with particular reference to the motion of air along pipes or ducts. A search for a simple, direct method yields disappointing results. Perhaps the most accurate method of measuring the mean rate of water flowing along a pipe is to weigh the quantity passing in a given time or to measure its volume. Weighing can obviously not be used for air; volume measurements, on the other hand, can, and they form the basis of the common gas meters described in Chapter XI. But unfortunately the use of gas meters is severely limited because both the two existing standard types can measure only rates of flow much smaller than those that usually concern engineers. Moreover, one of these types is not designed for high accuracy.

The only other direct method that suggests itself is to introduce some indicator, such as a small, light body or a puff of smoke, which will be carried along by the stream and can be timed over a measured distance. The *prima facie* attraction of this method – its simplicity – disappears on examination. In the first place, the duct or pipe must be transparent if the indicator is to be visible; in the second, the rate of flow varies from point to point across a pipe section, and this will make the determination of the mean rate of flow difficult even without the additional complication due to the diffusion across the stream of an indicator such as smoke. The first difficulty, invisibility, but not the second, can be overcome by the use of a radioactive tracer or a hot spot in the gas produced by an electric spark and subsequently made visible by an optical method such as Schlieren. These and other techniques (see, for example, refs. 1–3) have been tried, but so far have not been developed into practical engineering methods. Broadly, we can say that indicator methods are suitable only in particular cases, such as ventilation surveys in which the flow can be observed visually; and that most of them would usually be difficult to apply to the measurement of flowrates in pipes.

Thus, for most air-flow measurements they have to make, engineers cannot use direct methods, but have to resort to the measurement of some physical effect arising from the motion. Three such effects have been found by experience to be suitable: pressure changes associated with the motion; consequent mechanical effects, such as the rate of rotation induced in a rotor made up of light vanes mounted in the stream; and the rate of cooling of a hot body, such as an electrically heated wire, introduced into the air current. Of these, the first is of the greatest importance, since, as we shall see later, a properly designed instrument, suitably inserted in the stream, records a pressure difference which is entirely characteristic of the motion, and can be measured with a pressure gauge. If such an instrument is constructed in accordance with certain well-established principles, which are explained in Chapters III and IV, it may be used without calibration as a standard for the measurement of air speed. This is not true of anemometers† that depend for their action either on mechanical or on electrical effects; instruments of both these types are subject to individual variations difficult to control, and usually require calibration against a standard instrument of the pressure-measuring type.

Instruments that depend on the measurement of pressure can be subdivided into two distinct groups according to whether air does or does not flow through them. If there is no flow, we have what we shall term a pressure-tube anemometer, the characteristic feature of which is two independent and differently shaped tubes, each containing, in the end exposed to the air current, an orifice or a group of orifices at which a pressure is established by the motion of the stream. These tubes are connected at their other ends to opposite sides of a differential pressure gauge, which prevents all flow through the anemometer and measures the difference between the pressures at the two groups of orifices. Differential pressure is also measured in pressure anemometers through which there is flow. In both types the measured pressure difference depends on the speed and density of the air and also on the geometry of the instrument itself. It is obviously desirable that the geometry shall affect the readings as little as possible or, at any rate, with as little variation as possible over a wide speed range; and one variety of pressure-tube anemometer – the pitot–static tube

† The term anemometer, as its derivation implies, includes all types of instruments used for the measurement of air speed.

discussed in Chapters III and IV – has been devised which possesses this property more than any other type so far developed. The remarks made in the previous paragraph about the possibility of constructing anemometers that can be used to measure air speed without calibration apply to this type of instrument above all others; and it is by virtue of this property that the pitot–static tube has been adopted universally as the standard against which practically all other types of anemometer are ultimately checked.

Pressure anemometers through which there is a flow are more sensitive to small differences in shape than is a good pressure-tube anemometer. Nevertheless, within their limitations, such instruments have many useful applications. Commonly used examples are the plate orifice, the nozzle, and, to a smaller extent, the venturi tube. For accurate work they must be used under strictly controlled conditions.

Mechanical anemometers are of two main types. In the first, the working element is a rotor maintained in continuous rotation by the air at a rate depending on the air speed; in the second, the air deflects a plate or a vane controlled by gravity or a spring. The rotary type can be subdivided into the vane anemometer and the cup anemometer. The first of these is a most useful instrument which is widely used by engineers; it is discussed in some detail in Chapter VIII. The cup anemometer is used mainly by meteorologists and little by engineers, and, beyond the following brief description, will not be considered in this book.

In its original form, the cup anemometer consists of four hemispherical cups carried, with their bases vertical, at the outer ends of four light arms, which are symmetrically disposed in a horizontal plane and are attached to a central sleeve free to rotate about a vertical axis. The cups are arranged in pairs so that the concave side of one member of a pair is presented to the air current at the same time as the convex side of the diametrically opposite cup. Thus at any instant the air on one side of the median plane through the vertical axis of the instrument will be blowing into the interior of one or two cups and, on the other side of this plane, on the exterior of the opposite cup or cups. Hence, since the aerodynamic force on a cup with its concave side presented to the wind is greater than when the wind is blowing on its convex face, rotation ensues at a rate depending on the air speed. Instruments of this type are bulky and not easily portable.

In 1926 Patterson[(4)] claimed certain advantages for a three-cup type which was adopted to some extent in Canada. Later Sheppard[(5)] developed a much improved three-cup instrument with conical instead of hemispherical cups. Extreme lightness of construction and the use of an elegant indicating mechanism with little friction have resulted in a very sensitive instrument – it responds to an air speed as low as 0·2 m/sec – and one that behaves satisfactorily in a fluctuating wind,[(6)] in which the older types tend to overestimate the mean speed considerably. Sheppard's instrument has been used in investigations of atmospheric turbulence, and the Meteorological Office have adopted the three-cup type with conical cups as standard for their cup anemometers. These are, however, larger and heavier than the Sheppard instrument, as they are intended for use in stronger winds.

An early type of deflexion anemometer is the swinging-plate instrument, now rarely used. It consists of a vertical plate, suspended on knife edges, which deflects under the action of an air current to such an angle that the restoring torque due to the weight of the plate is equal to the air torque. The angle of inclination may therefore be taken as a measure of the air speed. In modern types such as the velometer and similar German instruments (see Chapter XI) the plate is replaced by a small vane under spring instead of gravity control, which carries a pointer moving over a scale of air speed. Yet again, the plate may be mounted on a torsion wire passing through an asymmetric vertical or horizontal axis, and there will be a relation between the air speed and the amount of twist of the wire necessary to hold the plate in normal presentation to the current. An anemometer of this kind was designed for low speed by J. P. Rees.[(7)] It indicated air speeds ranging from 0·05 to 1 m/sec (see also p. 199).

An example of another type of anemometer in which the airstream displaces an indicator giving an instantaneous reading of velocity or flowrate is the rotameter (see Chapter XI). The air flows upwards through a tapered vertical tube in which a float rises until its weight is balanced by the difference of pressure above and below it caused by the motion by the air through the restricted passage around the float.

Electrical or hot-wire anemometers are not extensively used in industrial situations, although certain types have been developed for this purpose. The chief limitation on their wider use is probably the more elaborate apparatus and manipulation that they require. As laboratory instruments, they may be made to give excellent results; and they are particularly well suited to the measurement of low air speeds,† a purpose for which they may be designed to have a considerably more open scale than is possible with pressure-tube anemometers. The hot-wire anemometer is essential for turbulence measurements, but this field is not discussed in this book.

A different type of electrical anemometer is worthy of mention: the ionization anemometer of Lovelock and Wasilewska.[(8)] It consisted essentially of a spherical openwork cage with a small spherical electrode at the centre coated with polonium which emitted α-particles and so ionized the air in its vicinity. The central sphere was connected to a 120-V electrical source, and the cage was earthed. Thus the ions were caused to flow between sphere and cage and an electric current was established, which could be measured. When the instrument was exposed to an airstream, the ion flow was disturbed and current

† Except when there is an appreciable vertical component of velocity, in which case the natural convection currents from the wire itself have a disturbing effect on the measurements.

changes occurred in terms of which the anemometer was calibrated. Its calibration was the same whatever the wind direction, and it was said to be able to measure velocities of less than 0·05 m/sec.

Another physical principle that has been used to measure fluid velocity is the effect of the motion on the velocity of acoustic-wave propagation; and flowrate in a pipe or duct has also been deduced from the displacement of an acoustic beam transverse to the flow. Although not used extensively, these methods are discussed briefly in Chapter XI. An account is also given in that Chapter of the measurement of flow velocity by observation of the Doppler frequency-shift of light reflected from minute solid particles (e.g. dust) moving with the fluid. This method has been made possible by the high-intensity narrow-beam illumination of lasers; it has received a great deal of attention during the past decade, and is still being developed.

References

1. A. B. Bauer, Direct measurement of velocity by hot-wire anemometry, *A.I.A.A. Jl* **3** (1965) 1189.
2. P. B. Earnshaw, The electric-spark technique applied to the measurement of velocity in a leading-edge vortex, R.A.E. Tech. Note Aero 2947 (1964).
3. C. Lahaye, E. G. Leger, and A. Lemay, Water velocity measurements using a sequence of sparks, *A.I.A.A. Jl* **5** (1967) 2274.
4. J. Patterson, The cup anemometer, *Trans. Roy. Soc. Canada* **20** (1926) 1.
5. P. A. Sheppard, An improved design of cup anemometer, *J. Scient. Instrum.* **17** (1940) 218. See also J. I. P. Jones, A portable sensitive anemometer with proportional d.c. output and a matching velocity-component resolver, *J. Scient. Instrum.* **42** (1965) 414.
6. F. J. Scrase and P. A. Sheppard, The errors of cup anemometers in fluctuating winds, *J. Scient. Instrum.* **21** (1944) 160.
7. J. P. Rees, A torsion anemometer, *J. Scient. Instrum.* **4** (1927) 311.
8. J. E. Lovelock and E. M. Wasilewska, An ionization anemometer, *J. Scient. Instrum.* **26** (1949) 367.

CHAPTER II

GENERAL PRINCIPLES OF THE PRESSURE-TUBE ANEMOMETER

As stated in Chapter I, a pressure-tube anemometer measures a pressure difference from which the speed of flow can be deduced. To understand the action of such an instrument, therefore, it is helpful to have some knowledge of the pressures that occur in a moving fluid. In the development of the argument we shall have to refer to a number of pressures arising from different causes. Although some of them will be defined again when they are first mentioned, it will be convenient to precede the discussion with the following summary of definitions of types of flow and pressures:

Incompressible flow – Flow in which changes in the fluid density due to the motion can be neglected.†

Compressible flow – Flow in which the motion produces appreciable changes in the fluid density.

Transonic flow – Compressible flow in which there are regions of both subsonic and supersonic speeds.

Static pressure – The pressure acting equally in all directions at a point in a stationary fluid, or normal to an element of surface parallel to the flow in the case of a fluid in motion.

Total pressure – The pressure that would arise if a moving fluid was brought to rest without change of total heat or entropy.

Velocity pressure – Half the product of the fluid density and the square of the speed: usually termed kinetic pressure in modern aerodynamics. In incompressible flow it is equal to the difference between the total pressure and the static pressure.

Dynamic pressure – The difference between the total pressure and the static pressure. Equal to the velocity pressure in incompressible flow.

Pitot pressure – The pressure measured by a correctly aligned facing (pitot) tube. When the local speed is subsonic, the pitot pressure is equal to the total pressure.

† Although air is a compressible fluid, many cases of air flow can be treated as incompressible flow (see p. 15).

Static and Velocity Pressures

The only pressure acting at any point in a stationary fluid is the hydrostatic pressure which acts equally in all directions. If now the fluid is set in uniform, unaccelerated motion, the hydrostatic pressure will still persist, although its magnitude may be changed; and if we imagine a very small, thin, flat disk placed at any point in the fluid, and moving with it, this static pressure, as it is called, will act on the two faces of the disk. Let us assume that there is a continuous stream of the fluid in movement, and that the motion is uniform and does not vary with time, so that across any section perpendicular to the stream the conditions are everywhere the same. If, instead of moving with the fluid, the small disk remains stationary at a point in the section under consideration, and if, moreover, the plane of the disk is parallel to the direction of motion, the pressure acting on the two faces will still be the static pressure existing at that point in the fluid provided that the presence of the disk does not disturb the flow.

Now consider what occurs if the disk is turned so that the stream impinges normally on one face: when the disk moves with the stream, it will again be acted upon by the static pressure only; but if it remains stationary, its front face will experience the static pressure and an additional dynamic pressure arising from the impact of the moving stream. It is important to remember that this pressure is entirely characteristic of the motion: it depends on the speed and density of the fluid only, and can therefore be used as a measure of the speed.

We see, therefore, that if the disk is stationary and normal to the stream, so that the forward motion of the fluid impinging on it is arrested, the pressure exerted will be the sum of the static and dynamic pressures at the point in the fluid at which the disk is situated; whereas if the disk is placed so that the fluid streams past it without disturbance the static pressure only is experienced. Suppose now that the imaginary disk is replaced by an open-ended tube of small bore bent to face the stream, and that the other end of the tube is connected to one limb of a simple U-tube pressure gauge (Fig. 2.1), the other limb of which is open to the atmosphere. The forward motion of the fluid impinging on the open end of the tube will again be arrested, and the pressure gauge will record the sum of the dynamic and static pressures,† or the total pressure as it is commonly called. It is found by experiment that the dynamic-pressure component of the total pressure measured by a facing tube is, within wide limits, unaffected by the shape and size of the tube, and, in incompressible flow, is in fact equal to the pressure that engineers commonly term velocity pressure (or velocity head, see p. 9). The relation that exists between the dynamic pressure and the density

† When pressures are measured by liquid-column manometers, it is often convenient to use as datum the prevailing atmospheric pressure, as in Fig. 2.1 (in which air is being moved along a pipe by the action of a fan). See page 9, however, for remarks about the use of pressure transducers and measurements in compressible flow.

and velocity of the fluid is derived later in this chapter; at this stage it is sufficient to remember that it is a function of the motion only.

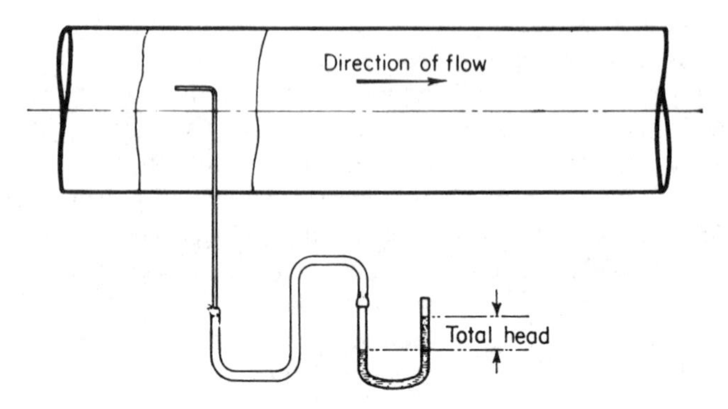

FIG. 2.1. Measurement of total head.

Thus, whereas the dynamic pressure is uniquely determined by the speed and density of the fluid, the static pressure is not. It follows that in two different sets of conditions the total pressures may be different although the dynamic pressures are the same. Therefore in order to determine the dynamic pressure, and hence the speed, it is necessary to measure both the total and static pressure since the dynamic pressure cannot be observed separately.

As an example of two instances in which a fluid moves with the same dynamic pressure but has different values of the total pressure, consider a stream of air that is being moved along a pipe by means of a fan. It is clear that the air may be made to travel with the same dynamic pressure whether the fan is blowing or sucking it down the pipe. On the other hand, the static pressure in the pipe near the fan will be higher than atmospheric when the fan is blowing and lower when it is sucking, so that the total pressure in the former case will be higher than in the latter.

Reverting now to the open-ended tube facing an airstream, we see that the pressure gauge to which it is connected will indicate the sum of the dynamic and static pressures of the air at the mouth of the tube. If, therefore, the other side of the gauge is connected to a tube or other device which measures the static pressure, the reading of the gauge will be equal to the difference between the pressures acting on the two sides, i.e. to the dynamic pressure, and will thus serve directly to determine the air speed.

Pressure-tube Anemometers

It is upon this principle that all pressure-tube anemometers depend. For the measurement of the total pressure, any open-ended tube facing upstream is reliable,† but some care is necessary in the design of the device for measuring the static pressure. The important condition to be observed is that, in the region where the static pressure is measured, the air should not acquire a velocity component perpendicular to its original direction of motion before the tube was inserted; in other words, any device for measuring static pressure will give erroneous results if it deflects the air appreciably. The design of static-pressure tubes is discussed in Chapters III and IV; at this stage it is sufficient to note that it is, in fact, possible to make a static tube that measures the static pressure with an accuracy well within practical requirements. Such a tube, suitably used in conjunction with a pitot tube, constitutes a complete pressure-tube anemometer which does not require calibration, save in exceptional cases, and therefore serves normally as a fundamental standard instrument for the measurement of air speed.

Anemometers of this nature are extensively used in practice. The only disadvantage inherent in this method of measurement is that the velocity pressures corresponding to low air speeds are small: a speed of 6 m/sec, for example, produces a pressure equivalent to only about 2 mm head of water. It follows that sensitive pressure gauges, or manometers, are necessary for the measurement of low speeds. In fact for air speeds below about 5 m/sec it is usually advisable to use some other measuring device, such as the vane anemometer, the orifice plate, or the venturi tube, or, for very low speeds, the hot-wire anemometer.

Before proceeding further, we shall find it necessary to understand the conventions used in the measurement of pressures. For convenience the pressure of the undisturbed atmosphere – the barometric pressure prevailing at the time and place of measurement – is often taken as the datum static pressure; all static pressures are then expressed in terms of the number of units of pressure by which they exceed or fall short of this datum. It follows that static pressures less than atmospheric will be recorded as negative pressures.

In practice, the gauges used to measure pressures in air-flow work are often based on the simple U-tube manometer (see Fig. 2.1) in which the pressure is balanced against the weight of a column of liquid, usually water or mercury. Hence it became customary to express a measured pressure in terms of the height or "head" of such a liquid column, i.e. as so many millimetres or centimetres rather than in terms of its proper unit of force per unit of area. It also became usual to use the terms total head and velocity head in preference to total pressure and velocity pressure. Now that various forms of pressure transducer are increasingly replacing liquid-column manometers, however, the practice of referring to pressures as heads, although often convenient, appears somewhat inappropriate. The use of atmospheric pressure as datum is likewise

† A facing tube of this kind is generally known as a pitot tube, after the French engineer H. Pitot who, in 1732, first described its use for the measurement of the speed of water flow and of ships.

often convenient; but in compressible flow pressure *ratios* are the important quantities (see Chapter IV), and, it is then essential to use "absolute" values (i.e. referred to vacuum as datum).

Bernoulli's Theorem and its Application to the Measurement of Fluid Velocity

Attempts to visualize conditions of flow are aided by the use of the conception of "streamlines". In any mass of moving fluid we assume the field to be mapped out by a system of lines such that the tangent to any one of these lines at any point coincides with the direction of motion of the particle of fluid situated at that point at the instant under consideration. The streamlines will thus give an instantaneous picture of the flow at any moment. If the flow is unsteady so that it varies with time, this picture will change from instant to instant; in periodic motion it will undergo a series of changes that recur at regular intervals. For steady flow, on the other hand, the streamlines will preserve their configuration unchanged at all instants, and will show the paths of individual particles of the fluid. This type of motion is known as streamline or laminar flow. It may occur in small pipes or at low rates of flow (see p. 77); but by far the more common is the type of flow known as turbulent, in which smaller random velocities, not confined to the general direction of the stream, are superposed on the main flow.

In general, the rigorous theoretical treatment of problems of fluid motion is difficult and often intractable, demanding mathematical analysis of a character beyond the scope of this book. By the introduction of simplifying assumptions, however, it is sometimes possible to obtain solutions of certain types of problems which serve a useful purpose by indicating how the various properties of the fluid may be expected to govern its motion. The value of such solutions is enhanced when, as not infrequently happens, experimental results agree more or less closely with the theoretical predictions. We must, however, avoid the danger of attaching to these solutions a significance that their nature does not warrant; it is essential to remember that they are only approximate and mainly of use as guides to thought.

The viscosity of a fluid (see p. 16) is the property that is largely responsible for the difficulty mathematicians find in solving problems of fluid motion; and most of the results of practical significance obtained from theory have been derived by neglecting viscosity altogether or by regarding its effects as localized within a narrow layer across which the flow velocity changes rapidly. Such a layer (Prandtl's "boundary layer") occurs adjacent to a solid boundary. Comparison with observation has shown that a certain theorem, which is of fundamental importance in the measurement of flow by pressure-tube anemometers, may be simply deduced on the former hypothesis, i.e. by assuming that no frictional forces occur between adjacent particles of the fluid, or between the fluid and any solid boundaries with which it is in contact. A fluid of this type is called an inviscid or ideal fluid, and the theorem is named after D. Bernoulli, by whom it was first stated in 1738. It may be derived in the following manner.

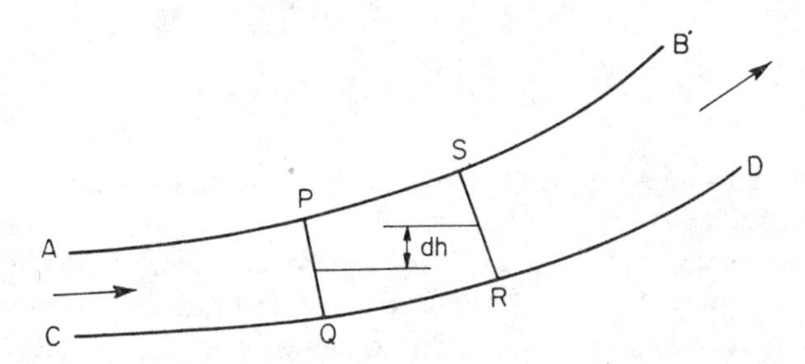

FIG. 2.2. Flow through a stream-tube.

Let *ABCD* (Fig. 2.2) represent a longitudinal median section through a tube of fluid bounded by streamlines. Such a tube is termed a stream-tube, and *AB* and *CD* are two of the bounding streamlines which, in steady motion, will preserve a constant shape.

Consider a small element *PQRS* of the tube of length $\mathrm{d}s$, which is so small that the elements *PS* and *QR* of the streamlines may be considered as straight lines. If v and p are respectively the speed and pressure across the section *PQ*, we may, since $\mathrm{d}s$ is small, write $v + \mathrm{d}v$ and $p + \mathrm{d}p$ for the corresponding quantities across section *RS*, where $\mathrm{d}v$ and $\mathrm{d}p$ are small. Let $a =$ area of section *PQ*; $\varrho =$ mass of unit volume of liquid; $g =$ acceleration due to gravity; and $\mathrm{d}h =$ difference of level between the centres of area of *PQ* and *RS*.

The relation between velocity and pressure is obtained by equating the total force acting along the stream-tube to the rate of change of momentum. This total force is equal to the sum of the forces due to the pressure difference at the two ends of the element and the component of the weight of the fluid in the element acting in the same direction. The force due to the pressure difference at the two ends is equal to $a\,\mathrm{d}p$, if second-order quantities are neglected; and the weight component is equal to $g\varrho a\,\mathrm{d}h$. Thus the total force acting on the element *PQRS* in the direction of motion is

$$-a\,\mathrm{d}p - g\varrho a\,\mathrm{d}h.$$

The mass of fluid entering per second across *PQ* is given by ϱav, which, since *PS* and *QR* are streamlines, is equal to the mass leaving at *RS*, so that the change of momentum per second is $a\varrho v\,\mathrm{d}v$ in the direction of motion.

We therefore have the equation

$$\varrho a v \, \mathrm{d}v = -a \, \mathrm{d}p - g \varrho a \, \mathrm{d}h,$$

or

$$v \, \mathrm{d}v + \frac{\mathrm{d}p}{\varrho} + g \, \mathrm{d}h = 0. \tag{1}$$

On integration along the stream-tube, this equation becomes

$$\frac{v^2}{2} + gh + \int \frac{\mathrm{d}p}{\varrho} = \text{constant} = C, \tag{2}$$

which applies to all sections of a given stream-tube. The quantity h may be taken as the height of the section measured above any arbitrary datum. An alteration in the datum level will merely result in an alteration in the value of C for the stream-tube. When the stream-tube becomes infinitely narrow it may be considered as forming a streamline, and we may therefore say that (2) gives the relation between pressure and velocity along a streamline in inviscid flow.

If the flow is horizontal, the datum level may be taken as being that of the streamline itself, and h becomes zero. In practice, when dealing with the flow of air we almost invariably find that the quantity gh is negligibly small in comparison with the sum of the other two terms of (2); so that the simplified form of this equation, namely

$$\frac{v^2}{2} + \int \frac{\mathrm{d}p}{\varrho} = C, \tag{3}$$

which is strictly valid only for horizontal flow, may generally be used without significant error.† In the following treatment we shall therefore not consider differences of level.

A further simplification can often be made, namely to treat the flow as incompressible, so that ϱ does not vary with pressure changes due to the motion and (3) becomes

$$\tfrac{1}{2}\varrho v^2 + p = C, \tag{4}$$

which is Bernoulli's equation in the form most commonly used in practical work. For air, the effect of compressibility can be neglected up to speeds of about 60 m/sec (see p. 15).

It will be seen that both terms occurring on the left-hand side of (4) have the dimensions of force per unit area, i.e. of pressure; the first term is the quantity we have called the velocity pressure, the second is the static pressure, and the sum of the two is the total pressure. Bernoulli's equation is therefore equivalent

† Occasionally, when two points of pressure measurement are at appreciably different levels, it may be necessary to allow for the difference in level. The difference at which this allowance becomes significant depends on the velocity head being measured. Approximately, the atmospheric pressure (i.e. the static-pressure datum) decreases by 9 mm of mercury (about 12 cm of water) per 100 m of height at low altitudes.

to the statement that the total pressure along a streamline is constant. This equation has been derived on the assumption that the flow is frictionless, but experience has shown that it can often be applied with good accuracy to the flow of real fluids.

The case of special interest in relation to the pressure measured by a pitot tube is that in which a streamline is brought to rest, resulting in what is called a "stagnation point" at which $v = 0$. If p_0 is the pressure at such a point, (4) becomes

$$p + \tfrac{1}{2}\varrho v^2 = p_0; \tag{5}$$

i.e. if the stagnation point is formed at the mouth of a pitot tube facing the stream, the pressure there, p_0, will be equal to the total pressure before the tube was inserted into the stream.†

Strictly, this will be true only if:

(a) the tube is so narrow that one streamline only is brought to rest;
(b) flow through the tube is prevented, as it is if the other end is connected to a manometer;
(c) the effects of viscosity can be neglected.

Of these conditions, only (b) can be fulfilled; and the fact that a tube of finite size facing the air current does indeed measure the total pressure rests not upon a rational theoretical basis but upon a large amount of reliable experimental evidence. This shows beyond doubt that the total pressure is accurately measured by a correctly aligned pitot tube, and further that variations in the size and shape of the tube within wide limits do not have any appreciable effect. As regards the effect of viscosity, experiments[1, 2] have shown that this can be neglected except for very small tubes working at very low speeds, outside the range ordinarily encountered in practice.‡ Thus the pitot tube gives an accurate reading of p_0 under practical conditions, although theoretically this result could only be expected for the streamline flow of an ideal inviscid fluid with an infinitely small tube.

Now we see from (5) that, in incompressible flow, p_0 is equal to the sum of the velocity pressure $\frac{1}{2}\varrho v^2$ and the static pressure p in the undisturbed flow. If, therefore, p is measured by a properly designed static-pressure tube connected to the other side of the same manometer to which the pitot tube is connected, the resultant pressure P indicated by the manometer is given by

$$P = p_0 - p = \tfrac{1}{2}\varrho v^2, \tag{6}$$

† Except in supersonic flow when the tube forms a shock wave (see p. 57).

‡ For air at ordinary temperatures and pressures, viscosity effects on the pitot tube are less than 1 per cent of $\frac{1}{2}\rho v^2$, provided that the product vd exceeds 1·5, where v is the speed in metres per second and d is the internal diameter of the tube in millimetres. In normal small tubing the internal diameter is about 0·6 times the external, so that, for such tubing, the criterion, based on external diameter, is about 2·5. Additional data are quoted on p. 44.

from which v can be calculated. Details of the method of calculation are given in Chapter VI. Values of ϱ for air are given in Table 5.1, p. 77.

Bernoulli's Equation in Compressible Flow

Equation (5) was derived on the assumption that the density ϱ of the fluid does not change however much the pressure changes. In a compressible fluid such as air, this assumption is obviously not strictly valid, but it is near enough to the truth in a large number of the problems of air-flow measurement that occur in practice. In many others, however, pressure and temperature changes produced by the flow are so great that the density of the air is no longer effectively constant throughout the flow field; and the pressure p_0 at a stagnation point has to be determined by integration of the general Bernoulli equation allowing for variation of density. Except in boundary layers, wakes, and jets, and when shock waves are present (see p. 57), the flow can usually be regarded as the isentropic flow of a perfect gas, for which the following well-known thermodynamic relations hold between pressure p, density ϱ, and absolute temperature T:

$$p = A\varrho^{\gamma}, \quad p = BT^{\gamma/(\gamma-1)}, \quad \varrho = KT^{1/(\gamma-1)}, \tag{7}$$

where A, B, and K are constants, γ is the ratio of the specific heat capacity of the gas at constant pressure to that at constant volume, and p is now *absolute* pressure (not relative to some arbitrary datum).

To determine the pressure at a stagnation point in isentropic compressible flow, we substitute $(p/A)^{1/\gamma}$ from (7) in the general Bernoulli equation (3) and obtain

$$\frac{v^2}{2} + A^{1/\gamma}\int p^{-1/\gamma}\,dp = C. \tag{8}$$

After integration and substitution of p/ϱ^{γ} for A, as given by (7), eqn. (8) becomes

$$\frac{v^2}{2} + \frac{\gamma}{\gamma-1}\frac{p}{\varrho} = C, \tag{9}$$

which is Bernoulli's equation for compressible flow.† It applies not only to subsonic flow but also to supersonic flow provided that there are no shock waves. If, as before, we denote conditions at the stagnation point by the suffix 0, we have

$$\frac{v^2}{2} + \frac{\gamma}{\gamma-1}\frac{p}{\varrho} = \frac{\gamma}{\gamma-1}\frac{p_0}{\varrho_0}, \tag{10}$$

† As in (5), gravitational terms have been neglected.

and the stagnation pressure p_0, i.e. the total pressure, is given by

$$p_0 = p\frac{\varrho_0}{\varrho}\left(1 + \frac{1}{2}\frac{\gamma - 1}{\gamma}\frac{\varrho v^2}{p}\right). \tag{11}$$

From (7),

$$\frac{\varrho_0}{\varrho} = \left(\frac{p_0}{p}\right)^{1/\gamma}.$$

Substituting this in (11), we obtain

$$p_0 = p\left(1 + \frac{1}{2}\frac{\gamma - 1}{\gamma}\frac{\varrho v^2}{p}\right)^{[\gamma/(\gamma-1)]}. \tag{12}$$

Now a, the speed of sound in air, is equal to $\sqrt{\gamma p/\varrho}$, and the ratio v/a is generally called the Mach number and denoted by M. Substitution of M in (12) gives

$$p_0 = p\left(1 + \frac{\gamma - 1}{2}M^2\right)^{[\gamma/(\gamma-1)]}. \tag{13}$$

Compressibility will begin to have a significant effect on the total pressure reading at some value of M less than 1. To find what this value is, we expand (13) by the binomial theorem and obtain

$$p_0 = p\left(1 + \frac{\gamma}{2}M^2 + \frac{\gamma}{8}M^4 + \frac{\gamma(2-\gamma)}{48}M^6 + \ldots\right)$$

$$= p + \frac{\gamma M^2}{2}p\left(1 + \frac{M^2}{4} + \frac{2-\gamma}{24}M^4 + \ldots\right);$$

i.e.

$$p_0 = p + \frac{1}{2}\varrho v^2\left(1 + \frac{M^2}{4} + \frac{2-\gamma}{24}M^4 + \ldots\right). \tag{14}$$

For air, γ is closely equal to 1·4, and (14) becomes

$$p_0 = p + \frac{1}{2}\varrho v^2\left(1 + \frac{M^2}{4} + \frac{M^4}{40} + \ldots\right). \tag{15}$$

At ordinary temperatures the velocity of sound in air is about 340 m/sec, so that v may be over 60 m/sec (i.e. $M \approx 0{\cdot}2$) before the second term in the expansion of (15) becomes equal to 0·01 of $\frac{1}{2}\varrho v^2$, while the third and subsequent terms are negligible. Hence for air speeds below about 60 m/sec the value of p_0 given by (13) differs from the value that would be measured in incompressible flow (5) by less than 1 per cent of $\frac{1}{2}\varrho v^2$. For such speeds, therefore, provided that an error in the measurement of v not exceeding 0·5 per cent (half the error on v^2) can be tolerated, compressibility effects can be ignored; and the low-speed form of (13), i.e. (5), can be used to obtain the total pressure.

For speeds higher than 60 m/sec the complete form of Bernoulli's equation, (13), must be used. Although this equation is valid also for supersonic speeds, it cannot be used to determine the total pressure when the pitot tube generates shock waves, which occurs at air speeds approaching the velocity of sound and above. The procedure then necessary is described in Chapter IV.

Viscosity

When air flows along a stationary solid boundary, such as a pipe wall, stresses are set up which exert a tangential force on the boundary as though the air were trying to adhere to it and drag it along. This is due to a property of air (and all real fluids) known as viscosity, which produces shearing stresses not only at the solid boundary but also within the moving air itself.

Although the viscosity of many common fluids such as air and water is quite small, it is nevertheless enough to produce marked effects in regions where the fluid velocity changes rapidly with distance normal to the flow direction: this occurs close to the surface of a body at rest relative to the adjacent fluid, in wakes and jets, and in pipe flow. Whilst many of the characteristics of pressure-tube anemometers can be predicted by the application of the theory of non-viscous fluids, some details of their behaviour cannot be explained unless the effects of viscosity are taken into account.

Consider a viscous fluid moving past a stationary plane surface, the flow being everywhere parallel to the surface. Theoretical arguments, strongly supported by experimental evidence, lead to the now generally accepted conclusion that the fluid particles in immediate contact with the surface are at rest. Farther out from the surface the speed increases, as the faster-moving layers of the fluid tend to accelerate the slower layers nearer the surface, while they themselves tend to be retarded. There is thus a tangential force on the surface in the direction of motion and a velocity gradient in the fluid in a direction normal to the surface. If v is the velocity parallel to the surface at a point distant y from it (Fig. 2.3), the velocity gradient at the surface (or the rate at which the velocity increases as the distance from the surface increases) is given by dv/dy measured at the surface; and it is found that for natural fluids, such as

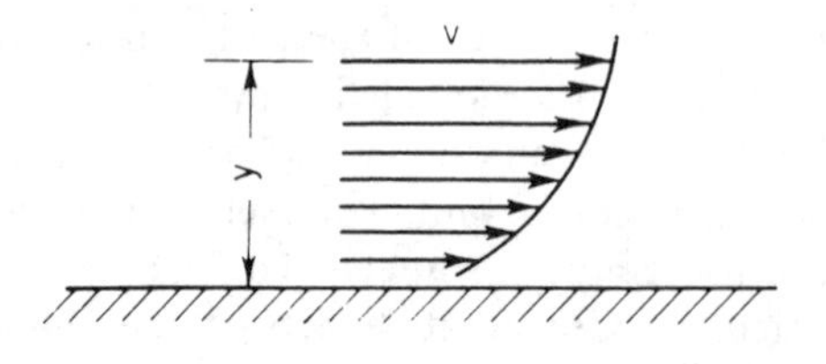

FIG. 2.3. Viscous flow near a solid boundary.

air and water, the tangential force F acting on unit area of the surface is proportional to the velocity gradient. Thus

$$F = \mu \frac{dv}{dy}, \tag{16}$$

where μ is a constant at any given temperature and pressure, and is called the coefficient of viscosity.

In general, if dv/dy is the transverse velocity gradient at any point in the fluid, not necessarily at a solid boundary, F in (16) represents the shearing stress at that point along the direction of v.

If the velocity is proportional to the distance from the surface or point, μ may be defined as the "tangential force on unit area of either of two horizontal planes of indefinite extent at unit distance apart, one of which is fixed while the other moves horizontally with unit velocity, the space between being filled with the viscous substance". This definition is due to Maxwell. Values of μ for air are given in Table 5.1, p. 77.

Fluids for which (16) holds are known as Newtonian fluids, because Newton first postulated this relationship, which has since been fully confirmed by experiment. However, it has been found[3, 4] that the addition of very small amounts – of the order of 1 per cent or less – of certain substances to water converts it into a fluid for which the relation between shearing stress and velocity gradient can no longer be expressed by the simple linear equation (16). In some cases the stress or the surface friction on a solid surface is found to be much less than for untreated water in the same flow conditions. Only Newtonian fluids are considered in this book.

The Boundary Layer

As already stated, when a fluid flows along a stationary solid surface, the velocity at the surface is zero. At a certain distance from the surface the full stream velocity is reached; and, in cases of practical importance with fluids of low viscosity such as air or water, the rise to this velocity occurs within a comparatively narrow layer of fluid known as the boundary layer (p. 10). The velocity gradient is rapid near the surface, and then falls away, the free-stream velocity being approached asymptotically.

Because the boundary layer is thin, pressures are transmitted across it without change (except in regions of rapid curvature), and considerable progress has been made in the analytical treatment of many problems of fluid motion by the assumption that the effects of viscosity are confined to the boundary layer and that outside it inviscid-flow conditions prevail. It is found, for example, that, provided that the boundary layer does not separate from the surface (see below), the pressures measured by pressure-tube anemometers can be related to the fluid velocity by inviscid-flow theory.

On the other hand, it is the viscous stresses within the boundary layer that are responsible for frictional resistance to fluid flow, as we shall see in Chapter V which deals with the flow in pipes. Moreover, if the boundary layer is confronted with too severe a rate of rise of pressure with distance along the direction of motion, it ceases to adhere to the surface and separates. Boundary-layer separation gives rise to undesirable effects such as loss of lift (stalling) on aircraft wings and turbine blades and increase of resistance. Such effects occur, for instance, in the blades of an overloaded fan and in too-sharply diverging diffusers (see pp. 106–9), with corresponding deterioration in performance. We shall see, too (Chapter III), that local separation of the boundary layer can affect the characteristics of pressure-tube anemometers, but not seriously in a properly designed instrument.

For these reasons, it will repay us to discuss the properties of the boundary layer in more detail.

Imagine that air is flowing steadily along a smooth flat plate which is so wide that, in the region we are considering, no disturbing transverse flow is introduced by the lateral edges. The thickness of the boundary layer will be zero at the leading edge of the plate and will grow progressively as the distance from the leading edge increases. Also, if no entry disturbances are introduced by the leading edge, the flow in the forward part of the boundary layer will be steady and streamline (or laminar, as it is more usually termed), i.e. everywhere parallel to the surface with no components of velocity perpendicular to the surface and no fluctuations of velocity along it.

This laminar boundary layer does not persist indefinitely. At a certain distance along the plate a different type of flow sets in; its characteristic feature is that, superimposed on the average flow along the surface, there are rapid and usually smaller velocity fluctuations both parallel and perpendicular to the surface, somewhat resembling the conditions in which, on a gusty day, the wind rises and falls intermittently above its average speed, with possibly up and down gusts as well. This turbulent boundary layer, as it is called, always occurs when the product of velocity and distance from the leading edge is high enough (see below); and, like the laminar layer, it thickens continuously as the flow proceeds along the plate.

Dimensional Homogeneity; Reynolds Number and Mach Number

No anemometer has yet been devised whose calibration can be predicted by theory alone: every type has to be calibrated, and it is of prime importance to know the precise conditions in which the calibration obtained for one particular instrument can be used for other geometrically similar instruments of the same type. The principle of dimensional homogeneity helps to elucidate this problem. Consider the pressure-tube anemometer as an example. It will record a pressure difference p, and the only properties of an incompressible fluid that affect this pressure are the velocity v, the mass density ϱ, and the viscosity μ. The pressure may also depend on the size of the anemometer, and if we consider a set of geometrically similar instruments, size will be defined by a characteristic linear dimension which we will denote by l. It follows that there will be an equation for p of the form

$$p = f(l, v, \varrho, \mu),$$

where f represents some function of the four variables within the bracket. This function can consist of any number of terms comprising these variables in combination, but each term must have the dimensions of a pressure.

The general form of one of these terms may be written $kl^\alpha v^\beta \varrho^\gamma \mu^\delta$ where the indices α, β, etc., may have any values, real or imaginary, and k is a numerical factor (non-dimensional). Expressed in relation to the three fundamental units of mass M, length L, and time T, the dimensions of such a term are $L^\alpha(LT^{-1})^\beta (ML^{-3})^\gamma (ML^{-1}T^{-1})^\delta$, and α, β, γ and δ must be such that these dimensions are those of pressure, i.e. $ML^{-1}T^{-2}$.

Hence, equating indices of like units, we have

$$1 = \gamma + \delta, \quad -1 = \alpha + \beta - 3\gamma - \delta, \quad \text{and} \quad -2 = -\beta - \delta.$$

From these three sets of relationships we obtain

$$\alpha = \beta - 2, \quad \gamma = \beta - 1, \quad \text{and} \quad \delta = 2 - \beta.$$

Thus the general term in the equation for pressure is of the form

$$kl^{\beta-2}v^\beta \varrho^{\beta-1}\mu^{2-\beta}$$

or

$$k\varrho v^2\left(\frac{\varrho vl}{\mu}\right)^{\beta-2}$$

Hence

$$p = \varrho v^2 f\left(\frac{vl}{\nu}\right), \tag{17}$$

where ν denotes the ratio μ/ϱ (usually termed the kinematic coefficient of viscosity – see Table 5.1) and f represents a function of the quantity vl/ν. The function can consist of any number of terms, but each one must involve the quantities v, l, and ν in the combination vl/ν, which, like $p/\varrho v^2$, is non-dimensional.

Consistently with (17), the relation between the velocity v and the differential pressure recorded by a pressure-tube anemometer is

$$p = K\varrho v^2, \tag{18}$$

where K is the (non-dimensional) calibration factor of the instrument which initially is determined by experiment.†

Comparing (17) with (18), we see that

$$K = f\left(\frac{vl}{\nu}\right), \tag{19}$$

i.e. the calibration factor K is a function of vl/ν.

Equation (19) is of great practical significance. From it we infer that once K has been determined experimentally by calibrating any one of a particular family of geometrically similar anemometers, this value can be applied, without the necessity for a new calibration, to any other member of the family, provided only that the value of vl/ν is the same in both cases, irrespective of the individual values of v, l, and ν. Hence if the calibration of the prototype is determined over a range of values of vl/ν, no other geometrically similar instrument need be calibrated within that range.

If we consider only air at a given temperature and pressure, ν will be constant, and the calibration factor K will depend only on the product vl, so that, for example, the value of K determined by calibration of the prototype at speed v will be the value of K applicable to a half-scale copy of the prototype at a speed $2v$. Further, the value of K determined for, say, air will be valid for any other fluid, e.g. water, provided that vl/ν is the same in both cases. As will be shown in Chapter III, this result of the application of the principle of similarity was used to extend the calibration range of a particular instrument in air from 20 to 75 m/sec by tests in water at speeds between about 1·5 and 6 m/sec.

This parameter vl/ν is of great importance in all branches of fluid motion, as was first pointed out in 1883 by Osborne Reynolds in investigations of the flow in pipes (see Chapter V); it is now always known as the Reynolds number. Using dimensional arguments similar to those we have applied to the pressure-tube anemometer, we can show that, for given conditions of smoothness and stream turbulence, the position at which the laminar boundary layer on a flat plate aligned with the flow direction changes to the turbulent type (see above) depends on a Reynolds number vx/ν, where x is the distance along the plate measured from the leading edge.

In the same way, we can use the principle of dimensional homogeneity to derive, for compressible fluids, a relationship between the readings of a pressure-tube anemometer and the relevant variables, similar to (17) which is valid only for incompressible fluids. To do this, we must include another property of the fluid that can affect the readings, namely the bulk modulus $\varkappa$. This is a measure of the compressibility of the fluid, and is defined as the ratio

† Provided that the quantities concerned are measured in a self-consistent system of units, the numerical values of non-dimensional products are independent of the particular units system used (see Appendix 1).

of a change in pressure to the volumetric strain (volume change per unit volume) it produces. $\varkappa$ has the dimensions of a pressure, i.e. $ML^{-1}T^{-2}$.

The results of the dimensional analysis, which we shall not repeat in detail, may be written

$$p = \varrho v^2 f\left(\frac{vl}{\nu}, \frac{\varrho v^2}{\varkappa}\right). \tag{20}$$

It is shown in textbooks on physics that $\varkappa = \varrho a^2$, where a is the velocity of sound in air. Hence $\varrho v^2/\varkappa = v^2/a^2$. This ratio of fluid speed v to the speed of sound a is usually known as the Mach number M, and we may therefore write (20) in the form

$$p = \varrho v^2 f(R, M), \tag{21}$$

where R denotes the Reynolds number.

Thus in compressible flow the calibration of a pressure-tube anemometer depends on both the Reynolds number and the Mach number, whereas in incompressible flow it depends only on the Reynolds number.

References

1. M. Barker, On the use of very small pitot-tubes for measuring wind velocity, *Proc. Roy. Soc.* A **101** (1922) 345.
2. F. A. MacMillan, Viscous effects on pitot tubes at low speeds, *J. Roy. Aero. Soc.* **58** (1954) 570.
3. J. F. Ripken and M. Pilch, Studies of the reduction of pipe friction with the non-newtonian additive CMC, University of Minnesota, St. Anthony Falls Hydraulic Laboratory Technical Paper 42, Series B (1963).
4. G. E. Gadd, Turbulence damping and drag reduction produced by certain additives in water, *Nature* **206** (1965) 463.

CHAPTER III

THE CHARACTERISTICS OF PITOT AND STATIC TUBES IN INCOMPRESSIBLE FLOW

It was shown in Chapter II that in incompressible flow the velocity pressure can be obtained from the difference between two pressure observations – the total pressure and the static pressure.

The measurement of total pressure is relatively simple. On the surface of any solid body, no matter what its shape, immersed in a stream of fluid there is some point (often the most forward point) at which the fluid is brought to rest and the pressure acting is the total pressure of the undisturbed flow.† Hence this pressure can be determined by providing an orifice at that point and connecting it to a manometer. This is the basis of the pitot tube‡ described in Chapter II, which has been universally adopted for the measurement of total pressure. There is abundant experimental evidence to confirm that the pressure indicated by a properly aligned open-ended tube facing the current† is accurately equal to the local total pressure (except for viscosity effects at very low Reynolds numbers, see p. 13); and that the shape of the tube may be varied within wide limits without sensibly affecting the observed pressure.

No such latitude in shape is possible in a tube designed to measure static pressure; for this purpose it is necessary to fulfil as closely as possible the conditions already stated in Chapter II, namely that the flow at the static orifices should be the same, both in speed and direction, as it was before the measuring instrument was introduced. Early attempts to meet these requirements were based on instruments such as that sketched in Fig. 3.1.

The head of the tube, which is introduced into the air current and faces upstream, is bent at right angles to the stem and its walls are parallel to the direction of flow. It terminates in a conical plug of gradual taper merging smoothly into the tube. This form of head deflects the air only slightly, and at a short distance back from the tapered portion a number of small holes is drilled. The other end of the stem is connected to a manometer. It used to be assumed

† Except in the supersonic flow of a compressible fluid (see Chapter IV).

‡ In this book, the term pitot tube means any open-ended tube facing the air stream. It does not include the combination of pitot tube surrounded by a second tube for static pressure measurement (see Fig. 3.3), which will be called a pitot–static tube.

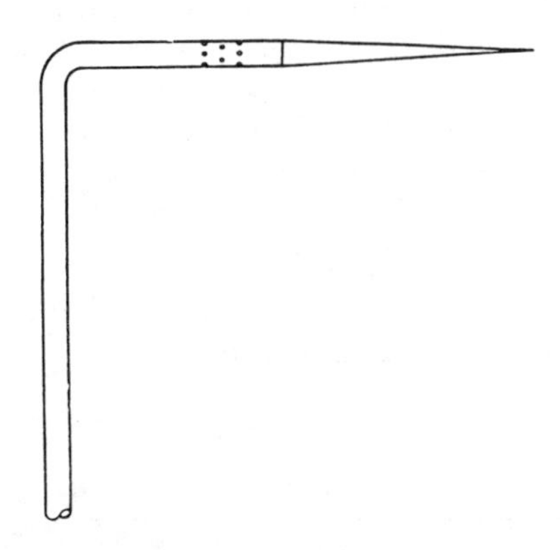

FIG. 3.1. Early static tube.

that the flow past the static orifices was sensibly parallel to the walls of the head, and that the stream static pressure would be recorded by the manometer provided that no external burrs had been left in drilling the holes.

Later experience has shown that it is necessary to take account of a number of factors which were ignored in early designs of static tubes, probably because they were then mostly unknown. The chief ones are the positions of the static orifices in relation to the rest of the instrument, particularly the stem; and the shape of the head which, besides disturbing the flow as little as possible, should be such that the same flow pattern is preserved over a wide speed range.

General Principles of Design; Pressure Distribution Along Surface

To explain how these requirements are met, it will be helpful to consider the salient features of the pressure distribution around a body immersed in a uniform stream of incompressible fluid. Since one fundamental requirement is the minimum disturbance of the flow, we shall consider the type of body generally known as "streamline".

Suppose that a solid of revolution of this type (Fig. 3.2) is immersed in a stream of incompressible fluid with its axis along the direction of flow, and that it is provided with means for enabling the pressures to be measured at a number of points along a generating line. We shall then obtain a pressure–distribution curve of the type sketched in the lower part of Fig. 3.2 in which the zero of pressure is the static pressure in the undisturbed flow.

As for the pitot tube, the pressure at the foremost (stagnation) point, where a filament of fluid is brought to rest, is equal to the velocity pressure, $\frac{1}{2}\varrho v^2$, plus p, the static pressure in the fluid. The pressure acting at any other point on the surface of the body will be the static pressure p plus some fraction or multiple f

(which may be either positive or negative) of the velocity pressure.† If, then, a small hole in the surface at the stagnation point and a similar hole elsewhere on the surface are connected to opposite sides of a differential manometer, the latter will indicate a pressure difference Δp given by

$$\Delta p = (1 - f)\,\tfrac{1}{2}\varrho v^2 = K\,\tfrac{1}{2}\varrho v^2. \tag{1}$$

If, therefore, we determine the value of the factor K by an experimental calibration, we can use such a streamline body, with the appropriate pressure connexions, as a pressure-tube anemometer.

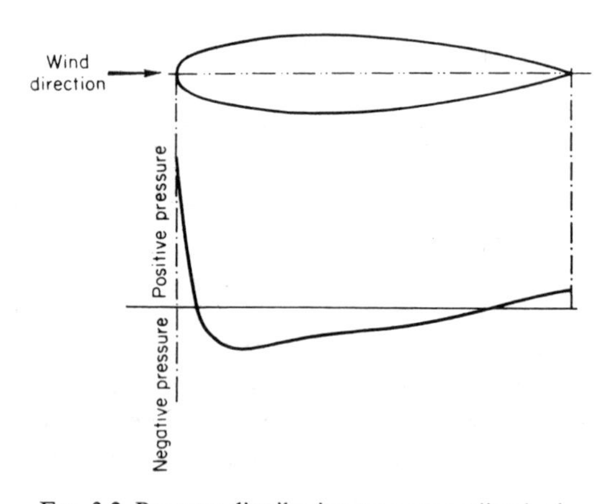

FIG. 3.2. Pressure distribution on a streamline body.

The Pitot–Static Tube

The combined pitot–static tube, an early U.K. example of which is shown in Fig. 3.3,‡ is a special case of such a device, in which the value of f in (1) is very nearly zero. Hence $K \approx 1$, and the differential pressure Δp is very nearly equal to the velocity pressure (in incompressible flow).

It will be seen from Fig. 3.3 that the head of the combined pitot–static tube consists of two co-axial tubes: the inner tube faces the flow and measures total pressure; the other tube is open to the stream only through the small static orifices a short distance back from the mouth. The head of the instrument is

† The maximum positive value of f, which occurs at the stagnation point in the nose, is 1. Its negative values, however, may be well below those indicated in Fig. 3.2, which relates to a typical streamline solid of revolution for which the negative value of f nowhere reaches 1 numerically.

‡ The circular disk shown near the end of the stem in Fig. 3.3 was provided to simplify mounting the whole instrument in a wind tunnel; it is generally omitted, without effect on the factor of the instrument, in modern designs for use in pipes or ducts. The value of D was 7·8 mm. Several of the leading dimensions were chosen to be round values in inches or fractions of an inch.

generally attached to a stem at right angles, which is held by a suitable streamline support or projects through the side of the pipe or duct in which measurements are being made, and carries the pressure connexions leading to the manometer.

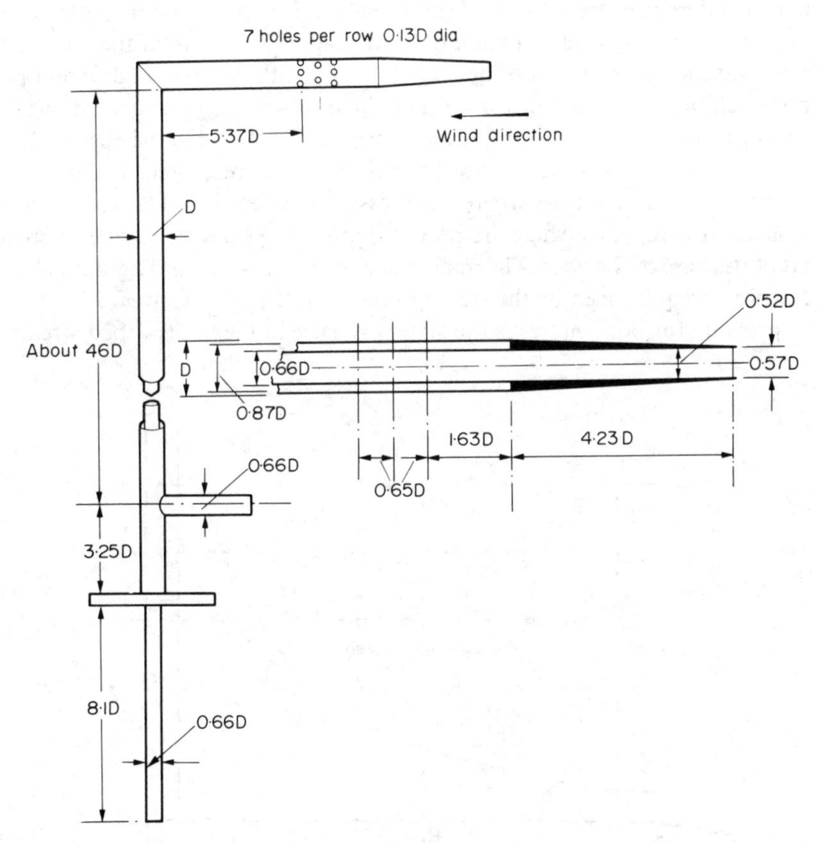

FIG. 3.3. First N.P.L. standard (tapered-nose) pitot–static tube.

The Characteristics of Pitot–Static Tubes

In the absence of the stem, the pressure distribution along the surface of the head is similar to that shown in Fig. 3.2 for the streamline solid of revolution, but with the forward region of positive pressure relatively rather shorter. If the value of K in (1) is to be 1, the static orifices must be placed in a region where the pressure acting on the surface of the head is equal to the stream static pressure. The pressure distribution, and hence the correct position of the static orifices, will depend partly on the shape of the front of the head (the nose), and partly on the position of the stem, which produces a pressure upstream of itself and so affects the pressure at the static orifices.

The effect of these variables – the shape of the nose, the effect of the stem, and the position of the static orifices – was first examined in an investigation of the characteristics of static tubes carried out in 1925.[1] Observations were made with four differently shaped noses, three of them tapered as in Fig. 3.3, but of different degrees of taper, and one hemispherical; the pressure distribution along the parallel portion of each head downstream of the base of the nose was measured both without the stem and with the stem at different positions behind the static orifices. With all the heads, the general character of the pressure distribution over the parallel portion of the head in the absence of the stem is shown by the lower curve of Fig. 3.4; there were appreciable differences near the base of the nose, where the pressure gradient is steep, but the curves tended to run together where the gradient flattens out some 5 or 6 tube-diameters aft of the base of the nose. The upper curve of Fig. 3.4 shows diagrammatically the pressure produced by the stem at points on the tube upstream.† The zero of pressure for both curves is the static pressure in the undisturbed stream. It

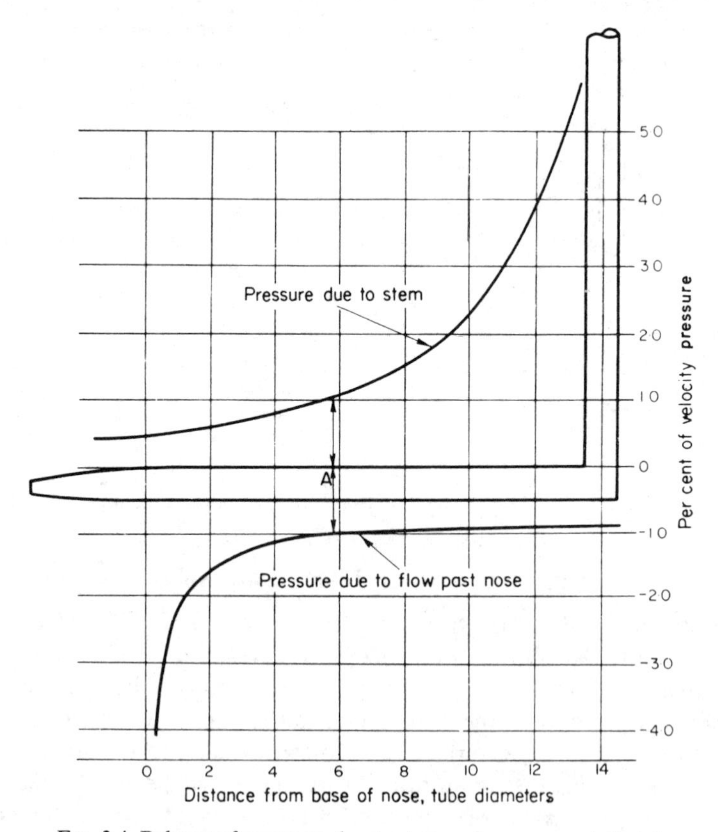

FIG. 3.4. Balance of pressures due to stem and nose on a static tube.

† See Fig. 3.9 for the actual values of these pressures.

will be seen that the pressure due to the flow around the nose is of opposite sign to that due to the stem; the correct position for the static holes is that at which the ordinates of the two curves are equal, as at the point A.

Obviously, this position of balance can be controlled by altering the distance between the stem and the base of the nose where the parallel portion of the stem begins; for each position of the stem there will be an intermediate position of the static holes at which the two opposing pressures will balance. If the static orifices are in a region of steep pressure gradient, slight errors in manufacture may lead to appreciable differences in the value of K from one instrument to another; on the other hand, if the static orifices are too far back, the head of the instrument will be undesirably long. From the results of this investigation it was recommended that the static orifices should be 6 tube-diameters back from the base of the nose, where the pressure gradient, as indicated by the lower curve of Fig. 3.4, has become small. The results showed also that if the stem was then 8 diameters behind the static holes, the pressure at these holes would be the stream static pressure. This work resulted in the pitot–static tube shown in Fig. 3.5, which was used for many years as the National Physical Laboratory (N.P.L.) standard hemispherical-nose pitot–static tube.

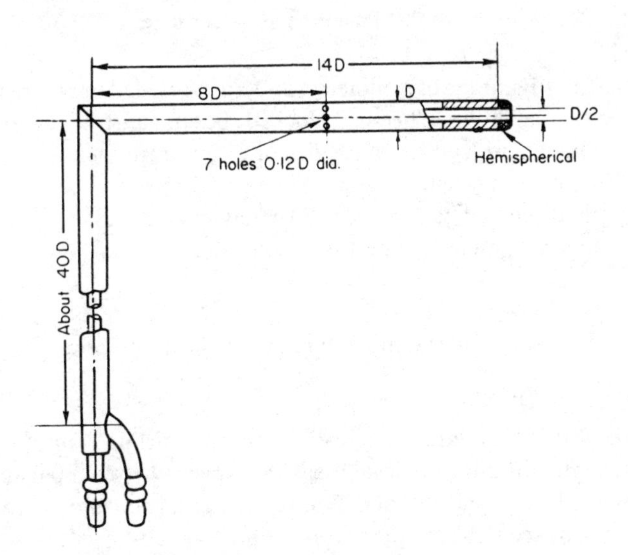

FIG. 3.5. N.P.L. standard pitot–static tube with hemispherical nose. (In the normal size $D \doteqdot 8$ mm)

The first N.P.L. standard pitot–static tube, shown in Fig. 3.3, was designed and calibrated at the Laboratory in 1912.[2] It is remarkable that, although little was then known about the properties of such instruments, so that the selected proportions were to a large extent arbitrary, nevertheless its factor was found to be closely equal to 1. It was used as the standard for many years; the only reason why the type shown in Fig. 3.5 came to be preferred for general work was that its hemispherical nose is less liable to mechanical damage than the sharp feather edge of the tapered nose of the original standard. A somewhat similar round-nose instrument is the German type, designed by Prandtl, with an annular slit instead of circular orifices for measuring static pressure (Fig. 3.6).

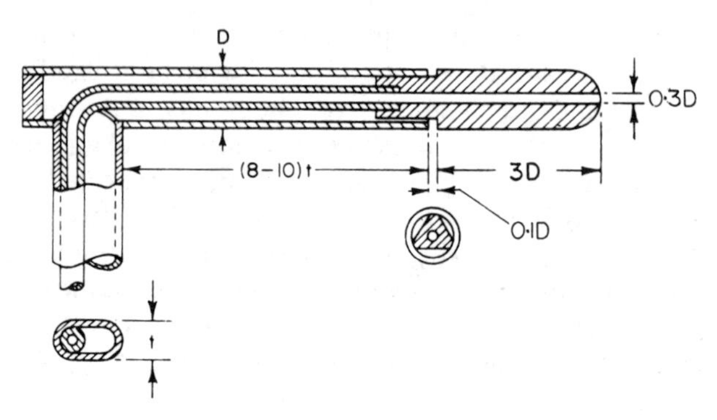

FIG. 3.6. Prandtl's pitot–static tube.

The slit has the advantage of symmetry, and, in a dust-laden airstream, it is less likely than are small holes to become blocked; but its sharp edges are a mechanical disadvantage and lead to difficulties of exact reproducibility in manufacture. The factor for this instrument also is very near to 1.

Later work, discussed below, has shown that an ellipsoidal nose is preferable to both the tapered and hemispherical types.

The Calibration Factors of Pitot–Static Tubes

(*a*) *Original Calibrations*

The two N.P.L. standards were calibrated on whirling arms, i.e. they were moved through initially still air, at measured speeds between 0·6 and 20 m/sec for the tapered nose and 0·6 and 7 m/sec for the hemispherical.[2, 3] At all speeds from 6 m/sec upwards the factor for the tapered-nose tube was found to be $1{\cdot}000 \pm 0{\cdot}001$, and the factor for the hemispherical-nose instrument, obtained from comparative tests in a wind tunnel with the tapered-nose instrument, was found to be the same within the accuracy of observation. Hence, for all practical purposes, K for both instruments at speeds above 6 m/sec was taken as 1·000.

At speeds between 0·6 and 6 m/sec, the observed values of K are given in Table 3.1. The values of K given in this table are mean values. At speeds above

TABLE 3.1. PITOT–STATIC FACTORS AT LOW REYNOLDS NUMBERS

Tapered-nose tube (3·8 mm dia)		Hemispherical-nose tube (3·9 mm dia)	
Reynolds No.[a]	K	Reynolds No.[a]	K
330	1·020	335	1·055
655	0·989	670	1·006
985	0·995	1000	1·001
1310	0·992	1335	0·996
1640	0·991	1670	0·992
1970	0·992	2005	0·991
2295	0·995	2340	0·992
2625	0·998	2675	0·996
2950	0·999	3005	0·999
3280	1·000	3340	1·001

[a] The Reynolds numbers were derived from integral values of air speed in feet per second at 15°C and 760 mm Hg. They differ slightly for the two instruments at the same air speed because the outside diameters were not exactly the same.

about 4 m/sec individual observations gave values within ±0·1 or 0·2 per cent of this mean. Over the speed range 1·8–3·7 m/sec the maximum variation from the mean was ±0·5 per cent, at 1·2 m/sec it was ±1 per cent, and at 0·6 m/sec ±2 per cent. The reason for this variation was probably the boundary-layer effect referred to below.

Soon after the tapered-nose instrument had been calibrated in air on the whirling arm, the calibration range for incompressible flow was extended to the equivalent of 75 m/sec in air by tests in water on the towing carriage of a ship-model testing tank up to speeds of about 6 m/sec. These tests were justified by the theory of dimensional homogeneity, which shows (see pp. 18–21) that the pitot-static factor K in incompressible flow can be expressed by the equation

$$K = f\left(\frac{vD}{\nu}\right), \tag{2}$$

where f represents some function of the quantity vD/ν, D is the external diameter of the tube, v is the velocity, and ν is the coefficient of kinematic viscosity.

At ordinary temperatures and pressures ν for water is about one-thirteenth that for air; therefore the value of K for any given pitot–static tube at any particular air speed can be obtained by moving it through water at about one-thirteenth of that speed.

If we consider only air at a given temperature and pressure, the Reynolds number vD/ν will vary only with the product vD; therefore, according to (2), the value of K for a given pitot–static tube will in general vary with air speed, or, if the speed remains constant, the value of K for geometrically similar tubes of different size will depend on their external diameter. Variation with Reynolds number is termed scale effect, and the fact that K is virtually constant for the standard instruments over such a wide speed range means that they suffer very little from scale effect. Absence of scale effect is a most desirable property in a device for measuring air speed.

Pitot–static tubes geometrically similar to the prototypes calibrated directly in air and water as already described are the only practical standards of reference for the measurement of air speed against which all other designs, and indeed all anemometers, have to be calibrated, either directly or indirectly. For many years the factors as determined by the initial calibrations were used: but as knowledge of fluid motion grew, and as methods of measurement became more refined, some doubts, supported by experimental evidence, arose about the precise values of the calibration factors to be used in research work where errors less than 1 per cent on velocity are desirable.

(*b*) *Modern Calibrations*

It was pointed out by Kettle[(4)] in 1953 that the tapered and hemispherical heads have pressure distributions with pronounced suction peaks near the nose, and that these could produce effects on the boundary layer along the tube which would vary with Reynolds number and with the turbulence in the air-stream. Corresponding changes in the value of K were therefore possible. One of the principal effects of a high suction peak would be to cause separation of the boundary layer from the walls of the tube because of the steep adverse pressure gradient downstream of the suction peak (see p. 18).

To overcome this defect of the hemispherical and tapered noses, Kettle designed an ellipsoidal head which had a much less pronounced suction peak, and found that its calibration remained constant over a wide speed range.[(4)]

A similar design (the modified ellipsoidal nose) was subsequently developed by Salter, Warsap, and Goodman[(5)] in the course of a comprehensive research during which they re-calibrated (but by a different method) instruments of the original tapered and hemispherical patterns. A diametral section of the modified ellipsoidal nose (Fig. 3.7) consists of two quarter ellipses separated by a distance equal to the pitot-hole diameter, the major and minor axes of the full ellipse ($2a$ and $2b$ in Fig. 3.7) being $4D$ and D–d respectively, with the length of the nose therefore equal to $2D$. Salter *et al.* found that this ellipsoidal nose, which has since been adopted in the new N.P.L. standard pitot–static tube (Fig. 3.8) specified by the British Standards Institution (B.S.I.),[(6)] is less sensitive to slight imperfections in shape, due, for example, to mechanical damage, than either of the two original patterns. It also has virtually no scale effect on K over the speed range covered by the tests (13–60 m/sec) for a tube of the usual diameter of about 8 mm, i.e. over a Reynolds-number range of about 6500 to 30,000.

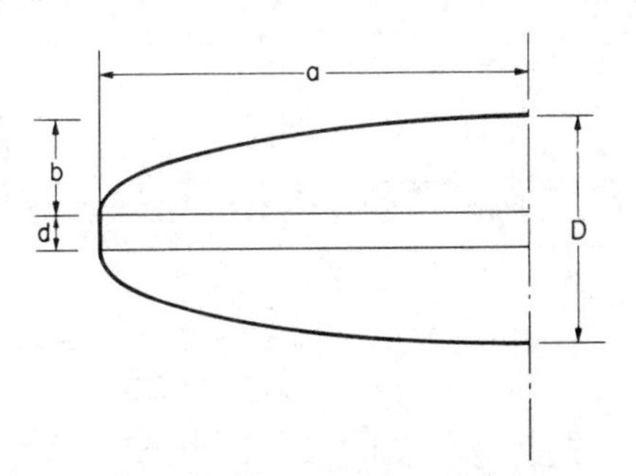

FIG. 3.7. Modified ellipsoidal nose.

To exclude the possibility of errors arising from scale effects† that would occur in the higher Reynolds-number range if larger pitot–static tubes were used, the B.S.I. specify 15 mm as the maximum permissible outside diameter of the head and stem.

Various types of the new head were tested,[5] all with the modified ellipsoidal nose, but with different distances from tip to static holes and from holes to stem. The difference between the usual right-angled mitred junction of stem and head and a rounded junction was also examined. It was found that rounding off the junction with a radius of 2·5–3·5D (see Fig. 3.8) increased the value of K by about 3 parts in 1000. It seems that the upstream effect of the stem is less with the curved junction.

The B.S.I. specify 8D (see Fig. 3.8) as the standard distance of the static holes from the tip, but allow for a range of distances between the holes and the axis of the stem. Values of the calibration factor K are given for distances ranging

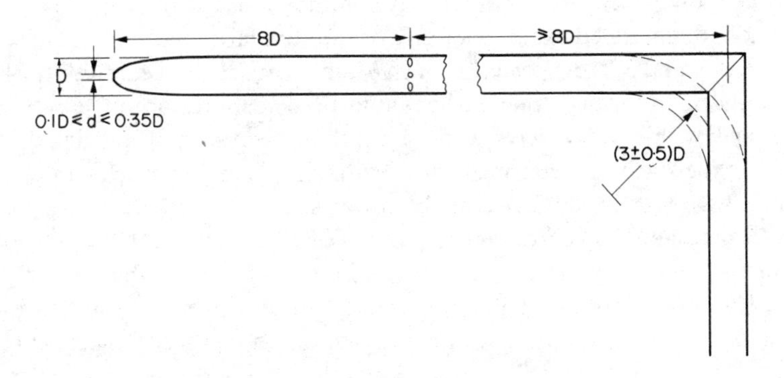

FIG. 3.8. N.P.L. standard "modified ellipsoidal" pitot–static tube.

† Due mainly to the change in the type of flow around the cylindrical stem.

TABLE 3.2. HEAD WITH ELLIPSOIDAL NOSE OF LENGTH $2D$, STATIC HOLES $8D$ FROM TIP ($6D$ FROM BASE OF NOSE), DISTANCE FROM STATIC HOLES TO STEM AXIS = $n\,D$; MITRED JUNCTION

n	39	28	20	17	14	12	10	9	8
K	1·002	1·001	1·000	0·999	0·998	0·997	0·996	0·995	0·994

from $8D$ (the minimum allowed) to over $50D$, in a graph on which the data of Table 3.2 are based.

For the radiused junction K is the same for all n values down to 17; for $n = 12$, 10, and 8 it is respectively 0·998, 0·9975, and 0·9975.

The recalibration of the copies of the two original N.P.L. standards ($D \simeq 8$ mm in both) gave values of K of 0·995 to 0·996 for the tapered nose and 0·993 to 0·995 for the hemispherical. Scale effect for the former was negligible over the speed range 13–43 m/sec and for the latter, although greater, was equivalent to only $1\frac{1}{2}$ parts in 1000 on the value of K over the same speed range.

All these results were obtained in a wind tunnel of low turbulence;† in a more turbulent tunnel‡ the factors over the speed range 12–24 m/sec were lower by about 0·002 for two tapered-nose instruments, and by rather less than this for a German design with hemispherical nose. These differences were due to boundary-layer changes induced by turbulence; they had no connexion with the effect of turbulent-velocity components on the pressures recorded by pitot and static tubes (see p. 41).

The Values of the Pitot–Static Factor to be Used in Practice

From the latest available information we can conclude that users who can tolerate errors in air-speed measurements that will most probably be less than 0·5 per cent on velocity§ can confidently continue to use a value of $K = 1$ for a well-made, undamaged example of any of the standard instruments shown in Figs. 3.3, 3.5, and 3.8 for speeds between 6 and 60 m/sec. Indeed, since the later investigation confirmed the earlier low-speed calibration of Ower and Johansen[3] within this order of accuracy in the speed range over which the two series of experiments overlapped, the results of the earlier work can also be accepted with some confidence at the lower speeds not covered in the later work. Consequently, according to the values of K given in Table 3.1, the lower limit of the speed range for which K can be taken as 1 for the two original N.P.L. standards can be extended to about 1 m/sec. Over most of this range below 6 m/sec the maximum error on air speed will not exceed 0·5 per cent, but

† Root mean square of the longitudinal turbulent component 0·1 per cent of v.
‡ Root mean square value 0·2 per cent.
§ Apart from possible errors due to turbulence, velocity gradient, etc. (see pp. 41–45).

at speeds below about 1·2 m/sec the assumption that $K = 1$ may lead to errors on speed of perhaps 1 per cent. For the ellipsoidal-nose instrument, with its lower scale effect, it seems reasonable to assume that the maximum error on velocity will not exceed 0·5 per cent over the entire speed range from 1 to 60 m/sec.

These speed ranges relate, of course, to air at ordinary temperatures and pressures, and to instruments of the usual external diameter D, which is about 8 mm for all three standards.† For geometrically similar copies of diameter D_1, the lower limits of the speed ranges quoted above will be altered in the ratio D/D_1 to preserve the same Reynolds number for air at the same temperature and pressure. Because of compressibility effects, however, the upper limit cannot be increased.

The results quoted on p. 32 show that for all three standard instruments – tapered, hemispherical, and ellipsoidal – the value of K is nearer to 0·995 than 1·000. Hence if slightly better accuracy, perhaps 0·25 per cent on velocity, is required, the observed dynamic pressure may be increased by 0·5 per cent before the velocity is calculated from it. Alternatively, still with the value of 1 for K, we can use a tube with a radiused junction between head and stem, radius not less than 2·5 D, or we can use an instrument with an increased distance between the stem and the static orifices. As shown in Table 3.2, an ellipsoidal-nose tube similar to that in Fig. 3.8, but with the stem some 20 tube-diameters instead of 8 behind the static orifices, will have a factor of 1 in a stream of low turbulence. This can be deduced from the relation between the pressure acting at the static orifices and their distance from the stem observed by Ower and Johansen[1] and reproduced in Fig. 3.9. This diagram can also be used to deduce the effect on K if, as sometimes happens, it is convenient to use a pitot–static tube with a standard head but without a stem, held by wires that produce very small upstream pressure fields.

Pitot–static tubes conforming to N.P.L. specifications are on the market with both hemispherical and ellipsoidal noses, and, subject to the margins of error quoted above, can be used without calibration. As a general rule, the engineer will prefer the latter type, whose calibration is less sensitive to small inaccuracies of profile, and so should be more easily reproducible from one instrument to another made to the same design. More important than exact compliance with dimensional specifications is reasonable care in manufacture to ensure smoothness of contour with no reversals of curvature anywhere along the nose.

On the other hand, the hemispherical nose has the advantage that it is less sensitive to errors of alignment with the wind direction up to angles of 30°, although up to angles of 15° the error with the ellipsoidal head is less (see p. 50).

† The corresponding Reynolds-number range for air at 15°C and 760 mm Hg is 500–33,000.

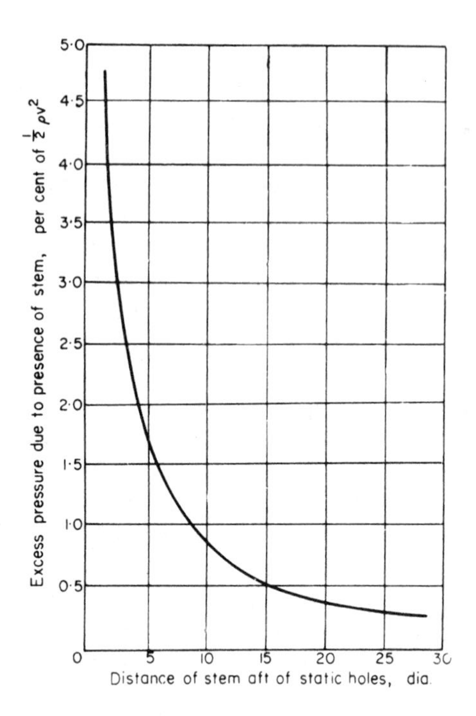

FIG. 3.9. Pressure due to stem.

Limitations of the Pitot–Static Tube for the Measurement of Low Speeds Under Industrial Conditions

The virtually constant value of the factor of the pitot–static tube over a wide range of air speeds – a range that will cover a great proportion of the cases with which the practical engineer has to deal – together with the ease of reproducibility of the factor from one instrument to another of the same geometrical pattern, constitute the main advantages that the pitot–static tube possesses over all other types of pressure-tube anemometer. Against this, however, the small differential pressure set up at low air speeds is a serious objection to the instrument; for it means that very sensitive manometers are needed if the necessary accuracy is to be achieved at such speeds, and under industrial conditions these manometers are most inconvenient to use, if, indeed, they can be used at all. For example, for an error limited to 1 per cent in the observation of the velocity head for an air speed of 0·6 m/sec we need a manometer sensitive to about 0·0002 mm of water-column; and an instrument of this sensitivity had to be specially designed† for the low-speed calibration of

† See Chapter X, p. 266.

the N.P.L. standard pitot–static tube.[3] Such an instrument can be used only in the laboratory; numerous precautions to ensure steady temperature conditions and firm foundations have to be taken if full use is to be made of its sensitivity. If we take 0·02 mm of water as a reasonable limiting sensitivity for a manometer that can conveniently be used in industrial work, see p. 251, then the lowest air speed that can be observed with an error not exceeding 1 per cent with the pitot–static tube is about 4·5 m/sec.

Pressure-tube Anemometers with Factors Greater than Unity

In order to overcome this disadvantage of the pitot–static tube, and to extend the practical range of pressure-tube anemometers in the low-speed region, a number of instruments have been designed having values of the factor K greater than 1. There is no practical difficulty about this; two of these instruments are illustrated in Fig. 3.10.

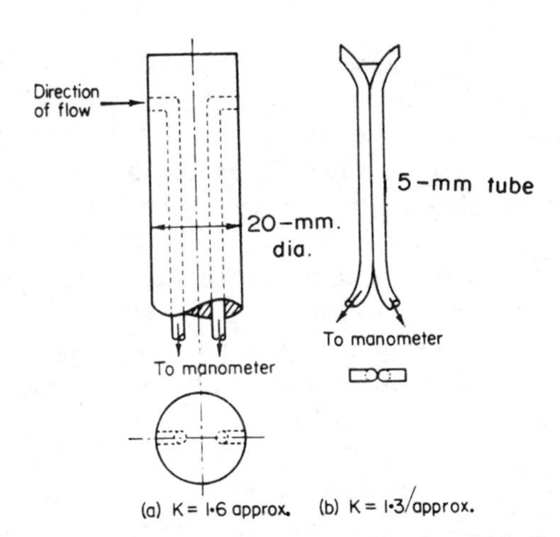

FIG. 3.10. Two pressure-tube anemometers for which $K > 1$.

But such instruments are unfortunately not to be recommended because they possess certain undesirable features. In the first place, experience shows that for most of those so far proposed the value of K varies appreciably with Reynolds number, particularly in the lower ranges for which the instrument is primarily intended. A typical calibration curve for an instrument of this type similar to, but not identical with, that shown in Fig. 3.10(b), is shown in Fig. 3.11, from which it can be seen that the factor drops from nearly 1·59 to 1·54 within the narrow speed range 12–20 m/sec, and that at the higher of these speeds it is still falling steeply. There is no need to stress the inconvenience of a

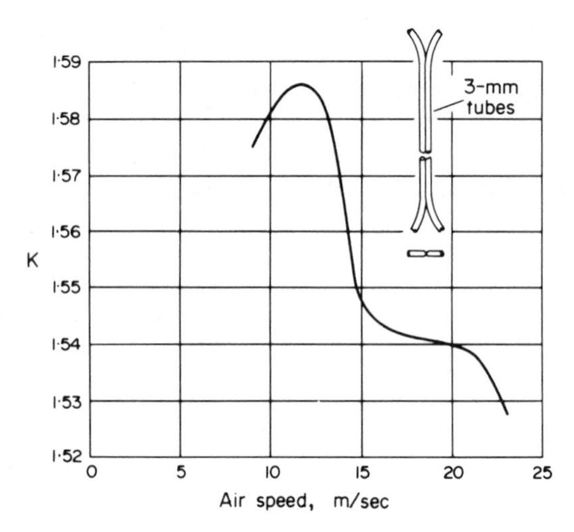

FIG. 3.11. Variation of factor with speed.

calibration that changes with speed, one result of which is that at least two calculations are needed to convert a pressure observation into velocity: the first has to be made with a guessed value of K, since the speed is initially unknown; and the approximate value of the speed thus obtained has to be used to obtain, from the calibration curve of the instrument, a nearer approach to the true value of K to be used in the second, more exact, calculation.

Another serious disadvantage of such instruments is that if K varies appreciably with Reynolds number, its value at a given Reynolds number is likely to depend much more than that of a standard pitot–static tube on the amount of turbulence present in the airstream, so that the calibration may be appreciably different, even at the same speed, in different pipe systems.

It must be remembered also that, since the velocity for a given pressure reading varies inversely as the square root of K, the extension of the lower range of speed that can be measured with a manometer of given sensitivity is not great. Thus, taking as before 4·5 m/sec as the lowest speed that can conveniently be measured with an error not exceeding 1 per cent with the pitot–static tube under industrial conditions, we find that, by using an instrument with a factor of 1·6, we can extend the lower limit of speed measurable to the same accuracy only to about 3·75 m/sec – not a significant extension.

Hence, even if the wandering calibration were overcome, little would be gained by using instruments having factors higher than that of the standard pitot–static tube, unless these factors could be increased to 2 or more. A number of such instruments are described in ref. 7. One, consisting of a straight cylindrical tube with one forward-facing and one backward-facing orifice near the tip, had a factor of 2·2; others had factors up to about 8. The latter type consists essentially of cylindrical tubes which are inserted into the airstream from one side with their axes perpendicular to the flow direction. Air entering through an upstream-facing orifice near the end of the tube passes through a venturi (see Chapter VII) or similar device inside the tube and co-axial with it, and returns to the main flow through the open end of the cylindrical casing, the other end of which is closed. The differential pressure between the venturi inlet and throat is measured: this is up to 8 times the velocity pressure, according to the particular design. Even higher factors, up to 14,[8] have been obtained with double-venturi probes, in which a small inner venturi is arranged with its outlet plane at the throat of a larger outer venturi.

All such instruments, however, seem to be subject to the above-mentioned disadvantage of calibration change with changing Reynolds number and therefore also, probably, with different degrees of turbulence. Moreover, they need calibration, and they cannot be used to measure the static pressure in a pipe; if a knowledge of the static pressure is required, it must be measured separately, e.g. at a hole in the pipe wall (see pp. 136–8).

Despite these drawbacks, when high accuracy is not important these high-factor instruments may be favoured when their overall dimensions are not too large in relation to the pipe diameter, because they can be used with cheaper, less sensitive manometers than the pitot–static tube needs at low speeds. The instruments described in ref. 7 have outside diameters of 3·8 cm and upwards, and should not be used in pipes of less than 75 cm diameter.

The Pitot Cylinder

In industrial pipe-flow measurements it is sometimes necessary to insert and withdraw a speed-measuring probe at frequent or fairly frequent intervals. Sometimes also, for example in investigations of the flow in turbo-machines, it is necessary to make total-head explorations in regions difficult or impossible to reach with the L-shaped pitot tube. Thus there is sometimes a use for a probe that can be inserted more easily into a pipe, and needs a smaller hole in the wall than the pitot tube, or is better suited for exploration in confined or difficult conditions. The so-called "pitot cylinder" may be used for such purposes. It is a straight cylindrical tube with one end closed, usually by a hemispherical cap, which is inserted into the pipe and, in its simplest form, has a single small hole drilled in its wall near the end exposed to the air stream (see Fig. 3.23(b) which shows a direction-finding pitot cylinder, with two pressure holes). When in use, the tube is orientated so that the orifice faces upstream and therefore, when connected to a manometer, registers the total pressure at the orifice. Usually there are two additional orifices in the same transverse plane as the forward-facing orifice and arranged symmetrically at 30–35° on either side of it. Their purpose is to enable the instrument to be accurately aligned: when the pressures at the two auxiliary orifices are equal, the total-pressure orifice will be normal to the direction of flow.

Static pressure is best measured separately. Theoretically, it could be measured by suitable location of the auxiliary orifices because the distribution of surface pressure around a transverse plane of the tube is such that there are two points on the periphery, symmetrically located with respect to the foremost point, at which the pressure is equal to the static pressure in the free stream. In practice, however, it is very difficult to measure static pressure accurately in this way for a number of reasons. In the first place, these points are located in the region where the surface pressure is varying most rapidly with angular position round the circumference, so that the holes must be accurately located within about 0·2° and, even so, the instrument will be sensitive to misalignment. Moreover, the position at which the surface pressure equals the static pressure varies with Reynolds number and the degree of turbulence in the airstream; it also tends to wander to some extent even in any particular set of normally "steady" conditions, so that the readings of a manometer connected to one of these orifices will fluctuate and be difficult to measure accurately. This difficulty and that due to sensitivity to alignment can be largely overcome by locating the two static-pressure orifices 36·4° on either side of the total-pressure orifice and connecting them together by a length of capillary from which a static-pressure lead to the manometer is taken.(9)

As regards the measurement of total pressure, unless the orifice is more than $2\frac{1}{2}$ tube diameters from the closed end of the pitot cylinder, effects due to the flow around the end cause the pressure at the orifice to differ from the total pressure at the same point in the free stream. The instrument factor is therefore not 1, but ranges from about 0·93 when the hole is 1 tube diameter from the end (and much less when it is nearer) to about 0·99 when the distance is 2 diameters.(10) The instrument therefore needs calibration, and its factor is found to depend to some extent upon Reynolds number and stream turbulence. The factor is also affected by inclinations of the flow direction to the axis of the cylinder: it decreases by about 1 per cent for a 5° inclination and by about 5 per cent for 10°.(10) Often, however, the flow direction will be known, at least approximately, and it will be possible to align the instrument sufficiently accurately by orientating it firstly about its axis until the pressures at the side holes are equal and then about a perpendicular axis passing through the total-pressure orifice until the total-pressure reading is a maximum.

Another fact that must be borne in mind in using this instrument is that, because of wall-proximity and pressure-gradient effects, its calibration, even for a specified distance of the total-pressure orifice from the end of the cylinder, is different when the orifice is near a wall. Hence, without an initial calibration at various positions along a diameter, the pitot cylinder cannot be used for accurate determinations of flowrates in pipes by one or other of the traverse methods described in Chapter VI. This difficulty cannot be overcome by using a pitot cylinder extending right across the pipe and sufficiently long to enable the total-pressure orifice to be located at any desired point on the diameter. Even for this continuous cylinder, pressure-gradient effects cause the total-pressure calibration to vary with orifice position.

Apart from pipe traverses, however, the convenience of the pitot cylinder for local explorations of the flow will favour its use for many purposes where high accuracy is not essential. Reference 10 shows that the instrument factor remains reasonably constant down to a Reynolds number (based on the cylinder diameter) certainly as low as 1 or 2×10^3, provided that the static pressure is determined by some other device. This was tacitly assumed in the above discussion of the variation of the factor with the distance of the total head orifice from the cylinder tip: for reasons already stated, it is difficult to obtain reliable measurements of the static pressure by means of side holes in the pitot cylinder itself.

A useful summary of information on the cantilever pitot cylinder will be found in ref. 11.

Disk Static Heads

In a dust-laden airstream the small holes in the static head of the tubular type are liable to become blocked; and for measurements in such conditions heads of the type shown in Fig. 3.12 are sometimes used. Experiments by Heenan and Gilbert[12] have shown that these heads can give satisfactory accuracy. It is

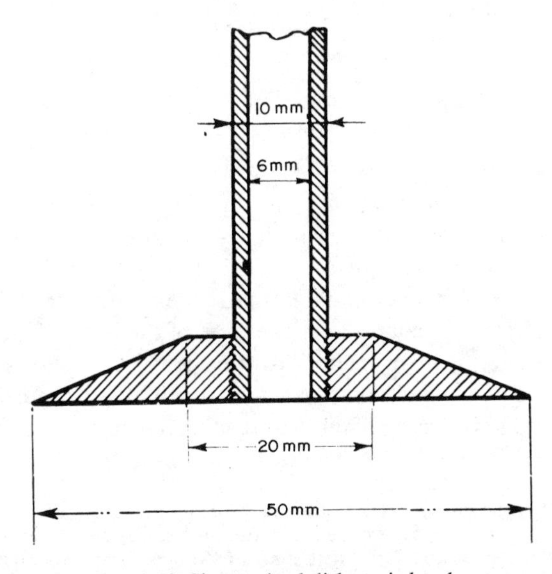

FIG. 3.12. Sharp-edged disk static head.

important that the face of the disk should be smooth and aligned with the flow direction,† and that the tube connecting the orifice to the manometer should be flush with the disk surface. Such heads may be used in conjunction with the usual form of pitot tube, as illustrated in Fig. 13.3, where one possible arrangement is shown; it is assumed that there are no appreciable differences between the flow conditions at the static face and the mouth of the pitot tube. Such combinations, however, may disturb the flow considerably more than a standard pitot–static tube does. Alternatively, the measurements of static pressure and pitot pressure may be made separately.

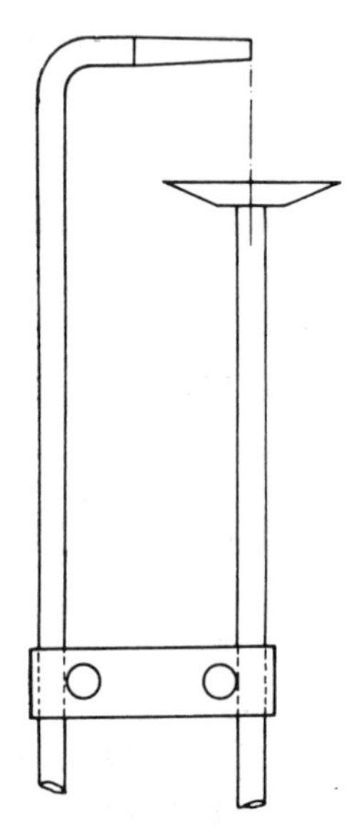

FIG. 3.13. Combination of total-head tube and disk static head.

By rounding the edges of the disk and suitably arranging the static orifice, we can considerably reduce the sensitivity of the disk static head to misalignment. One such instrument, shown in Fig. 3.14, is quite small, but there seems to be no reason why this tvpe should not be made to the overall dimensions of that in Fig. 3.12 and so retain the advantages of the latter in the presence of dust. Calibration of the instrument illustrated in Fig. 3.14 showed[13] that the reading was 0·115 of $\frac{1}{2}\varrho v^2$ below the static pressure of the undisturbed flow when the disk was set edgewise to the wind. This reading remained practically constant until the disk was inclined at an angle of $\pm 3°$ to the wind direction and then

† Heenan and Gilbert stated that this head "will indicate accurately even if the direction of flow makes a considerable angle with the face". But the reasons they gave for this statement are not clear or convincing, and in the light of present knowledge it is very difficult to accept. Their paper was written in 1894.

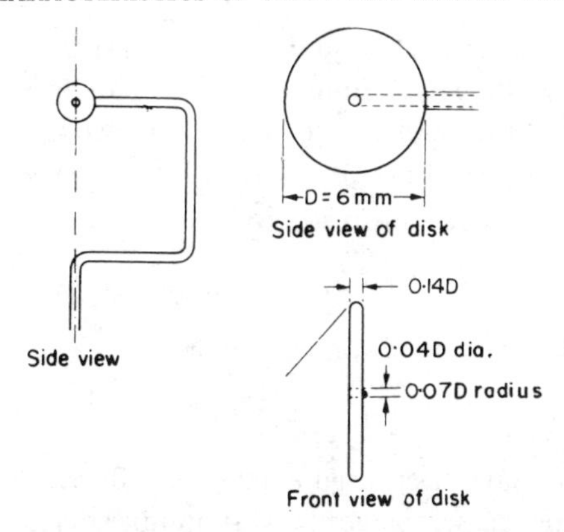

FIG. 3.14. Disk static head with rounded edges.

rose sharply for greater angles. The existence of this sharply defined minimum can be used to align the head to the wind direction. The minimum value was found to be insensitive to rotation about an axis normal to the surface of the disk to within 0·01 of $\frac{1}{2}\varrho v^2$ over $\pm 10°$, and to change of wind speed (to within 0·005 of $\frac{1}{2}\varrho v^2$) over the range 6–24 m/sec.

The Effect of Turbulence on Pitot and Static Tubes

The variation of K with Reynolds number (see p. 20) that may be caused by changes of turbulence in the airstream is a secondary effect of turbulence. A more direct effect is due to the pressure changes at the total-pressure and static orifices produced by the turbulent velocity fluctuations that practically always occur in moving streams of air. In general, we may regard the flow as having a steady velocity v on which is superimposed a random turbulent velocity which has a definite value at any instant but an average value of zero taken over a sufficiently long time interval. The turbulent velocity can be resolved into components v_x, v_y, and v_z parallel to the axes of a three-dimensional system of Cartesian co-ordinates with the x-axis along the direction of mean flow. At any instant, the velocity component in the x-direction is $v + v_x$; v_x may be positive or negative, but its average value is zero; and similarly for v_y and v_z. The associated pressure changes, however, depend on the squares of the velocities, and the mean value of these are *not* zero.

The turbulent velocity components affect the readings of both the pitot and the static tube. Goldstein[14] investigated these effects theoretically, neglecting possible effects of such factors as frequency, damping, resonance, and lag in the leads. Subject to these limitations, he showed that the pressure measured by the pitot tube in incompressible flow is not $p + \frac{1}{2}\varrho v^2$ but $p + \frac{1}{2}\varrho v^2 + \frac{1}{2}\varrho(\overline{v_x^2} + \overline{v_y^2} + \overline{v_z^2})$, where $\overline{v_x^2}$, etc., are the mean squares of the turbulent components

v_x, etc. This is consistent with the interpretation of total head as a measure of the energy per unit volume of the fluid (see p. 78). Goldstein's analysis showed also that the static tube might be expected to record a pressure equal to $p + \frac{1}{2}\varrho[\frac{1}{2}(\overline{v_y^2} + \overline{v_z^2})]$. In isotropic turbulence $\overline{v_x^2} = \overline{v_y^2} = \overline{v_z^2}$, and the readings of the pitot and static tubes become respectively

$$p + \tfrac{1}{2}\varrho(v^2 + 3\overline{v_x^2}) \quad \text{and} \quad p + \tfrac{1}{2}\varrho\overline{v_x^2},$$

so that the pitot–static combination will record

$$\tfrac{1}{2}\varrho v^2(1 + 2\overline{v_x^2}/v^2).$$

The above results have been used extensively. But turbulence can also be regarded as a statistical assemblage of a vast number of eddies of various sizes; and a later paper by Barat[(15)] points out that its effects depend not only on the turbulence intensity measured by $\overline{v_x^2}$, etc., but also on the geometrical scale of the turbulence, i.e. on the sizes of the component eddies. When the size of a typical eddy is small compared with the diameter of a static-pressure tube, the pressure fluctuations at the several orifices are not correlated with one another. In these circumstances the tube does read high, as already described, although a more detailed analysis by Toomre[(16)] concludes that the error is $\frac{1}{2}\varrho(\overline{v_y^2} + \overline{v_z^2})$, which is twice the value obtained by Goldstein. But when the turbulence scale becomes large compared with the tube diameter, the pressures at the several orifices become increasingly correlated, as if the flow direction at any instant were inclined to the tube, or the tube inclined to the flow, by an amount corresponding to the transverse velocity components. The tube would then be expected to read low, by an amount corresponding to some sort of time average of the fluctuating angle of yaw. Toomre shows that the error in the limit of indefinitely large turbulence scale is equal in magnitude (and opposite in sign) to the limit for indefinitely small scale. The way in which the error varies with scale between the two limits of $\pm \frac{1}{2}\varrho(\overline{v_y^2} + \overline{v_z^2})$ is not yet established. Early experimental results[(17)] are now suspect because of their intrinsic sensitivity to the exact value of the tube calibration factor K. An investigation subsequent to Toomre's[(18)] suggests that, with the tube sizes likely to be used in industrial applications, the static-pressure reading is likely to be nearer to the true static pressure than to either of the theoretical extreme values.

Usually, therefore, there will be little justification for attempting any correction at all to measurements made with a static tube unless definitive error-data become available.† For a pitot tube the error due to the fluctuating effective inclination of the flow is likely to be negligible in most cases, owing to its much greater yaw insensitivity, and Goldstein's result can accordingly be accepted

† For accurate pipe-flow measurements, it is preferable to measure static pressure by means of a hole in the wall, for which the turbulence corrections are known with greater certainty (see p. 109).

for most practical purposes. This is particularly true if the probe is square-ended and thin-walled,[19] in which case the degree of insensitivity to changes in flow direction is indeed large (see Fig. 3.18).†

The information at present available, therefore, suggests that a pitot–static combination in isotropic turbulent flow will record

$$\tfrac{1}{2}\varrho v^2(1 + \alpha \overline{v_x^2}/v^2), \tag{3}$$

where α varies between 1 (for turbulence of indefinitely small scale) and 5 (for indefinitely large scale). Thus even if r.m.s. turbulent velocity fluctuations are as great as 10 per cent of v, the error in velocity measurement, if the effects of turbulence are neglected, amounts only to 0·5–2·5 per cent. As a value of $\sqrt{\overline{v^2}}/v$ of 0·1 represents a fairly turbulent stream, we may conclude that turbulence effects on pitot–static readings are often negligible, although they may need to be taken into account in accurate measurements of the flow through a pipe (see Chapter V) or in a jet, in which the turbulence intensity rises to considerably greater values.

The Effect of a Velocity Gradient on the Reading of a Pitot Tube

Another small correction has to be applied in accurate work when there is a velocity gradient across the mouth of the pitot tube, as there is, for example, in measurements made in pipes. In such conditions, Young and Maas[20] have shown that the effective centre of the pitot tube is displaced from the geometric axis of the tube towards the region of higher velocity, and therefore the deduced local velocity at the true centre of the tube will be too high. Young and Maas confined their experiments to plain cylindrical tubes with square ends, such as the small tubes often used for total-pressure exploration in pipes. For such tubes of external diameter D and internal diameter d they found that the displacement Z was given by the equation

$$\frac{Z}{D} = 0{\cdot}13 + 0{\cdot}08\,\frac{d}{D}. \tag{4}$$

Later work by MacMillan[21] suggested that the value of Z given by this equation is a little too high. He used a tube in which $d/D = 0{\cdot}6$ and obtained a value of Z/D between 0·14 and 0·16, whereas (4) gives 0·18. For pitot tubes with sharp-edged conical noses, however, Livesey[22] found the effective displacement to be negligible. The effect of the Young and Maas correction on pipe-flow measurements is discussed in Chapter VI.

† On the other hand, the theory given in ref. 19 also indicated that round-nosed probes should (in large-scale turbulence) record only $p + \tfrac{1}{2}\varrho v^2$ when the internal diameter of the tube is less than 0·3 of the external diameter.

In boundary-layer investigations, use is often made of flattened pitot tubes because the region in which measurements have to be made is so thin. The displacement effect for such tubes, with h/H equal to 0·6 and W/H about 1·6 (where the symbols are defined in Fig. 3.15) was found by Quarmby and Das[23] to be given by the equation

$$\frac{Z}{H} = 0{\cdot}19 \pm 0{\cdot}01. \tag{4a}$$

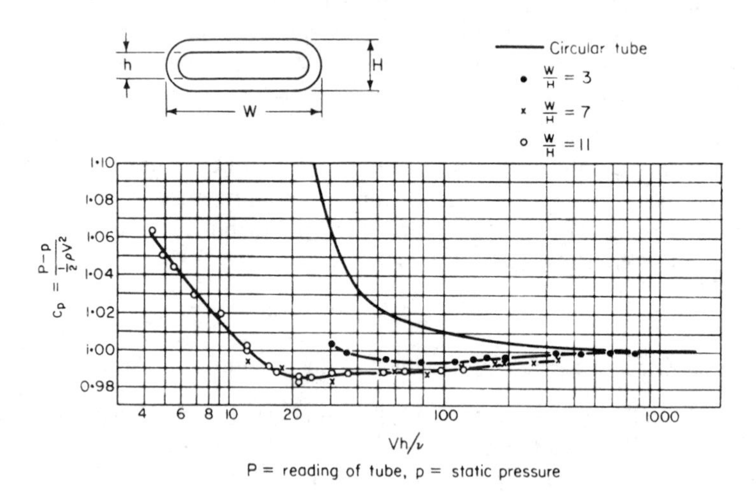

FIG. 3.15. Viscosity errors of small pitot tubes.

Because of the accompanying effects of turbulence, of viscosity when Reynolds number is low, and of wall proximity near a solid boundary (see below), the displacement effect due to velocity shear alone has proved difficult to determine with great precision. Fortunately, however, the displacement effect is usually small, and in practice a probe size can be chosen which reduces the uncertainty to acceptable proportions.

Effects of Viscosity and Wall Proximity

The readings of a very small pitot tube are subject to errors due to viscosity (see p. 13). Although their magnitude has been established (see Fig. 3.15) for both circular and flattened pitot tubes,[24] whenever possible such small tubes are best avoided because of their extremely slow rate of response. By using a tube large enough to avoid excessive lag, it will still usually be possible to keep the tube Reynolds number above that at which viscosity effects begin to be appreciable (typically about 250 based on internal diameter[25] or height;[24] see also ref. 5).

MacMillan[21] also found that if the geometric centre of a probe for which $d/D = 0{\cdot}6$ is nearer than $2D$ to a solid wall or boundary, as it may be in boundary-layer or pipe-flow measurements, a further correction is necessary: the measured velocity then has to be increased in accordance with curve A of Fig. 3.16. Curves B and C relate to results obtained subsequently by Quarmby and Das[23] for flattened tubes with h/H equal to 0·6. The effect of wall proximity is seen to be greater for the flattened tubes.

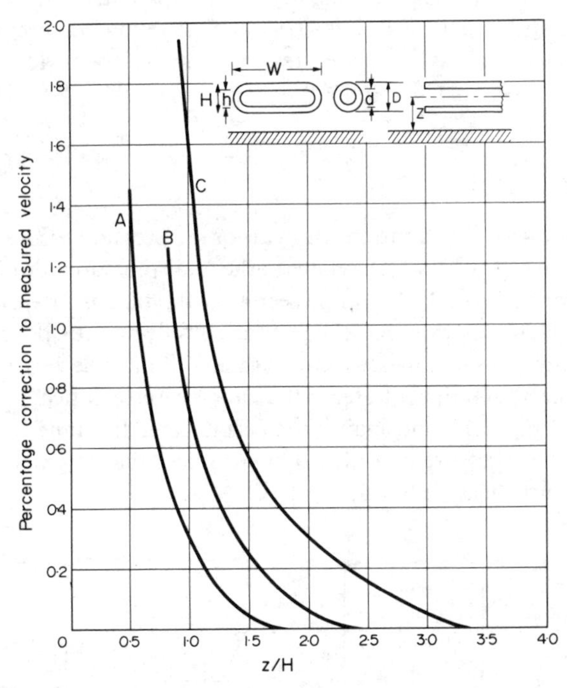

FIG. 3.16. Wall proximity correction for pitot tubes. Curve A: circular probe with $d/D = 0{\cdot}6$ (Macmillan). $R_p = 2{\cdot}4$ to $11{\cdot}8 \times 10^4$. Curves B and C: flattened tubes with $b/H = 0{\cdot}6$ (Quarmby and Das[23]). $R_p = 2{\cdot}8 \times 10^4$ for curve B and $5{\cdot}8 \times 10^4$ for curve C.

It should be noted that, in deriving curve A, MacMillan first corrected his observations for low-Reynolds-number viscosity effects (as previously determined[25] in uniform free-stream conditions), so that these effects were removed at the outset except in so far as they might themselves have been affected by velocity shear and wall proximity. Curves B and C (Quarmby and Das)[23] do not incorporate the corresponding correction for flattened tubes,[24] but display a marked dependence of wall-proximity effect on wall shear stress or (equivalently) on Reynolds number R_p of the test section (a circular pipe).

The values quoted for R_p were based on pipe diameter and the mean velocity at the test section.

The Effect of Misalignment

In general, errors will arise if a pitot head or static head is not accurately aligned with the direction of flow; but for small angles the errors are often small, particularly for pitot heads, which are much less sensitive to this effect than static heads. If there is a stem, angular deviations in the plane containing both head and stem (pitch) give slightly different results from corresponding deviations about the axis of the stem (yaw). The following discussion is confined to the effects of yaw in incompressible flow; the behaviour in compressible flow is discussed in Chapter IV.

(a) *Pitot Tubes*

A typical curve of the variation with yaw of the pressure recorded by a pitot tube is shown in Fig. 3.17, which relates to the total-pressure side of the original N.P.L. standard (Fig. 3.3). It will be seen that the tube is insensitive to quite large angles: at 20°, for example, the pressure is only about 1 per cent less than that at zero yaw. This feature seems to be general for a wide variety of different types of pitot head. A number tested in the Langley Aeronautical Laboratory[26] are sketched in Fig. 3.18, and against each is indicated the limiting angle of yaw below which the pressure recorded differs by less than 1 per cent from the dynamic pressure measured at zero yaw.

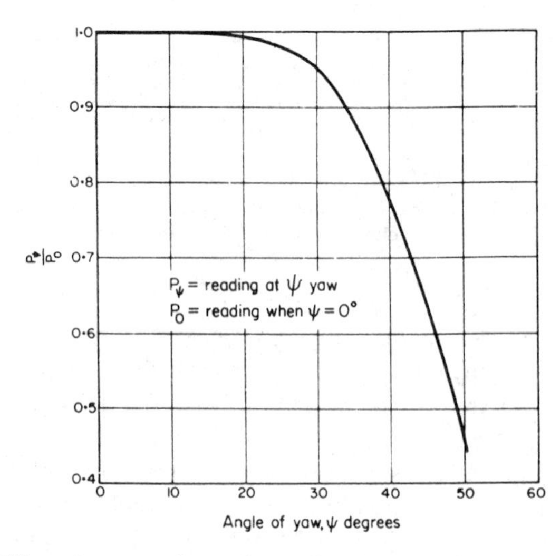

FIG. 3.17. Effect of yaw on pitot–tube readings relative to static pressure as datum.

From the results given for the simpler designs (a)–(d) of Fig. 3.18, it appears that for square-ended heads the sensitivity to yaw becomes greater as the ratio of the diameter of the total-pressure orifice *d* to the outside diameter of the tube *D* increases. The same is true for shaped heads, as is shown by the results[27] for heads with hemispherical noses of different *d*/*D* ratios (Fig. 3.19).

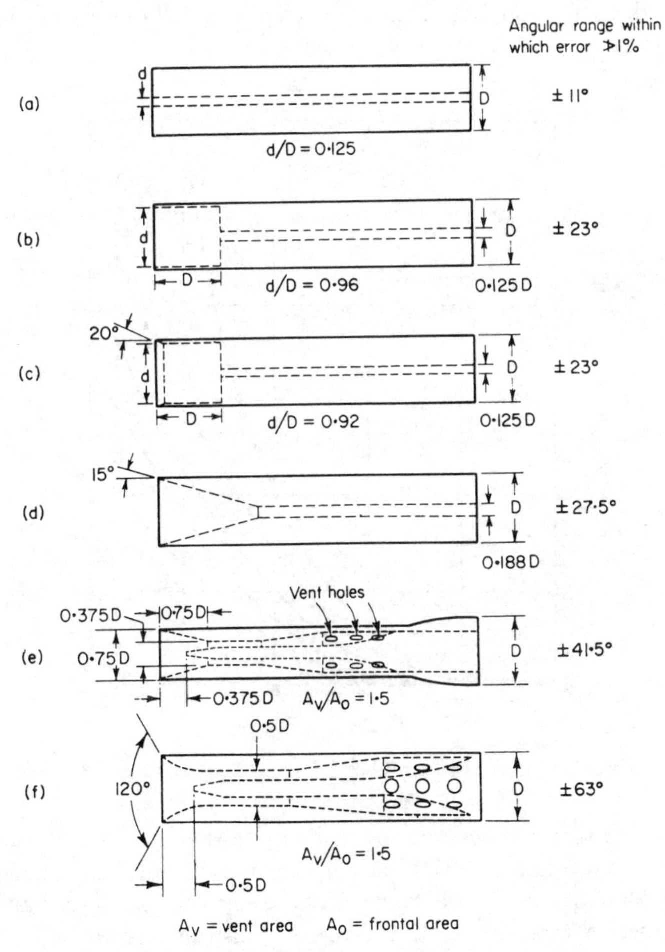

FIG. 3.18. Pitot heads of various degrees of yaw sensitivity.

The results given in Fig. 3.18 are typical of those for the numerous variants of each design that were tested. The shielded, vented type (Fig. 3.18(f)) was surprisingly insensitive to yaw, the error being less than 1 per cent of the dynamic pressure for all angles below 63°. The total vented area for this head was $1\frac{1}{2}$ times the area of the mouth of the shield.

Two rather simpler types of shielded head developed by Winternitz[28] are shown in Fig. 3.20. The one illustrated at (a) is insensitive to 45° misalignment (limiting error as before 1 per cent of $\frac{1}{2}\varrho v^2$), but rather difficult to manufacture to give consistent results from one instrument to another. The other instrument, in which the "sting" facing the airstream is replaced by a cylinder extending right across the tubular shield and having a small hole facing the airstream, is free from this objection, but the angular range in which the error is limited to 1 per cent of $\frac{1}{2}\varrho v^2$ is 28°.

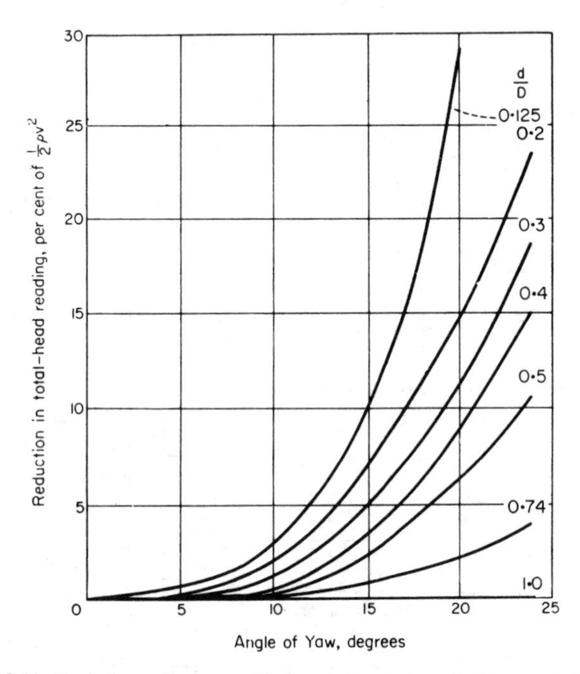

FIG. 3.19. Variation of yaw sensitivity of pitot tubes of different d/D ratios.

Instruments such as these can be made quite small. In that illustrated in Fig. 3.20(a), the length $L(=3D)$ was 7·6 mm, and the outside diameter of the sting was 0.8 mm; the outside diameter of the vertical central tube and of the facing orifice of type (b) was 0·6 mm. Such instruments are often used in investigations of the performance of turbo-machines, for which small probes of low sensitivity to direction are desirable (see also p. 37).

(b) *Static Tubes*

As already stated, static tubes are much more sensitive to angular deviations than pitot tubes. Examples in incompressible flow are shown in Fig. 3.21

which relates to the first two standard N.P.L. instruments, the tapered head of Fig. 3.3 and the hemispherical head of Fig. 3.5. A practically identical curve was obtained in similar tests with the new ellipsoidal standard of Fig. 3.8.

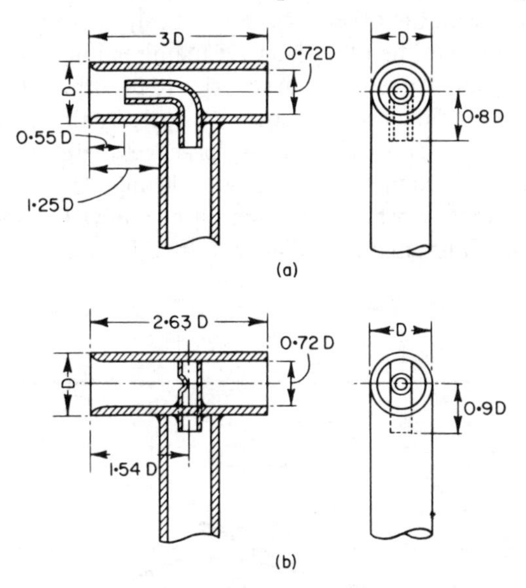

FIG. 3.20. Shielded pitot-heads.

(c) *The Pitot–Static Combination*

It may be convenient at times to know the effect of yaw on the readings of a pitot–static tube. These can, of course, be deduced if the effects on each of the components are known. The results for the three N.P.L. standards are shown in Fig. 3.22 From this diagram it is clear that, if the direction of flow is uncertain

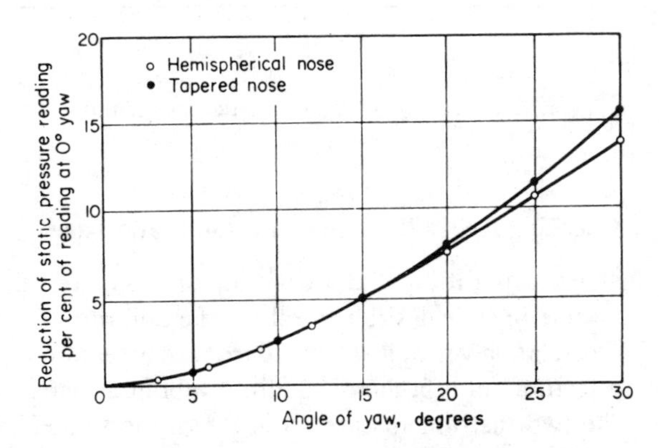

FIG. 3.21. Effect of yaw on static-tube readings.

to as much as 25° or 30°, the hemispherical nose is much to be preferred; for the maximum error that would be incurred due to misalignment within that angular range would be less than 5 per cent of $\frac{1}{2}\varrho v^2$, or about 2 per cent of v, whereas the errors with the other types could be 10 or 15 per cent of $\frac{1}{2}\varrho v^2$ at 25° and more at 30°. On the other hand, if the direction of flow is known to within 15°, the ellipsoidal nose is much superior to the other two.

It may be noted that in all three instruments correct alignment corresponds to a local minimum pressure difference. In practice, therefore, the instrument can often be correctly aligned by rotating it through small angles about an assumed direction of flow (which will generally be known to within ±20° or so) until a position of minimum differential pressure is found.

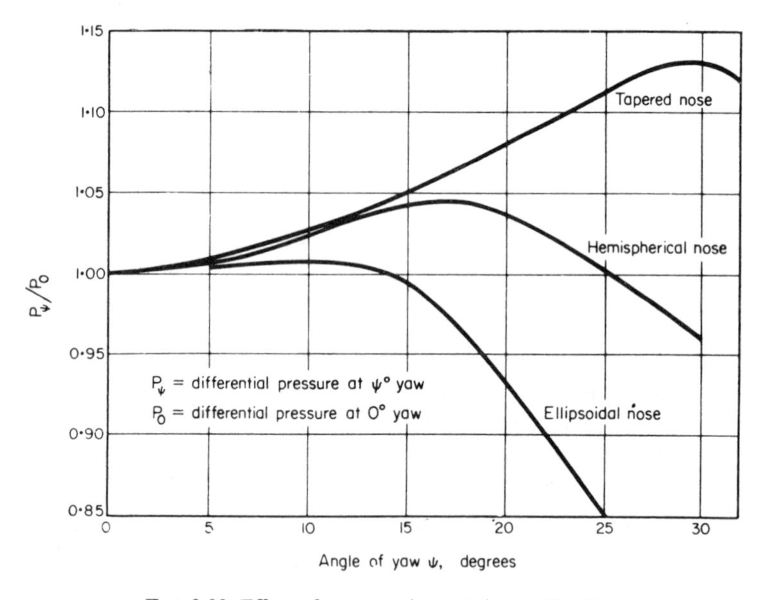

FIG. 3.22. Effect of yaw on pitot–static combinations.

Checking the Calibration of a Pitot–Static Tube

Although the basic calibration of a pitot–static tube is extremely difficult to carry out, it may sometimes be desirable to check the calibration of a particular instrument. This is relatively simple if a standard instrument and a small wind tunnel (working section about a metre square or about a metre in diameter) are available. The two instruments are mounted in the tunnel side by side, about 0·6 m apart for instruments of the usual size, and their static sides are connected to opposite sides of a sensitive manometer, which thus measures the difference between the static readings. This difference is observed for each of a number of air speeds; the two instruments are then interchanged, so that each now occupies the position previously occupied by the other, and the differences in the static readings are again observed through the same speed range. As a result of this change in position, the differences between the static readings of the two instruments at the various speeds are obtained free from the effect of any errors due to the possibility that the local static pressures in the two positions in the tunnel occupied by the instruments might have differed slightly. It is unnecessary to make a similar comparison of the total-pressure side unless it differs radically from the conventional pattern.

Effect of Vibration on the Readings of a Pitot–Static Tube

Vibration of a pitot–static tube both along the wind direction and in a transverse plane causes spurious velocity components at the measuring orifices, and so affects the pressures recorded. There is experimental evidence that, in adverse conditions, the errors due to this cause can be significant. Vibration can be aerodynamic or mechanical in origin; and the motion of the head, and hence presumably the magnitude of the errors as well, will be greatest when there is resonance between the exciting frequency and the natural frequency of the tube.

Aerodynamically excited vibrations are mainly of the transverse type, and are set up by the periodic shedding of vortices from alternate sides of the cylindrical stem of the instrument. The relation between the period of vortex shedding and the tube diameter and speed is given on p. 287. From this it follows that, for a cylinder of 8 mm diameter – the usual size of the standard pitot–static tube – the vortex frequency, which is very nearly proportional to the speed, is about 40 Hz at 1·5 m/sec and about 160 Hz at 6 m/sec. Thus, although resonance between the vortex frequency and the natural frequency of the pitot–static combination has been known to occur in water, it is unlikely to be encountered in air except at wind speeds that are in any case below the limit of useful range for this type of instrument.

Vibrations may also be transmitted to the instrument from mechanical sources, and these are probably more likely to cause trouble than those discussed in the previous paragraph, particularly if they cause the head to vibrate along the wind drection. If we assume that the vibration is simple harmonic, so that the displacement x of the head at any instant can be written $a \sin qt$ where a is the maximum amplitude of the vibration, the velocity due to the vibration is $aq \cos qt$, and the axial speed of the pitot head relative to the air is $v + aq \cos qt$ The corresponding dynamic pressure is

$$\tfrac{1}{2}\varrho(v + aq \cos qt)^2.$$

The average dynamic pressure p over a cycle is given by

$$p = \frac{q}{2\pi} \int_0^{2\pi/q} \tfrac{1}{2}\varrho(v + aq \cos qt)^2 \, dt,$$

which, on integration and insertion of limits, becomes

$$p = \tfrac{1}{2}\varrho(v^2 + a^2q^2).$$

If f is the frequency of vibration, $q = 2\pi/f$; and it follows that for high frequencies and not impossibly large amplitudes of vibration the term aq may not be negligible in relation to v, and errors on the velocity calculated from the observed value of p may be expected.† Damping and inertia in the manometer and its connexions may also affect the readings (see pp. 318–19).

If, therefore, there is any noticeable vibration of the pitot–static tube, it should either be stiffened, or – and this will probably be the more acceptable course as a rule – restrained by suitable wire or other bracing which does not seriously disturb the flow.

Determination of Flow Direction

To achieve the required accuracy in the measurement of wind speed, instruments such as the pitot–static tube have to be aligned with the local flow direction to within a few degrees.

In pipe flow it is often adequate to align the tube geometrically, and in conditions of two-dimensional flow a single rotation can suffice (about an axis perpendicular to the planes of two-dimensionality). In such cases it is often sufficient to use the instrument itself as direction indicator, e.g. by rotating a pitot–static tube until its reading is a minimum; but one or other of the more sensitive instruments mentioned below must be used if for some reason the flow direction is to be determined accurately in its own right. To specify flow direction in three-dimensional conditions it is necessary to determine two orientation angles, although in practice one of these angles is often known in advance.

Various designs of pressure-type yawmeters for measuring flow direction have been designed; some of these are shown in Fig. 3.23. Those built up from open-ended tubes are easier to make than those in which pressure holes are drilled into a body. Designs with only one pair of off-centre pressure holes (types (a), (b), (c), (d)) are intended for use in conditions of two-dimensional flow. The design shown in Fig. 3.23(b) is mounted with its axis normal to the

† An exploratory (unpublished) investigation has been made at the N.P.L. on pitot tubes undergoing forced oscillations in an airstream. Appreciable oscillation effects on the pitot reading were found; but their magnitude, and indeed their sign, varied with the type of oscillation. In general, therefore, the simple quasi-stationary theory given above is insufficient. These complications emphasize the importance of preventing probe vibration in practice.

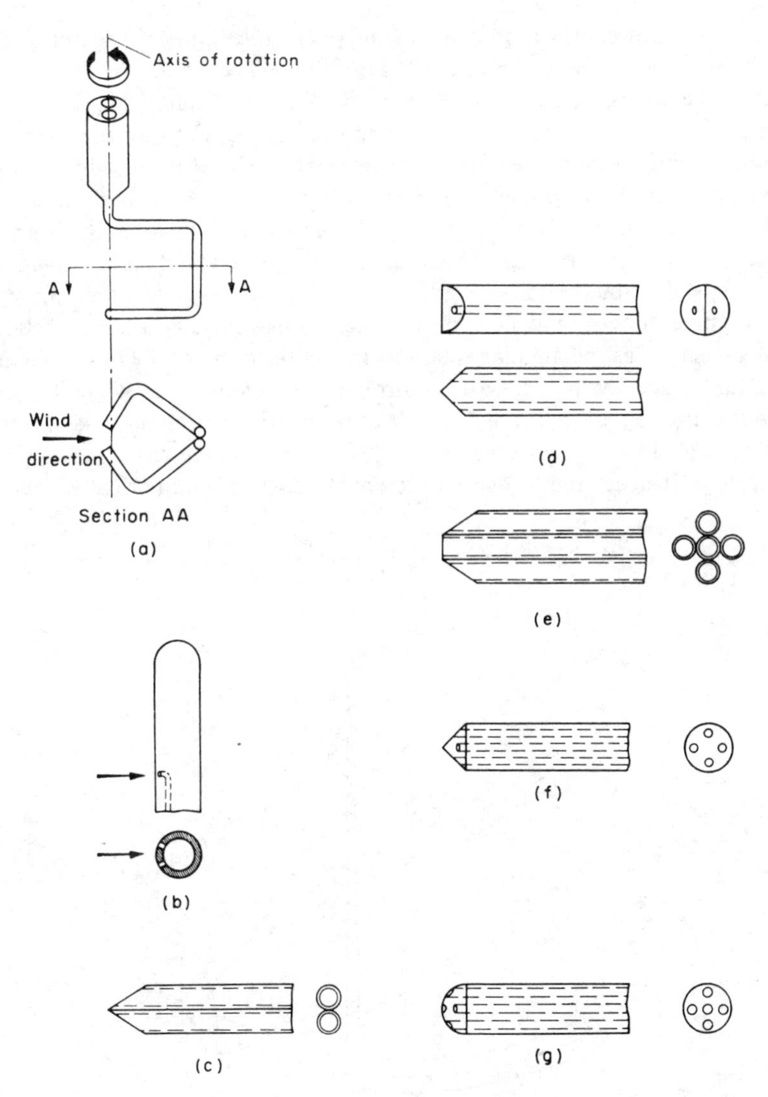

FIG. 3.23. Types of yawmeter.

planes of two-dimensionality, either cantilevered from a solid boundary or completely spanning the flow. Claw types such as (a) have now been largely superseded by the other more compact types which disturb the flow less and give a reading more nearly at a point. The overall transverse dimensions of all the instruments shown in Fig. 3.23 except type (a) can be 3 mm or even less. By adding another pair of off-centre pressure holes, as in types (e) and (f), the directions of three-dimensional flows can be determined; and the incorporation of a central pitot-orifice, as in types (e) and (g), enables an speed as well as flow direction to be measured. Pyramidal heads have been used for these three-dimensional instruments, as well as conical (f) and hemispherical (g).

Pressure-type yawmeters can be used in two distinct ways. In the alignment or null-reading method, use is made of the equality of pressure at symmetrically opposite points on a symmetrical probe: when the probe has been orientated until corresponding readings are all equal in pairs, the flow direction is simply related to the geometry of the instrument – ideally it would coincide with the probe axis. This method is especially convenient in two-dimensional flow; its use in three-dimensional flow necessitates somewhat elaborate orientation gear, especially as the angular rotations imparted to the probe in this method must take place about axes passing through the fixed point at which the local flow direction is required: the principle of one suitable mechanism is indicated in Fig. 3.24. In the second method, no attempt is made to align the probe accurately. Instead, the pressure differences between symmetrically opposite

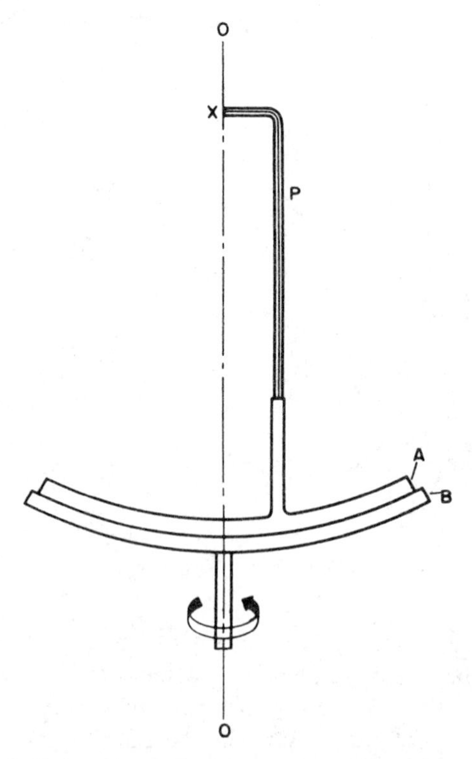

FIG. 3.24. Method of orientating three-dimensional yawmeter. A, circular-arc segment carrying probe P; B, circular-arc guide on which A slides under rack-and-pinion adjustment. Centres of A and B are at X, the mouth of the probe. The whole can be rotated about axis OO.

holes are measured (not equalized) and the flow direction is deduced from them by means of prior calibration. For convenience, the design of a probe used in this way should be such that the pressure differences are related linearly to the angles made by the probe axis to the flow direction. Further details may be found in ref. 13; and also in refs. 29 and 30, which include information about sensitivity and the effects of changes in Reynolds number and Mach number, and discuss the choice of method and procedure for particular circumstances. Usually the null-reading method is to be preferred.

Other methods for determining flow direction include the use of tufts or threads (fixed at one end) and vanes (rotating about an axis). These are subject to gravitational or inertial errors, and at best provide some sort of average indication over their length rather than a precise indication of the flow direction at a point; but they are simple to use, and adequate for many purposes.

References

1. E. Ower and F. C. Johansen, The design of pitot–static tubes, *R. & M.* 981 (1925).
2. F. H. Bramwell, E. F. Relf, and A. Fage, On the determination on the whirling arm of the pressure-velocity constant for a pitot (velocity head and static pressure) tube and on the absolute measurement of velocity in aeronautical work, *R. & M.* 71 (1912).
3. E. Ower and F. C. Johansen, On a determination of the pitot–static tube factor at low Reynolds numbers with special reference to the measurement of low air speeds, *Proc. Roy. Soc. A* **136** (1932) 153.
4. D. J. Kettle, The design of static and pitot–static tubes for subsonic speeds, *J. Roy. Aero. Soc.* **58** (1954) 835.
5. C. Salter, J. H. Warsap, and D. G. Goodman, A new design of pitot–static tube, with a discussion of pitot–static tubes and their calibration factors, *R. & M.* 3365 (1965).
6. B.S.1042:2A:1973, *Methods for the measurement of fluid flow in pipes. Part 2. Pitot tubes. Class A accuracy*, British Standards Institution, London (1973).
7. H. E. Dall, Withdrawable pitot devices for pipelines, *Instrum. Engr* **3** (1962) 135.
8. H. Peters, Einfluss der Zähigkeit bei Geschwindigkeitmessungen mit Staudruck Multiplikatoren, *Z. f. Flugtechnik u. Motorluftschiffahrt* **22** (1931) 321.
9. A. H. Glaser, The pitot cylinder as a static pressure probe in turbulent flow, *J. Scient. Instrum.* **29** (1952) 219.
10. F. A. L. Winternitz, Cantilevered pitot cylinder, *Engr* **199** (1955) 729.
11. W. Wuest, Eigenschaften von Zylindersonden zur Strömungsmessung, *Z. f. Instrumentenkunde* **71** (1963) 187.
12. H. Heenan and W. Gilbert, The design and testing of centrifugal fans, *Proc. Instn Civil Engrs* **123** (1895–6) 272.
13. D. W. Bryer, D. E. Walshe, and H. C. Garner, Pressure probes selected for mean-flow measurements. Exploration of turbulent boundary layers, *R. & M.* 3037 (1958).
14. S. Goldstein, A note on the measurement of total head and static pressure in a turbulent stream, *Proc. Roy. Soc. A* **155** (1936) 570.
15. M. Barat, Influence de la turbulence sur les prises de pression statique, *Comptes Rendus* **246** (1958) 1156.
16. A. Toomre, The effect of turbulence on static pressure measurements, Aero. Res. Council Rep. 22010 (1960) (unpublished).
17. A. Fage, On the static pressure in fully-developed turbulent flow, *Proc. Roy. Soc.* **A,155** (1936) 576.
18. P. Bradshaw and D. G. Goodman, The effect of turbulence on static pressure tubes, *R. & M.* 3527 (1968).

19. H. A. Becker and A. P. G. Brown, Response of pitot probes in turbulent streams, *J. Fluid Mech.* **62** (1) 85 (1974).
20. A. D. Young and J. N. Maas, The behaviour of a pitot tube in a transverse total-pressure gradient, *R. & M.* 1770 (1936).
21. F. A. MacMillan, Experiments on pitot tubes in shear flow, *R. & M.* 3028 (1956).
22. J. L. Livesey, The behaviour of transverse cylindrical and forward-facing total-head probes in transverse total-pressure gradients, *J. Aero. Sci.* **23** (1956) 949.
23. A. Quarmby and H. K. Das, Displacement effects on pitot tubes with rectangular mouths, *Aero. Quart.* **20** (1969) 129.
24. F. A. MacMillan, Viscous effects on flattened pitot tube at low speeds, *J. Roy. Aero. Soc.* **58** (1954) 837.
25. F. A. MacMillan, Viscous effects on pitot tubes at low speeds, *J. Roy. Aero. Soc.* **58** (1954) 570.
26. W. Gracey, Wind tunnel investigations of a number of total-pressure tubes at high angles of attack; subsonic, transonic and supersonic speeds, N.A.C.A. Tech. Rep. 1303 (1957).
27. W. B. Huston, Accuracy of airspeed measurements and flight calibration proceedings, N.A.C.A. Tech. Rep. 919 (1948).
28. F. A. L. Winternitz, Simple shielded total-pressure probes, *Aircr. Engng* **30** (1958) 313.
29. D. W. Bryer and R. C. Pankhurst, *Pressure Probe Methods for Determining Wind Speed and Flow Direction*, N.P.L. Monograph, H.M.S.O., London, 1971.
30. F. A. L. Winternitz, Probe measurements in three-dimensional flow, *Aircr. Engng* **28** (1956) 273.

CHAPTER IV

PITOT AND STATIC OBSERVATIONS IN COMPRESSIBLE FLOW

IN TRANSONIC and supersonic flow (see p. 6) the readings of pitot and static tubes are affected by shock waves formed by the measuring instruments introduced into the stream. A shock wave is a narrow zone, crossing the streamlines, through which there occur sharp increases in pressure, density, and temperature, accompanied by an abrupt degradation of energy (increase in entropy) and decrease of velocity. The component of velocity normal to the shock wave is supersonic upstream of the wave and subsonic immediately downstream. These shock waves affect the pressures acting both at the mouth of the pitot tube and at the orifice of the static tube.

Readings of Pitot Tubes in Compressible Flow

Bernoulli's general equation, equation (13) of Chapter II,† giving the relation between the static and total pressures in terms of the Mach number M, applies equally for subsonic flow ($M < 1$) and supersonic flow ($M > 1$), provided that the flow is isentropic; and this equation is used to calculate the Mach number from pitot observations in subsonic compressible flow. In supersonic flow, however, a shock wave always forms ahead of the mouth of a pitot tube, and the fluid is no longer brought to rest isentropically. In these conditions, the pitot pressure (see p. 6) is less than the total pressure which we wish to measure, and (1) cannot be applied to pitot observations.

Normal Shock Relations

The relationship between the total pressure p_0 and the pitot pressure p_0' in supersonic flow can be obtained with sufficient accuracy from the assumption

† We shall have to refer frequently to this equation in the following pages. It will therefore be convenient to repeat it here and call it equation (1) of the present chapter, viz.:

$$p_0 = p\left(1 + \frac{\gamma - 1}{2} M^2\right)^{[\gamma/(\gamma-1)]} \qquad (1)$$

that the shock wave formed ahead of the pitot tube is normal to the stagnation stream-tube; experimental evidence supporting this assumption may be found in ref. 1.

Let suffixes 1 and 2 refer to conditions immediately upstream and downstream of such a normal shock wave. The equations expressing the conservation of mass, momentum, and energy across the shock are respectively

$$\varrho_2 v_2 = \varrho_1 v_1, \tag{2}$$

$$p_2 - p_1 = \varrho_1 v_1^2 - \varrho_2 v_2^2, \tag{3}$$

and

$$\frac{v_2^2}{2} + C_p T_2 = \frac{v_1^2}{2} + C_p T_1, \tag{4}$$

where T is absolute temperature and C_p the specific heat capacity of air at constant pressure.

Since, from kinetic theory,

$$C_p = \frac{\gamma}{\gamma - 1} R,$$

and, for a perfect gas,

$$p = \varrho RT,$$

where R is the gas constant,† (4) may be written

$$\frac{v_2^2}{2} + \frac{\gamma}{\gamma - 1}\frac{p_2}{\varrho_2} = \frac{v_1^2}{2} + \frac{\gamma}{\gamma - 1}\frac{p_1}{\varrho_1}. \tag{5}$$

We thus have three equations, (2), (3), and (5), from which we can determine the three unknowns p_1, ϱ_1, and v_1 from p_2, ϱ_2, and v_2. As the algebra is tedious we shall simply quote the following results for $\gamma = 1{\cdot}4$, which is a very close approximation for air:

$$\frac{p_1}{p_2} = \frac{6}{7M_1^2 - 1} \tag{6}$$

and

$$M_2^2 = \frac{M_1^2 + 5}{7M_1^2 - 1}. \tag{7}$$

The recorded pitot pressure p_0' is the total pressure downstream of the normal shock, i.e. $(p_0)_2$. From this and the static pressure of the undisturbed flow $p(\equiv p_1)$ we determine the Mach number $M(\equiv M_1)$ as follows:

† When the so-called equation of state is expressed in the form $p = \varrho RT$, R relates to the particular gas concerned; for air, for instance, its value is approximately 287·1 J/K kg. The general form of the equation, valid for *any* perfect gas, is $p = \varrho RT/M$ where M denotes molar mass (approximately $28{\cdot}96 \times 10^{-3}$ kg/mol for air) and R is now the universal gas constant, approximately 8·314 J/K mol.

$$\frac{p}{p_0'} \equiv \frac{p_1}{(p_0)_2}$$

$$= \frac{p_1}{p_2} \frac{p_2}{(p_0)_2}$$

$$= \left(\frac{7M^2 - 1}{6}\right)^{5/2} \left(\frac{5}{6M^2}\right)^{7/2}, \tag{8}$$

using (6), (1), and (7).

If the total pressure p_0 of the undisturbed flow is required $[\equiv (p_0)_1]$, we write

$$\frac{p_0}{p_0'} \equiv \frac{(p_0)_1}{(p_0)_2}$$

$$= \frac{(p_0)_1}{p_1} \frac{p_1}{p_2} \frac{p_2}{(p_0)_2}$$

$$= \left(\frac{M_1^2 + 5}{5}\right)^{7/2} \left(\frac{6}{7M_1^2 - 1}\right) \left(\frac{5}{M_2^2 + 5}\right)^{7/2},$$

from (1) and (6). Substituting for M_2^2 from (7), we obtain finally

$$\frac{p_0}{p_0'} = \left(\frac{M^2 + 5}{6M^2}\right)^{7/2} \left(\frac{7M^2 - 1}{6}\right)^{5/2}. \tag{9}$$

To determine p_0 in these conditions, therefore, we need also to determine M: this is usually done by measuring also the static pressure p. Values of p_0'/p_0 $[\equiv (p_0)_2/(p_0)_1]$ are given in Table 4.1, from which a curve may be plotted: alternatively, use may be made of extensive tabulations available such as those of refs. 2 and 3, or of the curves already plotted in ref. 4.

Determination of M and $\frac{1}{2}\varrho v^2$ in Compressible Flow

To determine M and $\frac{1}{2}\varrho v^2$ in compressible flow, therefore, the procedure may be summarized as follows:

(1) Measure the stream static pressure, relative to some convenient datum pressure (often atmospheric).

(2) Measure the pitot pressure, similarly.

(3) Record the datum pressure(s).

(4) Express the static pressure and the pitot pressure relative to vacuum: let these absolute pressures be p and p_0' respectively. Form the ratio p/p_0'.

(5) Identify p_0' with the total pressure of the undisturbed stream (p_0) or with the total pressure downstream of a normal shock $[(p_0)_2]$ according as the flow is subsonic ($p/p_0' > 0{\cdot}528$) or supersonic ($p/p_0' < 0{\cdot}528$), and remember that in

TABLE 4.1. ISENTROPIC RELATIONS WITH $\gamma = 1{\cdot}4$
Notation: M = Mach number, p = static pressure, ϱ = density, v = velocity, T = absolute temperature. Suffix $_0$ denotes "total" conditions ($v = 0$)

M	$\frac{p}{p_0}$	$\frac{\frac{1}{2}\varrho v^2}{p_0}$	$\frac{\varrho}{\varrho_0}$	$\frac{T}{T_0}$
0	1	0	1	1
0·05	0·9983	0·0017	0·9988	0·9995
0·10	·9930	·0070	·9950	·9980
0·15	·9844	·0155	·9888	·9955
0·20	·9725	·0272	·9803	·9921
0·25	0·9575	0·0419	0·9694	0·9877
0·30	·9395	·0592	·9564	·9823
0·35	·9188	·0788	·9413	·9761
0·40	·8956	·1003	·9243	·9690
0·45	·8703	·1234	·9055	·9611
0·50	0·8430	0·1475	0·8852	0·9524
0·55	·8142	·1724	·8634	·9430
0·60	·7840	·1976	·8405	·9328
0·65	·7528	·2227	·8164	·9221
0·70	·7209	·2473	·7916	·9107
0·75	0·6886	0·2711	0·7660	0·8989
0·80	·6560	·2939	·7400	·8865
0·85	·6235	·3153	·7136	·8737
0·90	·5913	·3352	·6870	·8606
0·95	·5595	·3534	·6604	·8471
1·00	0·5283	0·3698	0·6339	0·8333
1·05	·4979	·3842	·6077	·8193
1·10	·4684	·3967	·5817	·8052
1·15	·4398	·4072	·5562	·7908
1·20	·4124	·4157	·5311	·7764
1·25	0·3861	0·4223	0·5067	0·7619
1·30	·3609	·4270	·4829	·7474
1·35	·3370	·4299	·4598	·7329
1·40	·3142	·4311	·4374	·7184
1·45	·2927	·4308	·4158	·7040
1·50	0·2724	0·4290	0·3950	0·6897
1·55	·2533	·4259	·3750	·6754
1·60	·2353	·4216	·3557	·6614
1·65	·2184	·4162	·3373	·6475
1·70	·2026	·4098	·3197	·6337
1·75	0·1878	0·4026	0·3029	0·6202
1·80	·1740	·3947	·2868	·6068
1·85	·1612	·3862	·2715	·5936
1·90	·1492	·3771	·2570	·5807
1·95	·1381	·3677	·2432	·5680
2·00	0·1278	0·3579	0·2300	0·5556

the latter case p is to be identified with the static pressure upstream of the normal shock (i.e. p_1). Hence deduce the Mach number of the undisturbed flow using (1) or (8) according as the flow is subsonic or supersonic. Curves relating M to pressure ratio may be prepared from Tables 4.1 and 4.2, or can be found already plotted in ref. 4. Alternatively, once again the tables in refs. 2 and 3 may be used.

(6) Deduce the kinetic pressure from the equation

$$\tfrac{1}{2}\varrho v^2 = \tfrac{1}{2}\varrho M^2 a^2 = \frac{\gamma}{2} M^2 p, \tag{10}$$

(since $a^2 = \gamma p/\varrho$), or from p_0 using the relation

$$\tfrac{1}{2}\varrho v^2 = \tfrac{7}{10} p_0 M^2 \left(\frac{5}{M^2 + 5}\right)^{7/2}, \tag{11}$$

(when $\gamma = 1{\cdot}4$): this follows from (1) and (10); numerical data are given in Table 4.1 and the references already cited.

TABLE 4.2. NORMAL-SHOCK RELATIONS WITH $\gamma = 1{\cdot}4$

Notation: M = Mach number, p = static pressure, p_0 = total pressure. Suffices $_1$, $_2$ refer respectively to conditions upstream and downstream of a normal shock

M_1	$\frac{p_1}{(p_0)_2}$	$\frac{(p_0)_2}{(p_0)_1}$	M_1	$\frac{p_1}{(p_0)_2}$	$\frac{(p_0)_2}{(p_0)_1}$
1·00	0·5283	1·0000	1·50	0·2930	0·9298
·05	·4980	0·9999	·55	·2773	·9132
·10	·4689	·9989	·60	·2628	·8952
·15	·4413	·9967	·65	·2493	·8760
·20	·4154	·9928	·70	·2368	·8557
1·25	0·3911	0·9871	1·75	0·2251	0·8346
·30	·3685	·9794	·80	·2142	·8127
·35	·3475	·9397	·85	·2040	·7902
·40	·3280	·9582	·90	·1945	·7674
·45	·3098	·9448	·95	·1856	·7442
			2·00	0·1773	0·7209

Yaw Effects in Compressible Flow

The marked insensitivity of the pitot-tube reading to misalignment in incompressible flow (see pp. 46–48) is retained in compressible flow right through to supersonic conditions, so that the tube can still be set with sufficient accuracy for most purposes by sighting by eye. Results for a hemispherical head[5] (Prandtl tube, Fig. 3.6) in subsonic flow are shown in Fig. 4.1.

Examples of how pitot tubes behave in supersonic flow are provided by data for the designs of pitot tube shown in Fig. 3.18 at a Mach number of 1·6. The

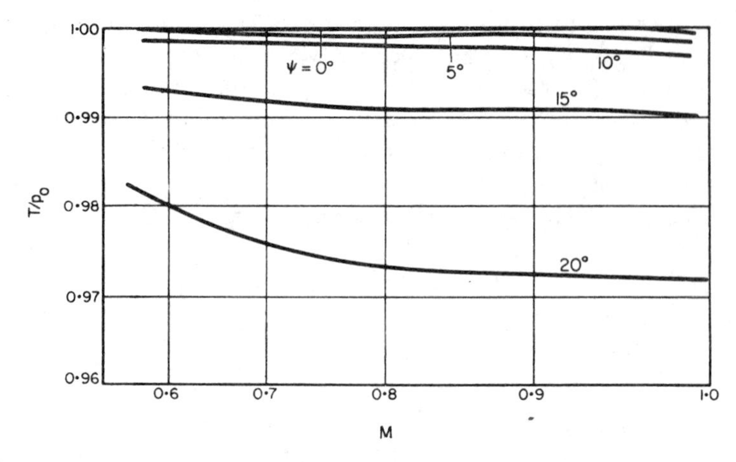

FIG. 4.1. Variation of total-pressure reading T with Mach number M and yaw angle ψ for Prandtl tube. p_0 = true total pressure.

yaw insensitivity can now conveniently be defined[(6)] as the angular misalignment at which the pitot reading has changed by 1 per cent of the difference between the pitot pressure (zero yaw) and the static pressure of the undisturbed flow. In incompressible flow this was equal to 1 per cent of the kinetic pressure of the undisturbed flow ($\frac{1}{2}\varrho v^2$); in compressible flow, however, its interpretation as an equivalent change in kinetic pressure or Mach number involves the use of the isentropic flow relations together with, in supersonic flow, the normal shock relations. With the above definition, the values of yaw insensitivity for the tube designs (a)–(f) of Fig. 3.18 at $M = 1{\cdot}6$ are: (a) $\pm 11°$; (b) $\pm 29°$; (c) $\pm 27\frac{1}{2}°$; (d) $\pm 30°$; (e) $\pm 38\frac{1}{2}°$; (f) $\pm 40°$.

Pitot Tubes in Supersonic Shear Flow

The effective displacement of the pitot-tube centre in a transverse gradient of total pressure in incompressible flow has been discussed on p. 43 on the basis of observations made in free shear layers (wakes) and in shear layers adjacent to solid surfaces (boundary layers). In compressible flow, no data appear to be available in wakes for probes which are small in comparison with the wake thickness. Measurements in (turbulent) boundary layers, however, have shown that the displacement Z is greater at supersonic speeds than in incompressible flow: when D, the external diameter of a circular probe† (or height of a flattened probe) is less than 0·14 times the boundary-layer thickness, Z is about $0{\cdot}4D$ at a free-stream Mach number of 2·0,[(7)] as compared with less than

† The ratio of internal to external diameter was 0·6.

$0{\cdot}2D$ in incompressible flow. Wall proximity reduces Z below $0{\cdot}4D$ when the probe centre is closer to the surface than D.

At a free-stream Mach number of 4·6,[8] the effective displacement was found to be far less nearly constant across the boundary layer than at $M = 2{\cdot}0$: Z/D increased with distance from the wall (or, equivalently, with the local Mach number within the boundary layer). In terms of local Mach number, Z/D was about 0·6 at $M = 2{\cdot}5$ and reached nearly 1·0 at $M = 4$.

At both free-stream Mach numbers (2·0 and 4·6), the use of too large a probe resulted in a distortion of the velocity profile (taking the form of a kink in the curve of M plotted against distance from the wall) superimposed on the displacement effect discussed above.

Readings of Static Tubes in Compressible Flow

In compressible flow a static tube aligned with the flow direction continues to record the free-stream static pressure correctly at first as the Mach number increases. Results obtained by Walchner[5] for a Prandtl tube, for example, show no appreciable error in static reading up to a Mach number of 0·8 at least (Fig. 4.2). At higher Mach numbers the recorded pressure increases steadily, probably because the effect of the pressure field of the tube support

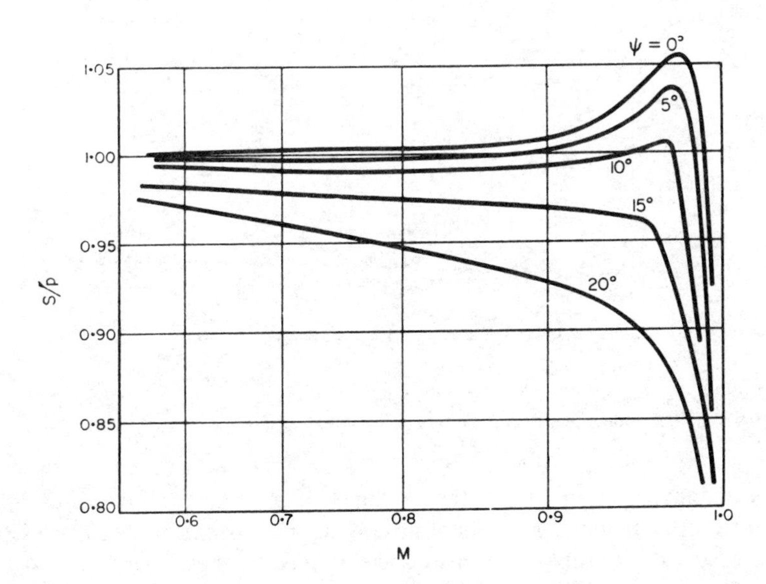

FIG. 4.2. Variation of static-pressure reading S with Mach number M and yaw angle ψ for Prandtl tube. p = true static pressure.

increases. Quite near a stream Mach number of unity, however, the reading falls steeply. This is due to the development of a local region of supersonic flow on the head of the instrument, terminating in a shock wave which moves aft and passes over the static hole as Mach number is raised: thereafter the static-pressure tapping is in the supersonic flow and the tube indicates the corresponding static pressure. A further consequence is that the pressure at the static hole can then no longer be influenced by the pressure field of the stem; hence one cannot still use the low-speed design principle of counteracting the pressure fall due to the head by the pressure rise due to the stem, and a tube designed on this principle to read correctly at low Mach numbers is bound to read erroneously in high-speed flow. These effects can be delayed to a Mach number very near unity, however, by using a finely tapered head: the conical head[9] shown in Fig. 4.3(a) and supported downstream on a long, tapered probe appears to read satisfactorily right through $M = 1$; in terms of deduced Mach number the corrections for this tube were estimated to be zero up to $M = 0{\cdot}98$, 0·002 for $0{\cdot}98 < M < 1{\cdot}01$, and 0·004 for $1{\cdot}01 < M < 1{\cdot}20$. Tests in America on a similar head[10] of larger physical size (Fig. 4.3(b)) showed that the error in static-pressure reading remained within 0·01 of the difference between the total pressure and the free-stream static pressure over the range of Mach number from 0·20 to 1·13 and for angles of incidence between $\pm 1°$, for values of x/D of both 5 and 7.

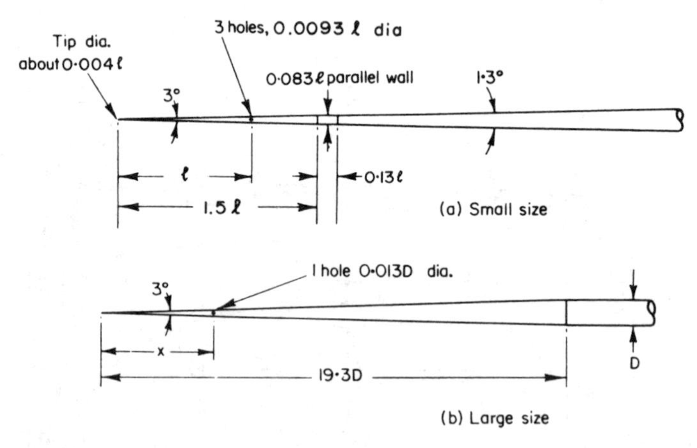

FIG. 4.3. Three-degree conical static probes. Sizes: (a) $l = 38$ mm; (b) $D = 25$ mm.

As in incompressible flow, the readings of a static tube continue to be sensitive to much smaller angles of misalignment with the flow direction than is the case for pitot tubes. This also is exemplified by the curves of Fig. 4.2. In some flow conditions it is advantageous to use a wedge-shaped head, as this is relatively insensitive to misalignment in its plane of symmetry. Calibrations for an instrument of this type over the range of Mach number 0·3–0·95 are given in ref. 11.

Since the error in the reading of a static-pressure tube in transonic flow depends on head shape and hole position as well as on Mach number and flow direction, any new design needs to be calibrated. This should be done in conditions which are closely representative of those in which the tube is subsequently to be used.

In supersonic flow, static tubes with a variety of head shapes can record the free-stream static pressure accurately provided that the static holes are sufficiently far downstream. Thus, for all the head shapes[1] shown in Fig. 4.4 the readings at $M = 1{\cdot}6$ lie within $\pm\frac{1}{2}$ per cent of the stream static if the pressure tappings are more than 10† tube diameters downstream of the shoulder. The yaw sensitivity of these heads is typically about 1 per cent reduction in measured pressure for 3°–5° of yaw. These figures relate to a probe with a sting support. If a stem support is used, this should not be closer to the pressure holes than about 13 times the tube diameter except for the square head, when a distance of 6 or 7 diameters is enough.

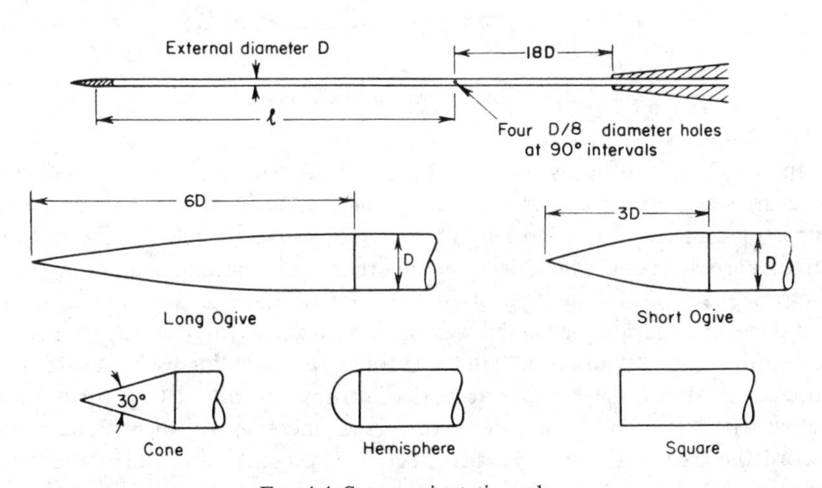

FIG. 4.4. Supersonic static probes.

A shorter probe, designed mathematically with its pressure tappings located where the local static pressure is theoretically equal to that of the undisturbed flow, is sketched in Fig. 4.5(a); calibration showed that this design[13] recorded the stream static pressure to within 3 per cent of its true value over the range of Mach number from 1·3 to 2·0, and that the error varied very little with M.

† This distance must be increased at higher Mach numbers: for a square-ended tube[12] 28 times the tube diameter is needed at $M = 2{\cdot}0$ and 22 times at $M = 2{\cdot}5$ and 3·0.

By incorporating a downstream shoulder as in Fig. 4.5(b), the speed range has been extended down to subsonic speeds: Hess, Smith, and Rivell[14] have confirmed the calibration of such instruments from $M = 0{\cdot}1$ to $M = 3{\cdot}4$.

Reduction of probe length is more commonly secured, however, by using a head shape in the form of a simple cone (usually with a 5° total included angle) with pressure tappings located conveniently near the apex; the static pressure of the undisturbed flow is then deduced from the reading by means of the known values of the pressure on an inclined surface in a supersonic stream.[4], [15–17]

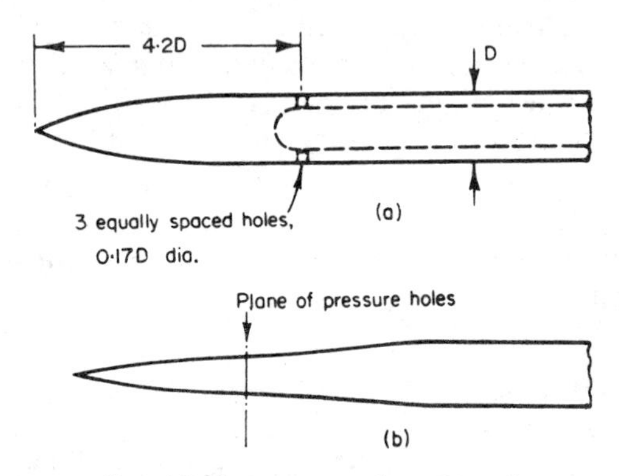

FIG. 4.5. Short supersonic static probes.

In some circumstances it might be convenient to use a probe that can measure static pressure, within an acceptable tolerance, at both subsonic and supersonic speeds. Hess and Smith[18] have shown that such a probe, with a circular cross-section, can be designed theoretically by suitable shaping in a diametrical plane, i.e. by an appropriate variation of diameter with downstream distance. A probe designed in this way, when tested in a wind-tunnel, was found to record static pressure to within 0·005 times $\frac{1}{2}\varrho v^2$ (reading high) at subsonic speeds; at $M = 1$ the reading decreased abruptly to about 0·02 times $\frac{1}{2}\varrho v^2$ below true static and then rose, with further increase in Mach number, to record the true static exactly when $M = 3$. The combination of this design principle with the use of a non-circular shape of tube cross-section[19] makes it possible to derive probe shapes that have been proved experimentally to have very little sensitivity to pitch and yaw over a range of about $\pm 5°$ (in pitch and yaw simultaneously) at $M = 0{\cdot}2$. The pitch and yaw sensitivity increased only slightly as M was increased to 3. There are, however, likely to be rather few occasions on which the advantages of such probes could be held to justify the additional effort involved in their manufacture.

Static pressure in supersonic flow has also been measured using the knife-edge flat plate shown in Fig. 4.6(a). The leading edge must be thin and sharp in order not to produce a detached shock wave, and the flat face must be aligned accurately with the flow. Further, the pressure tapping must lie upstream of the Mach waves (i.e. very weak shock waves) from the corners of the probe. This type of instrument does not greatly disturb the upstream flow, and is therefore useful for pressure explorations in the vicinity of shock waves. Misalignment errors are serious, but they can be reduced by the use of two such probes, symmetrically disposed. A hollow cylindrical design can also be used (Fig. 4.6(b)).

Static-pressure readings can be affected by variations in the diameter of the circular pressure holes and in their surface shape and cross-sectional shape. These effects also vary with Mach number, but they are usually small: some examples are quoted in ref. 10. The diameter of the circular pressure holes is typically about a twentieth of the external diameter of the tube.

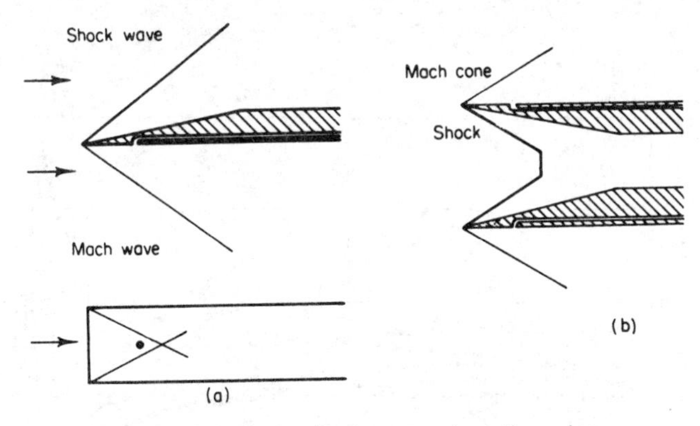

FIG. 4.6. Sharp-edged supersonic static probes.

Other precautions must be observed in making measurements of static pressure in compressible flow, in addition to avoiding unacceptably large misalignment angles and, in the transonic region, the passage of a shock wave over the pressure tapping as already described. Thus, care must be taken with regard to possible changes with Mach number of the upstream effects of the stem of the tube, even in the absence of shock waves; in supersonic flow there is the further risk that the detached bow wave of the tube support, if this is bulky or too close to the pressure tapping, may affect the static reading. Care must also be taken that shock waves from other sources do not impinge on the tube near the pressure tapping and thus affect the reading, e.g. by provoking boundary layer separation. One possible situation in which the reading is thus affected can arise from the reflection of the bow shock of the tube from a nearby solid boundary such as the wall of a duct. Nor can static pressure be measured directly by a static tube near strong shocks.

In some circumstances, difficulties of static-pressure measurement in transonic flow can be avoided by using a tube which extends upstream so that its nose is situated in a region where the flow is well subsonic.

Pitot–Static Tubes

Although separate static-pressure and total-pressure probes are often to be preferred, the combined pitot–static tube may be used in compressible flow. The necessarily blunt nose of such an instrument, however, accentuates the effects of compressibility on the readings in transonic flow. Experimental data from ref. 5 on the variation of the pitot–static difference $(T-S)$ with Mach number (and angle of yaw) for a Prandtl tube are shown in Fig. 4.7. Similar tests on a standard aircraft pitot–static tube showed that the velocity error $\delta v/v$ remained within $\pm 0{\cdot}005$ over the range of Mach number from 0·4 to 0·8, and for angles of incidence up to at least 10°; $\delta(\varrho v^2)/\varrho v^2$ remained within $\pm 0{\cdot}01$ up to 15° at $M = 0{\cdot}4$, 12° at 0·6 and 0·7, and 8° at $M = 0{\cdot}8$.

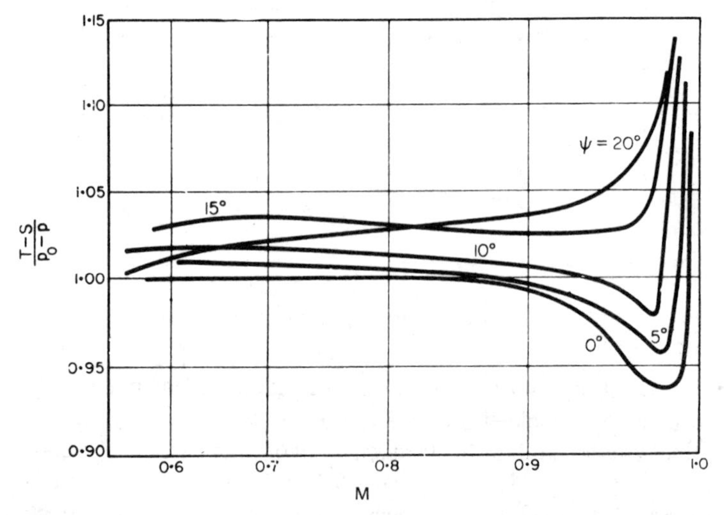

FIG. 4.7. Variation of dynamic-pressure reading T–S with Mach number M and yaw angle ψ for Prandtl tube. $p_0 - p$ = true dynamic pressure.

The errors in Mach number, as well as in kinetic pressure and flow velocity, due to given errors in static-pressure and total-pressure readings, are discussed on pp. 72–73. For a correctly aligned probe, the error due to the tube itself resides entirely in the static-pressure reading. The error in Mach number consequent upon a given error in static pressure is shown plotted in Fig. 4.8 with the pressure error expressed in the form $\delta p/(p_0 - p)$ (for subsonic flow) or $\delta p/(p_0' - p)$ (for supersonic flow) as well as in the form $\delta p/p$; curves relating errors expressed in these alternative forms are shown in Fig. 4.9.

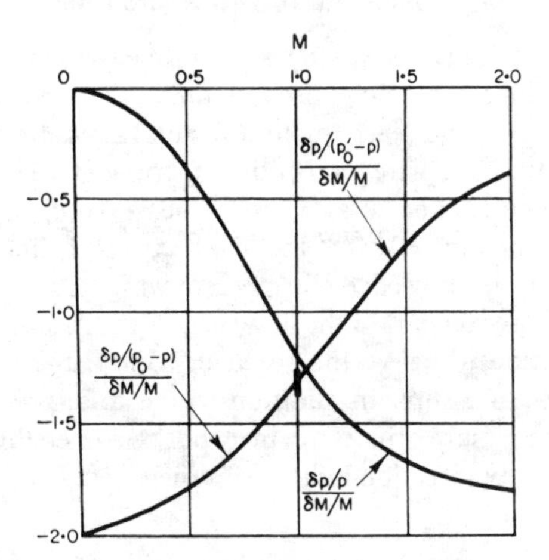

FIG. 4.8. Relationship between static-pressure error and consequent error in Mach number.

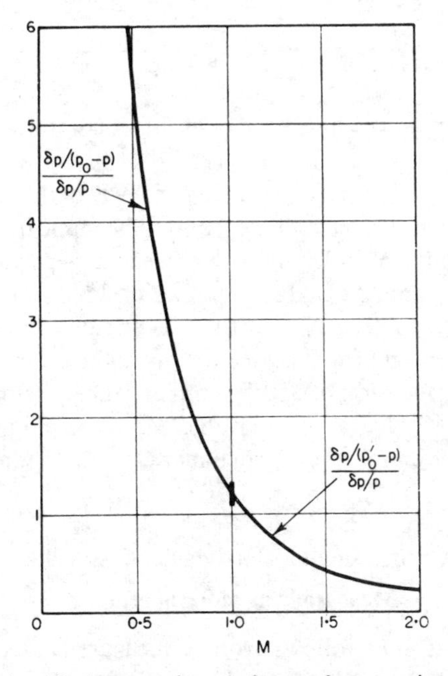

FIG. 4.9. Relationship between the two forms of error-ratio used in Fig. 4.8.

A Supersonic-flow Total-pressure Probe

As has already been explained (p. 57), a pitot tube is unable to measure total pressure directly in supersonic flow, because fluid in the stream-tube reaching the mouth of the tube suffers a reduction in total pressure as it crosses the shock wave that forms upstream of the instrument. This total-pressure reduction can be avoided, and total pressure measured directly, by means of the device sketched in Fig. 4.10.[20] In this, a curved cylinder C, whose tip resembles in shape the leading edge of a crescent wing, provides a compression surface S which imposes on part of the flow field – the "compression fan" – a shock-free deceleration to below the speed of sound: the pitot tube P then records the full total-pressure. In addition to the total-pressure sensor (the pitot tube P), static-pressure orifices are incorporated to enable the instrument to be used also as a flow-direction indicator. When tested at Mach numbers of

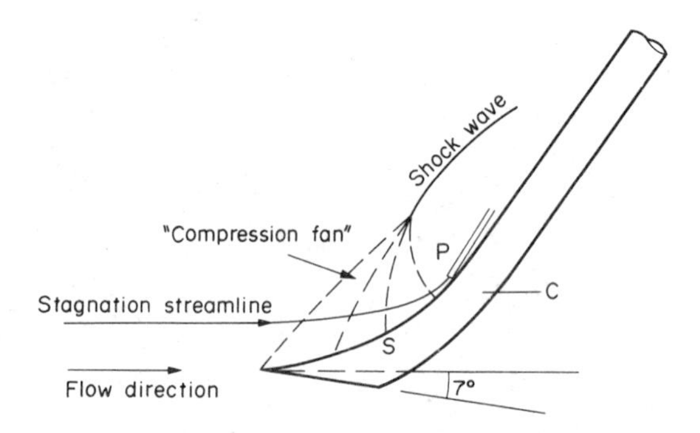

FIG. 4.10. Supersonic-flow total-pressure probe.

1·5, 1·9, and 2·1, the device was found to record 0·999 of the total pressure of the undisturbed flow when correctly aligned, and to be insensitive to misalignment over a range of about ±6° of leading-edge sweep at zero yaw and about ±8° of yaw at constant sweep. At a Mach number of 2·5, however, the maximum total-pressure recovery fell to 0·99 of the free-stream total pressure. The full capabilities and potential of this device remain to be established, particularly as regards upwards extension of Mach-number range, accurate determination of misalignment sensitivity, and achievable degree of miniaturization: in the instrument tested, the diameter of the cylinder C was 9·45 mm.

Determination of Velocity and Mass Velocity: Measurement of Temperature

We have seen that $\frac{1}{2}\varrho v^2$ follows from a measurement of the pitot–static pressure difference alone in incompressible flow, and from a pressure ratio such as p/p_0 (to determine M) together with p or p_0 in compressible flow [see (10) and (11)]. The determination of the velocity v or the so-called mass velocity ϱv (rate of mass flow per unit cross-sectional area) requires additional information: for instance, we need ϱ in order to derive v from $\frac{1}{2}\varrho v^2$. Alternatively, in compressible flow, we need the speed of sound if we choose to derive v from M. In practice, both routes require the measurement of temperature.

In incompressible flow, temperature can be measured with sufficient accuracy by means of an ordinary thermometer: the fluid density ϱ then follows from the measured temperature and the (absolute) static pressure through the equation of state for a perfect gas.

In compressible flow it is difficult to measure the temperature that would be recorded by a thermometer at rest relative to the fluid (and shielded from radiation): this is termed the "static" temperature T, by analogy with the corresponding terminology for pressures, in order to distinguish it from the more easily measured "total" or "stagnation" temperature T_0, defined as the temperature that would be attained if the fluid were brought to rest adiabatically. (In this case it is unnecessary to stipulate isentropic stagnation: "total" and "stagnation" temperatures are identical, as all the kinetic energy is converted into heat whether the stagnation process is isentropic or not.)

Instruments for determining T_0 usually take the form of shielded thermocouples as shown in Fig. 4.11. Radiation losses are reduced by the shielding, and heat-insulating material is used in the construction in order to reduce conduction losses. Vents provide a flow past the thermocouple: the shorter the time than an element of fluid spends within the probe, the lower the heat-loss effects. But this advantage has to be weighed against the fact that the temperature recovery is necessarily incomplete because the fluid is not brought fully to

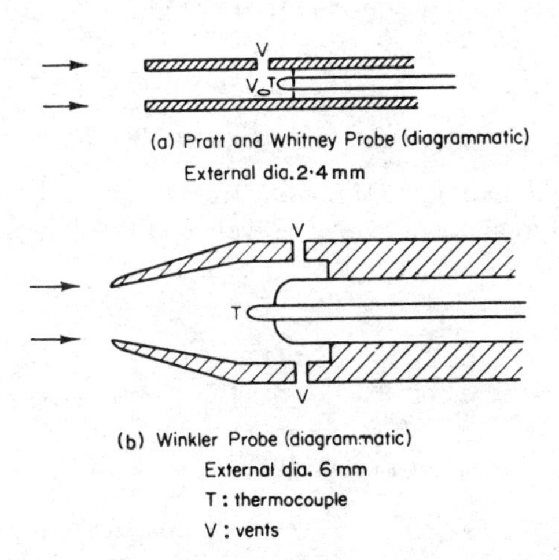

FIG. 4.11. Temperature-measuring probes for compressible flow.

rest. Calibration shows that a suitably designed instrument may record a temperature T' close to T_0, so that the *recovery factor* defined by the equation

$$r = \frac{T' - T}{T_0 - T}$$

is very close to unity. The degree of insensitivity of these instruments to misalignment with the flow direction is much the same as for pitot tubes of similar external shape.

As their performance depends on geometrical scale, flow velocities, fluid properties, and temperature ranges, shielded-thermocouple probes are best designed to suit the particular conditions in which they are to be used. An early diffuser-type design(21) gave a recovery factor very close to unity at speeds up to the speed of sound in the conditions in which it was tested; but subsequent investigators have failed to reproduce this result at reduced geometrical scale. The small Pratt and Whitney probe(22) shown in Fig. 4.11(a) was designed for supercharger testing and gave excellent results, whilst the design(23) shown in Fig. 4.11(b) has been calibrated, over a range of Reynolds number, at supersonic speeds up to $M = 7{\cdot}6$.

Having determined T_0, we can obtain the static temperature from (7) and (13) of Chapter II, which give

$$\frac{T}{T_0} = \left(1 + \frac{\gamma - 1}{2} M^2\right)^{-1}.$$

Numerical values of this function may be found in Table 4.1 (when $\gamma = 1{\cdot}4$) and in the volumes already cited. Hence, finally,

$$v = Ma = M\sqrt{(\gamma RT)}.$$

Error Equations in Compressible Flow

The error in Mach number (δM) arising from errors in the measurement of static pressure and total pressure may be evaluated as follows, when γ is equal to 1·4.

In subsonic flow we use (1) directly:

$$\frac{p}{p_0} = \left(1 + \frac{M^2}{5}\right)^{-7/2},$$

and differentiation with respect to M gives

$$\frac{\delta M}{M} = -\frac{5 + M^2}{7M^2}\left(\frac{\delta p}{p} - \frac{\delta p_0}{p_0}\right).$$

In supersonic flow, differentiation of (8) gives

$$\frac{\delta M}{M} = -\frac{7M^2 - 1}{7(2M^2 - 1)}\left(\frac{\delta p}{p} - \frac{\delta p_0'}{p_0'}\right).$$

The error in *kinetic pressure* may be obtained similarly. If this quantity is obtained by using (10), we have

$$\frac{\delta(\frac{1}{2}\varrho v^2)}{\frac{1}{2}\varrho v^2} = \frac{\delta p}{p} + 2\frac{\delta M}{M}.$$

If (11) is used instead, the error is given by

$$\frac{\delta(\frac{1}{2}\varrho v^2)}{\frac{1}{2}\varrho v^2} = \frac{\delta p_0}{p_0} - \frac{5(M^2 - 2)}{M^2 + 5}\frac{\delta M}{M}.$$

The error in *velocity* is given by

$$\frac{\delta v}{v} = \frac{\delta M}{M} + \frac{1}{2}\frac{\delta T}{T};$$

and if the temperature is obtained from an instrument whose recovery factor is r (see p. 71), we have also

$$T' = T\left(1 + \frac{\gamma - 1}{2} rM^2\right),$$

where T' denotes the measured temperature. Thus

$$\frac{\delta T}{T} = \frac{\delta T'}{T'} + \frac{\gamma - 1}{2}\frac{rM^2}{1 + \frac{\gamma - 1}{2} rM^2}\left(\frac{\delta r}{r} + 2\frac{\delta M}{M}\right).$$

When $\gamma = 1{\cdot}4$, therefore,

$$\frac{\delta T}{T} = \frac{\delta T'}{T'} + \frac{0{\cdot}2rM^2}{1 + 0{\cdot}2rM^2}\left(\frac{\delta r}{r} + 2\frac{\delta M}{M}\right).$$

References

1. D. W. Holder, R. J. North, and A. Chinneck, Experiments with static tubes in a supersonic airstream, Parts I and II, *R. & M.* 2782 (1953).
2. E. L. Houghton and A. E. Brock, *Tables for the Compressible Flow of Dry Air*, Arnold, London, 1961.
3. L. Rosenhead (Ed.), *Compressible Airflow Tables*. Clarendon Press, Oxford, 1952.
4. L. Rosenhead (Ed.), *Compressible Airflow Graphs*, Clarendon Press, Oxford, 1954.
5. O. Walchner, Über den Einfluss der Kompressibilität auf die Druckanzeige eines Prandtl-Rohres bei Strömungen mit Unterschallgeschwindigkeit, *Jb. der deutschen Luftfahrtforschung* (1938) I, 578.
6. W. Gracey, Wind tunnel investigations of a number of total-pressure tubes at high angles of attack, N.A.C.A. Tech. Rep. 1303 (1957).

7. J. M. Allen, Pitot-tube displacement in a supersonic turbulent boundary layer, NASA TN D-6759 (1972).
8. J. M. Allen, Effects of Mach number on pitot-probe displacement in a turbulent boundary layer, NASA TN D-7466 (1974).
9. E. P. Sutton, The development of slotted working-section liners for transonic operation of the R.A.E. Bedford 3-ft wind tunnel, *R & M.* 3085 (1958).
10. W. Gracey, Measurement of aircraft static pressure, N.A.C.A. Tech. Rep. 1364 (1958).
11. C. G. Gatelman and L. N. Krause, *Characteristics of a wedge with various holder configurations for static-pressure measurements in subsonic gas streams*, N.A.C.A. paper R.M. E51 G09 (1951).
12. J. F. Hahn, The effect of Mach number, incidence and hole position on the static pressure measured by a square-ended probe, *J. Roy. Aero. Soc.* **68** (1964) 54.
13. E. L. Mollo-Christensen, M. T. Landahl, and J. R. Martucelli, A short static-pressure probe independent of Mach number. *J. Aero. Sci.* **24** (1957) 625.
14. J. L. Hess, A. M. O. Smith, and T. L. Rivell, *Development of an Improved Static-Pressure Sensing Probe for all Mach Numbers: Phase I*, Douglas Aircraft Company Rep. ES40336 (1961).
15. J. L. Sims, Tables for supersonic flow around circular cones at zero angle of attack, NASA SP-3004 (1964).
16. R. S. Crabbe and G. S. Campbell, *Tables of Derivations to Extend the Zero Incidence Cone Table to Intermediate Mach Numbers and Cone Angles*, Nat. Res. Council of Canada Aero. Report LR451, NRC 9136 (1966).
17. Z. Kopal, *Tables of Supersonic Flow Around Cones*, Massachusetts Institute of Technology Tech. Report No. 1 (1947).
18. J. L. Hess and A. M. O. Smith, Static-pressure probes derived from supersonic slender-body theory, *J. Aircraft* **4**, 409 (1967).
19. A. M. O. Smith and A. B. Bauer, Static-pressure probes that are theoretically insensitive to pitch, yaw, and Mach number, *J. Fluid Mech.* **44** (3), 513 (1970).
20. M. J. Goodyer, A stagnation-pressure probe for supersonic and subsonic flow, *Aeronaut. Quart.* **25** (2), 91 (1974).
21. A. Franz, Messtechnische Fragen bei Laderuntersuchungen, *Jb. der deutschen Luftfahrtforschung* (1938) II, 215.
22. H. C. Hottel and A. Kalitinsky, Temperature measurements in high-velocity airstreams *J. App. Mechs* **12** (1945) A. 25.
23. E. M. Winkler, *Stagnation Temperature Probes for Use at High Subsonic Speeds and Elevated Temperatures*, NAVORD Rep. 3834 (1954).

CHAPTER V

THE FLOW OF AIR IN PIPES OF CIRCULAR CROSS-SECTION

MANY of the problems of air flow that concern the engineer relate to the flow in pipes, and the present chapter will be devoted to consideration of the characteristics of that type of flow. This will be useful also in disclosing many features of which account must be taken in all methods of measurement.

A large amount of work has been done in this important field, mainly in Great Britain, the United States, and Germany, with the result that the characteristics of pipe flow, at least in circular pipes, are now well established. The flow in non-circular pipes has not been so extensively studied.

Pipe flow can be regarded as a special case of boundary-layer flow. In Chapter II we saw that the boundary layer along a flat surface is at first laminar, then turbulent, and increases in thickness as the distance from the leading edge increases. Now suppose that the flat surface is rolled up to form a hollow cylinder with the air flowing along the inside. We then have a pipe, and the conditions of flow are similar in many ways to those for a surface. Near the inlet, when entry disturbances have been damped out, we have a length of streamline or laminar flow, which is succeeded by a breakdown to turbulent flow at a certain distance along the pipe.

If there is a suitably shaped flared inlet to the pipe, so that entry disturbances are reduced as far as possible, we shall find that, quite close to the inlet, the velocities are equal at all points in a transverse section practically right up to the walls – the boundary layer is just beginning to form and is very thin. At sections farther downstream, as wall friction makes its effect felt, the boundary layer becomes progressively thicker and spreads farther into the interior of the pipe, as is shown by the velocity profiles (curves showing the distribution of velocity across the section) of Fig. 5.1. Along a flat surface there is no limit to the thickness the boundary layer can develop, but in a pipe the limit is obviously half the diameter. When this thickness is reached, the whole of the flow across the section is boundary-layer flow, and we find that, if the inlet length is long enough, the velocity profile remains the same at all sections farther downstream – a most important fact from the practical point of view, to which we shall refer later. This condition is termed fully developed pipe flow.

It was Osborne Reynolds[(1)] who first studied in detail the two types of flow

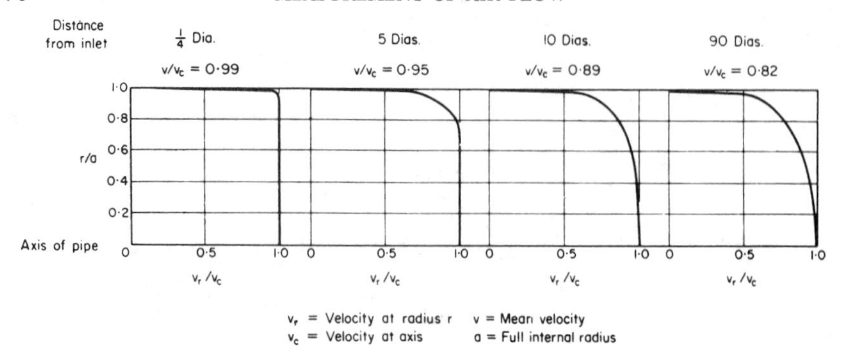

FIG. 5.1. Velocity distribution in a straight pipe at various distances from flared inlet.

(laminar and turbulent) in pipes, and showed the conditions in which each exists. His experiments were carried out with water flowing along straight horizontal glass tubes; and he was able to distinguish between laminar and turbulent flow by watching the behaviour of a small quantity of dye introduced into the inlet by a small jet on the axis. At low speeds of flow, the dye was drawn out into a steady, well-defined filament along the tube. As the speed was increased, there came a stage when the filament assumed a wavy appearance and showed clear evidence of the beginning of eddy formation in the water. With further increase of speed, these disturbances grew, until finally the filament completely lost its individuality and became diffused throughout the body of the water, in which turbulent motion had now become general. In a given size of tube and with water at a constant temperature, the speed at which eddy formation began was more or less constant; but when the experiment was reversed, the speed being gradually reduced from an initial condition of turbulent flow, laminar motion began at a speed that was much more definite and rather lower than that at which eddy motion started in the experiment with ascending speeds. The transition from turbulent to steady laminar flow in this reversed experiment was detected not by the colour method, which now could obviously not be applied, but by observation of the change in the relation between velocity and resistance to flow that occurs when the type of flow changes: for laminar flow the resistance† is proportional to the first power of the velocity; for turbulent flow approximately to the square.

The speed at which the change of flow occurs is termed the critical speed; and, from experiments with water at different temperatures and flowing at various speeds. Reynolds found that when the speed was gradually reduced, starting from turbulent flow, laminar motion always set in at a definite value of the Reynolds number vd/ν (see p. 20), where v‡ is the mean velocity in the pipe

† For an explanation of the term resistance, see p. 78.

‡ Throughout this chapter, unless otherwise stated, the velocity in a pipe is to be taken as the mean velocity, i.e. the volume of fluid passing in a given time divided by the cross-sectional area of the pipe.

and d is the diameter. We have already noted the widespread significance of the Reynolds number in fluid-motion problems. One of its chief merits, which, since Reynolds's time, has been abundantly confirmed by experiment, is that it enables laws determined from experiments with a particular fluid to be generalized for all other fluids. Thus, Reynolds found from his experiments with water that the critical velocity v_{crit} is reached when

$$\frac{vd}{\nu} = 2000,$$

so that

$$v_{\text{crit}} = \frac{2000\nu}{d}.$$

For a 2·5 cm diameter pipe, this gives the critical velocity as about 9 cm/sec for water at a temperature of 15°C. But the same expression holds for the flow of air, and indeed for any other fluid that can be assumed incompressible. Putting in the appropriate values of ν from Table 5.1,† we find that for air at atmospheric pressure and 15°C the critical velocity is $2{\cdot}9/d$ m/sec where d is in centimetres; i.e. about 10 cm/sec in a 30-cm-diameter pipe or about 30 cm/sec in a 10-cm pipe, and is proportionally less in larger pipes. From an engineering standpoint, therefore, the critical velocity for air is low; and in the great majority of cases the engineer meets he will be dealing with turbulent flow. Fortunately, this causes no serious difficulties of measurement: with all ordinary degrees of turbulence, the airstream, if generated by a reasonably

TABLE 5.1. VALUES OF μ, ϱ, AND ν FOR AIR AT 760 mm PRESSURE

Temperature (°C)	$10^5 \times \mu$ (kg/m sec)	ϱ (kg/m³)	$10^5 \times \mu/\varrho$ ($= 10^5 \times \nu$) m²/sec
0	1·72	1·292	1·33
5	1·74	1·269	1·37
10	1·76	1·247	1·42
15	1·79	1·225	1·46
20	1·82	1·204	1·51
25	1·84	1·184	1·55
30	1·86	1·164	1·60
40	1·91	1·127	1·70
60	2·00	1·060	1·88
80	2·08	1·000	2·09
100	2·17	0·946	2·30
200	2·57	0·746	3·44
300	2·93	0·616	4·75
400	3·25	0·524	6·20
500	3·54	0·457	7·78

N.B. The viscosity μ is independent of pressure except at very low and very high pressures.

† Based on the data of the *ICAO Standard Atmosphere* (see Appendix 3).

steadily running actuator, behaves, as far as the instruments for measuring average speeds and pressure are concerned, virtually as though the motion were steady. The turbulent components superimposed on the average motion are usually small enough to have little effect (see below and Chapter VI).

It should be remarked that the value 2000 ν/d for the critical velocity in a pipe is approximate only: it represents the average value under ordinary conditions. If special precautions are taken to avoid all disturbances, laminar flow may be made to persist up to much higher Reynolds numbers. We must also bear in mind that all the discussion so far relates to smooth pipes, that is, pipes for which the average surface irregularities are small in relation to the pipe diameter. Further remarks on the subject of roughness will be found later (p. 83), where an approximate quantitative definition of roughness is given.

The Resistance of a Pipe to the Motion of a Fluid Along it

There has been much discussion among engineers of the nature of the resistance of a pipe to the flow of fluid along it, and how this resistance should be measured. When a viscous fluid such as air flows along a straight length of pipe of constant diameter it loses energy owing to the effects of viscosity. If the pipe system includes lengths of different diameter, and also bends and obstructions such as valves, there will be additional energy losses. To maintain the flow against these energy losses there must be a difference between the pressures at the two ends of the system, i.e. the pressure at the upstream end must be higher than that at the outlet; and it has become customary to express the resistance of a length or system of piping in terms of the pressure drop along it. In the general case, this must be the drop in the mean total pressure for, as is shown below, the loss of total pressure is equal to the loss of energy per unit volume of the air in flowing through the system. This is a useful concept of resistance since the engineer is primarily concerned with the work the fan or other actuator has to do to force unit volume of the air along the pipe in unit time. We shall therefore use energy considerations to derive formulae for the pipe resistance.

Consider firstly the simplest (but hypothetical) case of flow along a horizontal length of parallel pipe at a speed v which is the same at all points in a cross-section. Let p be the static pressure above some chosen datum and ϱ the mass density of the air. A mass m of the air has kinetic energy $\frac{1}{2}mv^2$; and it has also potential energy by virtue of the work that has been done upon it (e.g. by the fan maintaining the motion) in raising its pressure to p. If V is the volume of the mass m of the air, this potential energy is pV, and the total energy referred to the mean level of the pipe is therefore $\frac{1}{2}mv^2 + pV$. Hence, since $m = \varrho V$ by definition,

$$\text{total energy} = V(\tfrac{1}{2}\varrho v^2 + p),$$

and

$$\text{energy per unit volume} = \tfrac{1}{2}\varrho v^2 + p.$$

It will be noted that each of the two terms in the latter expression has the dimensions of a pressure, and that their sum is, in fact, the total pressure of the fluid in incompressible flow.

It is thus apparent that the resistance offered by a length of pipe to the passage of a volume of air (or other fluid) along it may be measured in terms of the drop in total pressure experienced by the air in passing along this length; and this method of expressing the resistance is the only practicable one for general engineering use.

So far we have considered only flow in a parallel-walled straight pipe in which both the velocity and the static pressure are constant across a section, so that the mean total pressure at that section can be obtained from a single measurement at any point in its plane. Such flow, however, never occurs in practice; the velocities are never the same at all radii in a given section of a parallel pipe along which a fluid is moving steadily.† It is found that the velocity is usually a maximum at the axis of the pipe and falls away towards the walls (see Fig. 5.1). Therefore, to measure resistance in an actual case we have to determine the mean total pressure across a section where the velocity is not uniform.

Before discussing the general case we may consider fully developed flow in a straight parallel pipe, which presents no difficulties. In the first place, if the flow is parallel to the walls the static pressure across any cross-section is constant.‡ Secondly, once the flow has settled down to the normal distribution in a straight pipe (p. 76), that is at a sufficient distance from the inlet, we know that the distribution of velocity at all sections is the same. Now the mean total pressure at any section will be equal to the velocity pressure corresponding to the mean kinetic energy of unit volume of the air passing the section, together with the static pressure p at that section which, being uniform, may be obtained by a single measurement anywhere in the section. If we denote the velocity corresponding to this mean kinetic energy by $\bar{v}$, the velocity pressure is $\frac{1}{2}\varrho\bar{v}^2$. We need not concern ourselves here with the question of how $\bar{v}$ is to be measured because, as we shall show immediately, it cancels out in the final result.

Consider now a length of pipe whose end sections are denoted by A and B (Fig. 5.2). The mean total pressure at A will be greater than at B by a pressure P, which is equal to the work done on unit volume of the fluid in moving from A to B, i.e. to the resistance of the length AB of the pipe. We therefore have the equation

$$\tfrac{1}{2}\varrho_A\bar{v}_A^2 + p_A = \tfrac{1}{2}\varrho_B\bar{v}_B^2 + p_B + P. \tag{1}$$

† Steadily here is used in the sense that the motion is unaffected by external influences such as the effect of the inlet to the pipe; and that there are no appreciable fluctuations in the speed of the fan or other actuator producing the motion. It is not to be taken as implying non-turbulent motion.

‡ Except for small variations due to turbulence (see p. 109).

By assumption, the velocity distributions at A and B are identical, so that $\bar{v}_A = \bar{v}_B$. Further, we shall assume that the flow is incompressible, so that $\varrho_A = \varrho_B$. For this assumption to be valid, the static-pressure difference $p_A - p_B$ must be small compared with the absolute values of p_A and p_B, as it is in most practical cases. Equation (1) therefore reduces to

$$P = p_A - p_B. \tag{2}$$

The physical significance of (2) is that, *in the special case of flow in a straight length of parallel pipe* in which the velocity distributions at the two ends are the same, the resistance is measured by the difference in static pressure at the two ends.

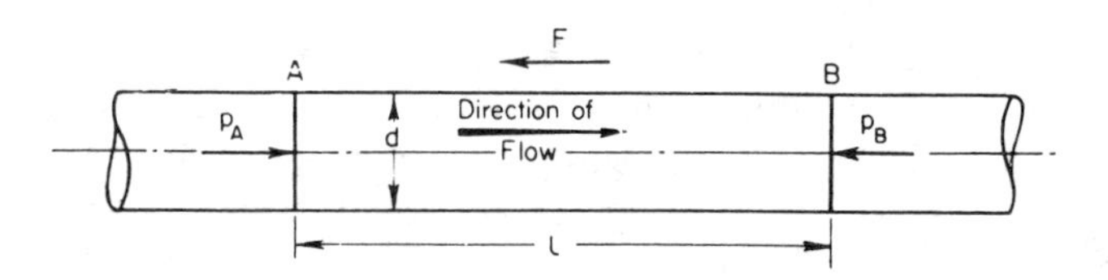

FIG. 5.2. Frictional resistance of a straight, parallel length of pipe.

The fact that this is a special case cannot be too strongly emphasized. It is the failure to realize this that has led to much of the confusion that is often associated with the measurement of pipe resistance. In the general case, one can measure resistance only by determining the loss of mean total pressure; and statements such as the one often made that resistance is measured by "a total-head gauge on the inlet side of a fan and by a static gauge on the outlet" are extremely misleading to anyone who accepts them without realizing their limitations. Further reference to this particular statement will be made later (see p. 340).

The Resistance of Smooth, Straight, Cylindrical Pipes

It has just been shown that when fully developed flow is established in a length of straight, parallel pipe, the resistance, as defined from energy considerations, can be measured in terms of the drop in static pressure instead of the drop in mean total pressure. The same conclusion follows for this particular case from what is perhaps a more natural approach to the concept of resistance, namely that it is due to frictional forces produced by viscosity. We assume that the effects of viscosity are equivalent to a shearing or frictional force at the pipe walls. Let τ_0 be the value of this force per unit area of the internal surface of the pipe, whose diameter is d (Fig. 5.2), and consider the forces acting on the cylinder of fluid within the length l of the pipe. This fluid experiences a retarding force $\tau_0 \pi dl$, equal and opposite to the axial force exerted by the fluid on the pipe due to the friction at the walls; and, to maintain the motion, the retarding force must be overcome by an equal and opposite force due to the difference in the pressures at the ends A and B. This force is $(p_A - p_B)\pi d^2/4$, and equating it to the frictional force $\tau_0 dl$ we obtain, after some rearrangement,

$$\tau_0 = (p_A - p_B)\frac{d}{4l}. \tag{3}$$

As already explained, the static-pressure drop along a length of straight parallel pipe in fully developed pipe flow is equal to the total-pressure drop and can therefore be used as a measure of the resistance. It is customary to express this resistance in terms of a non-dimensional coefficient γ which is equal to $\tau_0/\frac{1}{2}\varrho v^2$, where v is the mean velocity of flow. In terms of this coefficient, (3) becomes

$$\gamma = \frac{p_A - p_B}{\varrho v^2}\frac{d}{2l}, \tag{4}$$

from which, if γ is known, we can calculate the pressure drop for any given set of conditions.

Values of the resistance coefficient γ at various rates of flow in smooth,† straight pipes have been established largely as a result of two major experimental researches by Stanton and Pannell[2] and by Nikuradse,[3] supplemented by extensive theoretical analysis (see ref. 4). The results relate to pipe lengths sufficiently far downstream of the inlet to ensure that fully developed pipe flow has been established, so that results from one pipe will be applicable to the same type of flow along all other pipes of the same smoothness (expressed as a ratio of the size of the average surface excrescence to the pipe diameter). The requisite inlet length seems to be about 100 pipe diameters for the highest accuracy, but can be considerably shorter, say 40 diameters, without affecting the results seriously, particularly if a shaped, flared entry is fitted to reduce inlet disturbances. In general, the higher the Reynolds number R, which for pipes is defined as vd/ν, the longer the inlet length required, because the boundary layer becomes thinner in relation to the pipe diameter as the Reynolds number increases.

The results of Stanton and Pannell are of much interest and importance. They were obtained in smooth, straight pipes of drawn brass of diameters ranging from 0·3 to 12·7 cm, at rates of flow between about 0·3 and 52 m/sec, and with three different fluids – air, water, and thick oil. Values of the resistance coefficient γ as defined by (4) were calculated; and when these were plotted against the Reynolds number all the observations, irrespective of the fluid, the size of the pipe, or the speed, could be represented by the right-hand of the two

† For a discussion of what constitutes a smooth pipe, see pp. 83–85.

full curves shown in Fig. 5.3 provided that the Reynolds number was above about 3000, i.e. that the critical velocity had been exceeded, so that the flow was turbulent. In no case was the departure from the mean curve greater than could be accounted for by the limits of the experimental accuracy. Similarly, for laminar flow at Reynolds numbers below the critical, the results could again be represented by a single curve† – the left-hand curve in Fig. 5.3. The curves for the two different regimes were connected by a band of instability marking the transition from one type of flow to the other; within this band the observations were widely scattered between the two curves.

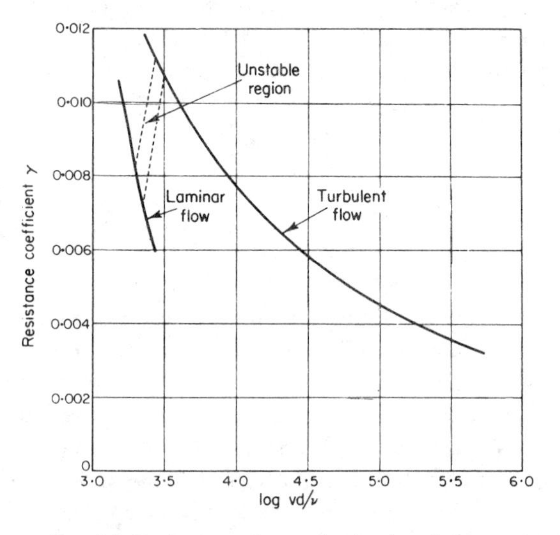

FIG. 5.3. Resistance of smooth, circular pipes.

The maximum Reynolds number reached by Stanton and Pannell was about 400,000. Nikuradse extended this to over 1,300,000, and, in their common range, the two sets of measurements of resistance are in excellent agreement. Boundary-layer theory[(4)] shows that the resistance coefficient γ and the Reynolds number R should be related by an equation of the form

$$\gamma^{-1/2} = A + B \log R\gamma^{1/2}, \tag{5}$$

and it is found that the two sets of results can indeed be represented by an equation of this form with values of the constants A and B of $-0{\cdot}40$ and $4{\cdot}00$ respectively. This equation can be used with good accuracy to calculate the resistance coefficient of smooth pipes. As it is rather troublesome to evaluate, however, a set of values is given in Table 5.2 for a range of Reynolds numbers.

† For the equation of this curve, see p. 92.

If required, values of γ at other Reynolds numbers, or vice versa, can be obtained from ref. 5.

The drop in static pressure along a smooth, straight pipe, a quantity that is often required in engineering calculations, can be obtained from the value of γ by the use of (4).

TABLE 5.2. RESISTANCE COEFFICIENT γ FOR SMOOTH, CIRCULAR PIPES CALCULATED FROM (5) WITH $A = -0{\cdot}40$ AND $B = 4{\cdot}00$

γ	$\log R$	$R \times 10^{-3}$	γ	$\log R$	$R \times 10^{-3}$
0·011	3·463	2·90	0·006	4·438	27·44
0·010	3·600	3·98	0·005	4·788	61·38
0·009	3·758	5·73	0·004	5·252	178·5
0·008	3·944	8·78	0·003	5·926	842·9
0·007	4·168	14·71	0·0026	6·295	1972

It may be noted here that to calculate the resistance of pipes in which the rate of flow is so high (but still subsonic) that compressibility effects have to be taken into account, the values of γ given by Fig. 5.3 are still valid,[5] but the equations for calculating the pressure drop from these values of γ are more complicated. Full details of the method of calculation for compressible subsonic flow will be found in refs. 6 and 7. In the present chapter only incompressible flow is considered.

Rough Pipes

The results discussed above are applicable strictly only to smooth-walled pipes. It is not easy to define smoothness; later work has shown that what can be considered smooth for a certain Reynolds number may be rough at higher Reynolds numbers. Stanton and Pannell used smooth, drawn brass pipes; but their results can probably be applied as a basis for resistance estimations in practice to all pipes that feel smooth to the touch, provided, of course, that there is an entry length of at least 40 diameters upstream of the length of pipe under consideration (see also p. 81).

Some experiments carried out by Nikuradse[8] in the turbulent range, with artificial roughness formed by closely spaced grains of sand stuck to the inside of the pipe, show the effect of different degrees of roughness; and give an indication of the limiting size of excrescence that can be permitted at different Reynolds numbers without causing a "smooth" pipe to become "rough". The effect of roughness is to cause the descending curve for turbulent flow in Fig. 5.3 to rise and approach a horizontal line, showing that the resistance tends to vary as the square of the speed. Nikuradse's results for different degrees of roughness formed the family of curves shown in Fig. 5.4. The values of the resistance all fell on the curve for smooth pipes at the lower Reynolds numbers, but departed from it as the rate of flow increased; the greater the roughness,† the earlier the departure from the smooth-pipe law and the higher the ultimate value of the resistance.

For the type of roughness used by Nikuradse, his results showed that a pipe can be regarded as smooth provided that the quantity $\varepsilon\sqrt{(\tau_0/\varrho)}/\nu$, i.e. $v\varepsilon\sqrt{(\gamma/2)}/\nu$, is less than 4. Thus $v\varepsilon\sqrt{(\gamma/2)}/\nu = 4$ represents the condition at which the curves of Fig. 5.4 begin to rise above the smooth-pipe value of the resistance. Nikuradse also found that they become horizontal when $v\varepsilon\sqrt{(\gamma/2)}/\nu$ has a value of about 100, and remain so for all higher values. The resistance then varies as the square of the speed, its value at any speed depending on the degree of roughness, i.e. the size of the excrescences.

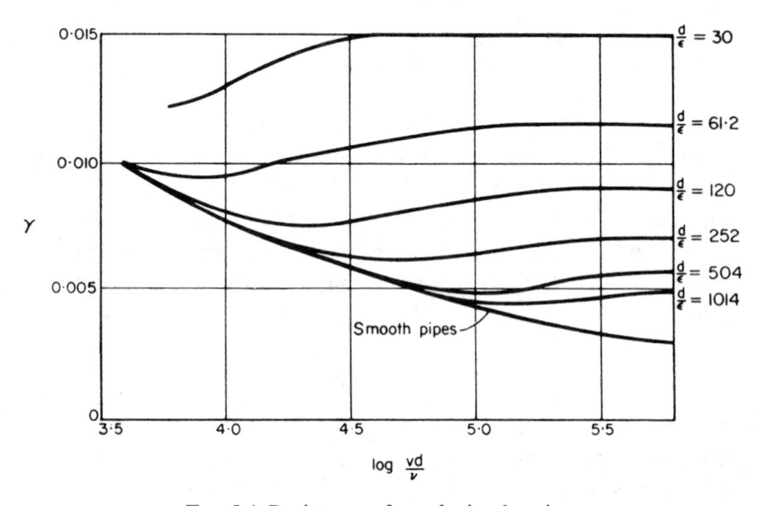

FIG. 5.4. Resistance of rough circular pipes.

The limiting height of excrescences of this type beyond which the pipe ceases to behave as smooth is obtained from the equation

$$\frac{v\varepsilon\sqrt{(\gamma/2)}}{\nu} = 4,$$

from which

$$\varepsilon = 4\nu v\sqrt{\left(\frac{\gamma}{2}\right)}.$$

Hence we can calculate the limiting value of ε for any value of the Reynolds number vd/ν by using (5) or Table 5.2 to get the appropriate value of γ. Thus we find that, for air at ordinary temperatures and pressures, the limiting size of

† Defined by the ratio ε/d, where ε denotes the average height of the sand grains.

excrescence for a 30-cm diameter pipe is about 0·4 mm at a mean rate of flow of about 3 m/sec, and about 0·05 mm when the mean rate of flow is 30 m/sec. Obviously, in smaller pipes the maximum allowable size of excrescence is less, and vice versa.

It should be noted that these remarks apply only to roughnesses consisting of closely spaced excrescences; the effects of a few isolated excrescences would, of course, be much less.

Experiments by Colebrook and White[9] have provided data on the resistance coefficients of commercial pipes. For smooth pipes, they gave (5) in the form

$$\gamma^{-1/2} = 4 \log \frac{R\gamma^{1/2}}{1{\cdot}255} \tag{6}$$

and their equation for rough pipes is

$$\gamma^{-1/2} = 4 \log 3{\cdot}7 \frac{d}{\varepsilon}, \tag{7}$$

where ε, as before, is the height of the average roughness.

In ordinary practice, pipes are rarely as smooth as those used by Stanton and Pannell or Nikuradse; and, according to Colebrook,[10] the following equation enables the resistance coefficient to be calculated with sufficient accuracy:

$$\gamma^{-1/2} = -4 \log\left(\frac{1{\cdot}255}{R\gamma^{1/2}} + \frac{\varepsilon}{3{\cdot}7d}\right). \tag{8}$$

It will be seen that this expression tends towards (6) when $\varepsilon \to 0$, i.e. for smooth pipes, and towards (7) at high values of R when, as already stated, small excrescences produce relatively larger effects. The values given in Table 5.3 of ε are from ref. 9.

TABLE 5.3. AVERAGE HEIGHT OF EXCRESCENCES IN COMMERCIAL PIPES

Material	ε (mm)
Non-ferrous drawn tubing, including plastics	0·0015
Black steel pipes	0·045
Aluminium pipes	0·05
Galvanized steel pipes	0·15
Cast-iron pipes	0·20

Values for ε for a number of commonly used surfaces are also given in ref. 5.

Velocity Distribution Across a Section of a Smooth Circular Pipe

As already stated on p. 75, when fully developed pipe flow has been established at a distance sufficiently far from the mouth for inlet disturbances to have died out, the velocity profile is the same at all sections farther downstream, and changes only if the rate of flow changes. More generally, we may say that the velocity distribution in fully developed pipe flow depends only on the Reynolds number, i.e. if we restrict ourselves to a particular fluid at constant temperature and pressure, only on the product vd. For example, if the velocity at a certain radius r of a pipe along which air is flowing is, say, half the velocity at the axis, then it will be half this velocity at the same radius at all sections farther downstream. Moreover, for air at the same temperature and pressure, the velocity in a pipe double the diameter of the first will be half the velocity at the axis at the same fraction of the full pipe radius as in the first case, *provided that the value of vd is the same, i.e. that the mean velocity is half that of the first case.*

This statement, which can be predicted from theoretical considerations, was confirmed by Stanton[11] by some experiments in smooth pipes in fully developed turbulent flow. His curve of velocity distribution for a Reynolds number of $5{\cdot}2 \times 10^4$ is shown in Fig. 5.5, which is based on observations in two pipes, one of 4·93-cm diameter and the other 7·4-cm, at the appropriate rates of flow. The velocity distribution is approximately parabolic from the axis outwards to a radius of about $0{\cdot}8a$, a being the full pipe radius. Thus if v_r is the velocity at a radius r, where $r < 0{\cdot}8a$, we can calculate v_r from the equation

$$v_r = v_c\left(1 - C\frac{r^2}{a^2}\right), \tag{9}$$

where v_c is the velocity at the axis and C is some non-dimensional constant.

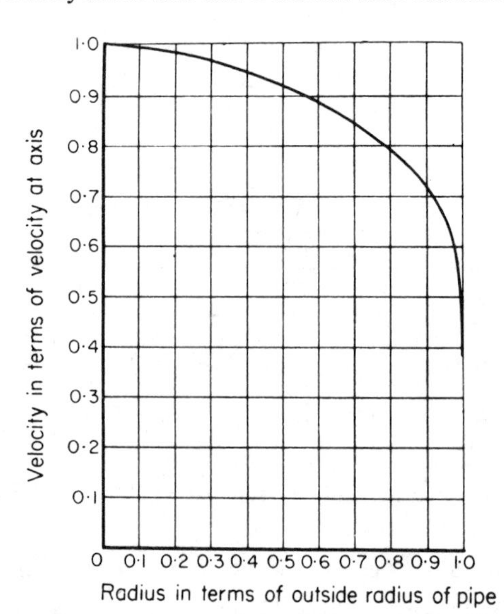

FIG. 5.5. Velocity distribution across a section of a smooth, circular pipe (Reynolds number $5{\cdot}2 \times 10^4$).

For the Reynolds number quoted, viz. $5{\cdot}2 \times 10^4$, Stanton's results show that C is equal to about 0·35. Although C will vary with the Reynolds number, the variation is not great except at low Reynolds numbers; Stanton found no distinguishable difference in the value of C from another set of experiments made at a Reynolds number of $1{\cdot}1 \times 10^5$.

Since this work by Stanton, however, many more measurements of the velocity distribution in straight pipes have been made; and analysis[12] shows that v may be expressed in terms of the axial velocity v_c, over a wide range of Reynolds number, by the power law

$$\frac{v_r}{v_c} = \left(1 - \frac{r}{a}\right)^m. \tag{10}$$

The value of m varies with Reynolds number: for turbulent flow the data given by Nikuradse[3] lead to the values shown in Table 5.4. It should be noted, however, that (10) cannot be strictly valid close to the pipe axis ($r = 0$), because the velocity profile it predicts is not flat in this region as it must be in practice if the flow is symmetrical. The same remark applies to the more accurate formulation

$$\frac{v}{v_*} = \alpha + \beta \log \frac{v_* y}{\nu},$$

where y denotes distance from the wall and

$$v_* = \sqrt{(\tau_0/\varrho)}$$

(known as the "friction velocity"). Again, in the very thin "laminar sub-layer" adjacent to the wall, v increases linearly with y. For flowrate determinations,

TABLE 5.4. DATA BASED ON RESULTS OF NIKURADSE[3]

vd/ν	m	$v_c d/\nu$
10^4	0·168	$1{\cdot}27 \times 10^4$
2×10^4	0·156	$2{\cdot}49 \times 10^4$
5×10^4	0·141	$6{\cdot}10 \times 10^4$
10^5	0·130	$1{\cdot}20 \times 10^5$
2×10^5	0·120	$2{\cdot}37 \times 10^5$
5×10^5	0·109	$5{\cdot}84 \times 10^5$
10^6	0·101	$1{\cdot}16 \times 10^6$

extensive use has been made of the so-called "log-linear" velocity distribution given by an equation of the form

$$v = A + B \log \frac{y}{d} + C\frac{y}{d}$$

as is described on p. 115. Nevertheless, the power-law approximation is convenient and adequate for many engineering purposes.

Ratio of Mean to Axial Velocity

It is often convenient in practice to determine the mean rate of flow in a smooth pipe from a single measurement of the velocity at the axis.† Since there is a characteristic velocity distribution for every Reynolds number, it follows that there must be a relation between v_c and the mean velocity v which depends only on the Reynolds number, provided that we are working sufficiently far downstream for normal pipe flow, unaffected by inlet disturbances, to have become established. In 1914, Stanton and Pannell[(2)] determined the ratio v/v_c for smooth pipes for Reynolds numbers between about 2400 and 100,000, and their results are given by the lower curve in Fig. 5.6. At the left, the Reynolds numbers are approaching that for the critical velocity at which the flow becomes laminar; and we see that the value of v/v_c is tending towards $\frac{1}{2}$, which is its theoretical value for laminar flow (see p. 91). Many years afterwards, Nikuradse[(3)] published the results of similar measurements, also in smooth pipes, in which he extended the range of Reynolds numbers to considerably higher values. His results are shown in the upper curve of Fig. 5.6, and will be seen to be consistently higher than those of Stanton and Pannell, although, where they overlap, the difference is small – about 1·5 per cent – except near the critical régime. The reason for this discrepancy has not been established; Nikuradse made no attempt to explain it in his paper. The value of v/v_c given by the power-law approximation of (10) is $2/(1 + m)(2 + m)$.

Because of this uncertainty about the true value of v/v_c, the curve given in Fig. 5.6 cannot be used in practice, if accuracy better than 98 per cent is required, in order to estimate the mean rate of flow from a single reading at the axis. Anyone wishing to use this convenient method – practical details will be found in Chapter XIII – should make his own initial calibration, i.e. construct his own curve of v/v_c against rate of flow for subsequent use; this will in any event be desirable in most practical cases, as there will generally be some uncertainty about possible departures from fully developed pipe-flow conditions at the section of measurement due to inlet disturbances or other causes, particularly axial asymmetry and swirl.

Velocity Distribution in Rough Pipes

In rough pipes, for which the frictional resistance varied as the square of the velocity, Stanton[(11)] found that the velocity distribution was parabolic up to a radius very nearly equal to the full radius of the pipe, and that this distribution did not, as it did in the smooth pipes, vary with the Reynolds number. The ratio v/v_c for these rough pipes had the value of about 0·76. It should be noted that the pipes in question were artificially roughened by cutting double screwthreads

† See Chapter XI.

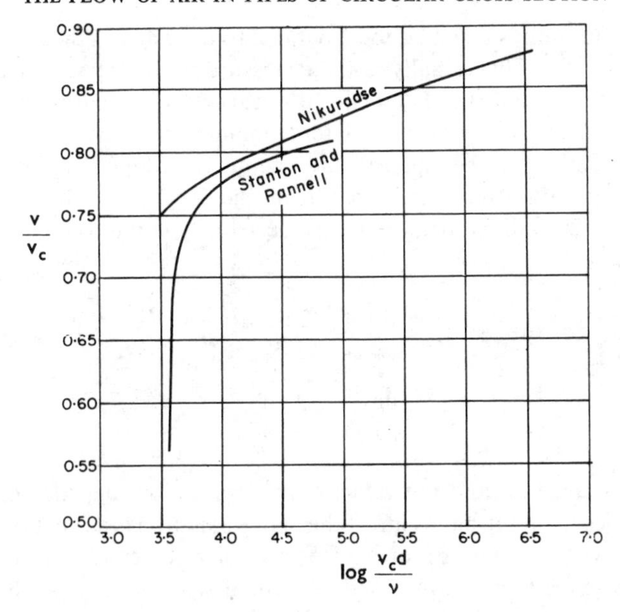

FIG. 5.6. Relation between mean and axial velocity in smooth, circular pipes.

on the internal surfaces, in order that the frictional resistance should vary as the square of the speed – a point that was verified experimentally. The pipes were therefore much rougher than those commonly used in practice.

Streamline or Laminar Flow in Pipes

Let us imagine a fluid flowing steadily along a straight, parallel pipe under the influence of viscous forces and pressure differences only, and let the flow be the same across all sections. Acting on any cylindrical element AB (Fig. 5.7) of the fluid co-axial with the pipe there will be a viscous drag which opposes the motion; and for the motion to proceed steadily this drag must be balanced by the difference in the pressures acting on the two ends A and B.

Let the length of the cylindrical element be $\mathrm{d}l$ and its radius r, and let a be the radius of the pipe.

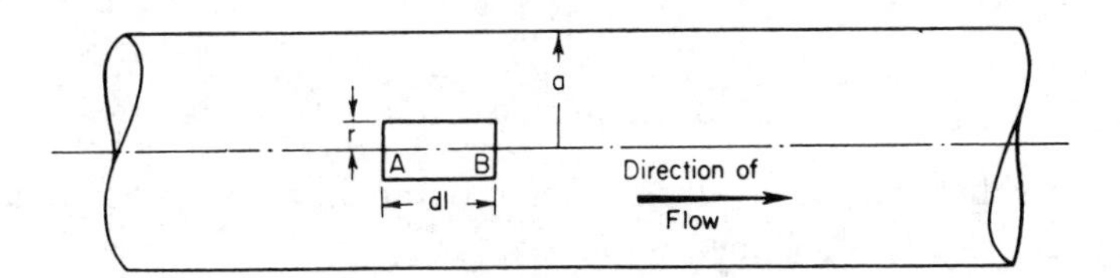

FIG. 5.7. Laminar flow in a pipe.

Since the motion is assumed to be parallel to the walls of the pipe, there are no viscous forces acting radially, and the pressure across any section perpendicular to the axis is uniform. Let p and $(p - \mathrm{d}p)$ be the pressures on the faces A and B respectively. The pressure difference maintaining the motion is therefore $\pi r^2\,\mathrm{d}p$; and this must be equal to the viscous retarding force acting on the curved surface of the element, which, from the definition of the coefficient of viscosity μ ,is $2\pi r\mu\,\mathrm{d}l\,\mathrm{d}v_r/\mathrm{d}r$, where v_r is the velocity at radius r.

Thus

$$\pi r^2\,\mathrm{d}p = -2\pi r\mu\,\mathrm{d}l\frac{\mathrm{d}v_r}{\mathrm{d}r},$$

or

$$\frac{\mathrm{d}p}{\mathrm{d}l} = -\frac{2\mu}{r}\frac{\mathrm{d}v_r}{\mathrm{d}r}.$$

Now μ is constant and the velocity, by assumption, depends only on the radius. Hence the right-hand side of the above equation is a function of r only; and it follows that $\mathrm{d}p/\mathrm{d}l$ is independent of l for given conditions of flow. We may therefore write $(p_1 - p_2)/l$ for $\mathrm{d}p/\mathrm{d}l$, where p_1 and p_2 are now the pressures at two sections a finite distance l apart; i.e.

$$\frac{p_1 - p_2}{l}\,r\,\mathrm{d}r = -2\mu\,\mathrm{d}v_r; \tag{11}$$

and, on integration between $r = 0$ and $r = a$, we have

$$\frac{p_1 - p_2}{l}\frac{a^2}{2} = -2\mu(v_a - v_c), \tag{12}$$

where v_c is the velocity at the axis.

If we assume that the velocity is zero at the walls,

$$v_a = 0$$

and

$$\frac{p_1 - p_2}{l} = \frac{4\mu v_c}{a^2}. \tag{13}$$

The velocity at any radius r is obtained by integrating (11) between the limits 0 and r and, after substitution of the value of v_c given by (13), is found to be

$$v_r = \frac{p_1 - p_2}{4\mu l}(a^2 - r^2); \tag{14}$$

which shows that the velocity distribution is parabolic across the section.

The mean velocity is obtained as follows:

Imagine an annulus of the pipe at a radius r and of width $\mathrm{d}r$; the volume of fluid flowing through the annulus is given by $2\pi r\, v_r\, dr$, where v_r is obtained from (14). The total volume flowing is therefore

$$V = \int_0^a 2\pi r v_r \, dr = \frac{\pi(p_1 - p_2)}{2\mu l} \int_0^a r(a^2 - r^2) \, dr$$

$$= \frac{\pi(p_1 - p_2)}{2\mu l} \left(\frac{r^2 a^2}{2} - \frac{r^4}{4} \right)_0^a$$

$$= \frac{\pi(p_1 - p_2) a^4}{8\mu l}. \qquad (15)$$

Now the mass of fluid passing in unit time is obviously constant; and if the pressure drop is small compared with either p_1 or p_2 (which are usually approximately atmospheric) we may neglect changes of density even in a gas and regard the volume flowing in unit time as constant. For liquids, which are practically incompressible, this will be so even if the pressure drop is large. In either case, therefore, the volume flowing in unit time may be taken as constant and equal to $\pi a^2 v$, where v is the mean velocity of flow.

Hence (15) may be written

$$\pi a^2 v = \frac{\pi a^4 (p_1 - p_2)}{8\mu l},$$

from which

$$v = \frac{p_1 - p_2}{8\mu l} a^2; \qquad (16)$$

i.e. in terms of the pipe diameter d

$$v = \frac{p_1 - p_2}{32\mu l} d^2. \qquad (17)$$

From (13),

$$v_c = \frac{p_1 - p_2}{16\mu l} d^2. \qquad (18)$$

In laminar flow, therefore, the mean velocity of flow is half the velocity at the axis.

The resistance coefficient in laminar flow is derived as follows:

This coefficient has already been defined as $\tau_0 / \frac{1}{2}\varrho v^2$, where τ_0 is the retarding force per unit area of pipe wall.

From (3)

$$\frac{p_1 - p_2}{l} = \frac{4}{d} \tau_0, \qquad (19)$$

and from (17)

$$\frac{p_1 - p_2}{l} = \frac{32\mu v}{d^2}. \qquad (20)$$

Hence, combining (19) and (20), we obtain

$$\tau_0 = \frac{8\mu v}{d},$$

and the resistance coefficient $$\frac{\tau_0}{\frac{1}{2}\varrho v^2} = \frac{16\mu}{\varrho v d} = \frac{16}{R}, \tag{21}$$

where R is the Reynolds number.

Numerous experimental checks of the equation have been made (see ref. 12), and have shown that the law is closely followed in laminar flow. It is the equation of the left-hand curve in Fig. 5.3.

The Resistance of a Pipe of Varying Cross-section

It has been shown that the resistance of any given length of pipe is equal to the loss in energy suffered by unit volume of the fluid in passing along this length. When the pipe is parallel between the two sections considered and the velocity distributions at these sections are the same, we have seen that the resistance is given by the difference between the static pressures at the sections; but if the diameters of the two ends of the length of pipe are not the same, the resistance is measured by the difference between the mean energy of unit volume of the fluid at the two sections. The same is true for a parallel pipe in which the velocity distributions at the two ends are not identical. In the general case we have therefore to measure the mean energy per unit volume, i.e. the mean total pressure, at the two sections, and to subtract one from the other. This may be done as follows, on the assumption that the sections at which measurements are made are in lengths of parallel pipe (but not necessarily of the same diameter); i.e. measurements are made in regions of axial flow. Otherwise the problem becomes more complicated because of the need to allow for the effects of flow direction on the rate of transport of axial momentum and for the variation of static pressure over the duct cross-section.

Consider any section of the pipe of radius a, and let v be the mean velocity at radius r through an annular element of width dr. Also let p be the static pressure at this radius referred to some convenient datum, usually atmospheric pressure. The volume of fluid passing through the annulus in unit time is $2\pi r\, \mathrm{d}r v$; its mass is therefore $2\pi r\, \mathrm{d}r \varrho v$, its kinetic energy $2\pi r\, \mathrm{d}r \varrho v v^2/2$, and its potential energy $2\pi r\, \mathrm{d}r v p$. The total energy of the fluid flowing across the section per second is therefore

$$\pi\varrho \int_0^a v^3 r\, \mathrm{d}r + 2\pi \int_0^a p v r\, \mathrm{d}r.$$

Also the total volume flowing per second is $2\pi \int_0^a vr\,\mathrm{d}r$, so that energy per unit volume, or the mean total pressure, is equal to

$$\tfrac{1}{2}\varrho \frac{\int_0^a v^3 r\,\mathrm{d}r}{\int_0^a vr\,\mathrm{d}r} + \frac{\int_0^a pvr\,\mathrm{d}r}{\int_0^a vr\,\mathrm{d}r}.$$

Therefore, to obtain the mean total pressure at a section, the values of the above integrals must be determined by taking sufficient measurements of v and p across the section. Thus, to determine, for example, $\int_0^a pvr\,\mathrm{d}r$, p and v must be measured at a number of values of r; a curve of the product pvr is then plotted on a base of r, and its area gives the value of the integral. The total pressures at each of the two sections forming the ends of the length of piping whose resistance is required must be obtained in this way, and their difference will be the resistance.

Usually it will be found that, although the pipe diameters at the two sections may be different, the sections themselves nevertheless form parts of parallel-walled pipes, so that the flow is axial and p may be taken as constant across the section. The integral $\int_0^a pvr\,\mathrm{d}r$ is then equivalent to $p\int_0^a vr\,\mathrm{d}r$, so that the expression for the mean total pressure at a section reduces to

$$\tfrac{1}{2}\varrho \frac{\int_0^a v^3 r\,\mathrm{d}r}{\int_0^a vr\,\mathrm{d}r} + p. \tag{22}$$

It may be noted that if v were constant across the section it could be taken outside the integrals in which it occurs, and the expression would reduce to its familiar form of $\frac{1}{2}\varrho v^2 + p$.

Now in fan testing and similar work, resistance has frequently to be determined in addition to the volume of air flowing per second. The latter is deduced by measuring the mean velocity v_m across the section by one of the methods described in subsequent chapters. It is then often assumed that the mean total pressure at the section, for the purpose of computing resistance, is equal to $\frac{1}{2}\varrho v_m^2 + p$. This assumption is incorrect, and in view of its wide acceptance we now examine the possible errors to which it may lead.

As a concrete example, we may take the case of the smooth, parallel pipe for which Stanton measured the velocity distribution as given in Fig. 5.5. Table 5.5 shows the values from which this curve was obtained; as before, v is the velocity at radius r, v_c is the velocity at the axis, and a is the internal radius

TABLE 5.5. VELOCITY DISTRIBUTION IN A SMOOTH PIPE ($R = 5{\cdot}2 \times 10^4$)

r/a	v/v_c	r/a	v/v_c
0	1·000	0·799	0·789
0·183	0·987	0·852	0·753
0·284	0·967	0·905	0·711
0·389	0·944	0·925	0·688
0·491	0·915	0·954	0·645
0·596	0·883	0·978	0·596
0·698	0·838	0·990	0·388

of the pipe. The integrals occurring in (22) can now be evaluated graphically thus:

For $\int_0^a v^3 r \, dr$ we may substitute its equivalent

$$v_c^3 a^2 \int_0^1 \left(\frac{v}{v_c}\right)^3 \frac{r}{a} \, d\left(\frac{r}{a}\right);$$

similarly $\int_0^a vr \, dr$ may be replaced by

$$v_c a^2 \int_0^1 \frac{v}{v_c} \frac{r}{a} \, d\left(\frac{r}{a}\right).$$

If we now plot on a base of r/a the values of

$$\left(\frac{v}{v_c}\right)^3 \frac{r}{a} \quad \text{and} \quad \frac{v}{v_c} \frac{r}{a}$$

obtained from Table 5.5, we shall derive two curves whose areas are respectively the two integrals required. Thus we shall find that

$$v_c^3 a^2 \int_0^1 \left(\frac{v}{v_c}\right)^3 \frac{r}{a} \, d\left(\frac{r}{a}\right) = 0{\cdot}291 v_c^3 a^2 \tag{23a}$$

and

$$v_c a^2 \int_0^1 \frac{v}{v_c} \frac{r}{a} \, d\left(\frac{r}{a}\right) = 0{\cdot}405 v_c a^2, \tag{23b}$$

so that the true mean total pressure from (22) becomes

$$\tfrac{1}{2}\varrho \times 0{\cdot}718 v_c^2 + p.$$

Let us compare this with the mean total pressure incorrectly, but frequently, stated to be $\frac{1}{2}\varrho v_m^2 + p$. In this case

$$v_m = \frac{2\pi \int_0^a vr \, dr}{\pi a^2} = \frac{2 v_c a^2 \int_0^1 \frac{v}{v_c} \frac{r}{a} \, d\left(\frac{r}{a}\right)}{a^2} = 0{\cdot}81 v_c.$$

Hence

$$v_m^2 = 0{\cdot}656v_c^2$$

and the mean total pressure calculated in this way is

$$\tfrac{1}{2}\varrho \times 0{\cdot}656v_c^2 + p.$$

The true mean total pressure therefore exceeds that calculated by the erroneous method by $0{\cdot}062 \times \frac{1}{2}\varrho v_c^2$, i.e. by $0{\cdot}095 \times \frac{1}{2}\varrho v_m^2$, so that the error incurred in estimating the mean total pressure by this method is, in this particular case, $9\frac{1}{2}$ per cent of the pressure corresponding to the mean velocity. The smaller the velocity at a section in relation to the static pressure, the less will be the error in total pressure. If we take the case, which is quite common in practice, in which the mean velocity pressure is about equal to the static pressure at a section, we see that the mean total pressure deduced by the usual method may be about 5 per cent too low – an error that is by no means negligible. On the other hand, if the resistance of a pipe system is obtained by applying this incorrect method to measure the *difference* between the mean total pressure at the two ends of the system, the error will probably be less than this unless the velocity distributions at the two ends are markedly dissimilar. For fully developed pipe flow the error is zero.

Methods of measuring the true mean total pressure at any section in practice are described in Chapter VI.

The Resistance of Bends and Sharp Corners in Incompressible Flow

Any appreciable departure from straightness in a pipeline is accompanied by an increase in resistance, which becomes larger the sharper the bend. When air flows round a bend, the centrifugal action, combined with the effects of viscosity, may cause the flow to leave the pipe walls at certain places. Thus at and slightly upstream of the bend, the centrifugal action gives rise to a continuous increase of pressure across the section from the inside towards the outside of the bend. If the curvature of the bend is sharp, the pressure at the outer wall may be raised considerably, and the retarded flow near the wall may not have sufficient kinetic energy to enable it to force its way into this high-pressure zone. In that case, the air particles at the outer wall are brought to rest, or may even move in an upstream direction against the main flow, which then separates from the wall leaving a dead-air region in which there is more or less intense eddying motion. The energy lost in this eddying movement causes an increase of resistance (see Chapter II). Separation may also occur at the inner wall slightly downstream of a sharp bend. This happens because the flow cannot change its direction so abruptly as to follow the bend, and it gives rise to a further increase of resistance.

Much work has been done on the resistance of bends, but it is difficult to correlate the results because many of the investigators have not realized all the factors involved, so that their experimental conditions are not comparable. The greatest source of discrepancy, apart from differences of Reynolds number and of surface roughness, is the different lengths of straight pipe upstream and downstream of the bend, particularly the latter. Experiments have shown that the disturbance produced by a bend may persist for some 50 pipe diameters downstream, and that the loss of pressure due to the bend, as measured, say, 3 diameters downstream of the bend, may be less than half that measured 50 diameters downstream.

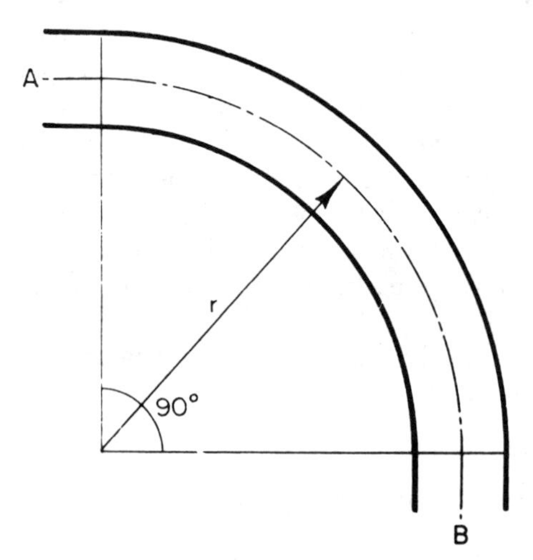

FIG. 5.8. Definition of bend radius r.

A comprehensive analysis of published data on the resistance of bends in incompressible flow has been made by Ward Smith,[13] on which the following résumé of the resistance of 90° bends is mainly based.†

Figure 5.8 shows a 90° bend AB in a circular-section pipe, the radius r of the bend being measured to the pipe axis. The axial length of the bend is $\pi r/2$, and it is usual to define the resistance of the bend as the pressure drop Δp that takes place within the bend in excess of the pressure drop that would occur in the same length ($\pi r/2$) of straight pipe. We shall first assume that the pipe is smooth, so that values of the resistance coefficient γ for straight lengths of constant cross-section can be taken from Fig. 5.3. Then, if p_A and p_B are the

† Ward Smith[7] has also studied the resistance of bends in compressible flow, and has derived equations that yield results in good agreement with published data for 90° bends in rectangular and circular ducts. As for incompressible flow, the resistance is greatly increased by a bend.

mean static pressures at entry A and exit B of the bend, the pressure drop Δp due to the bend, as above defined, is

$$\Delta p = p_A - p_B - \left(\frac{\pi \gamma r}{d} \varrho v^2\right), \tag{24}$$

where the term in brackets is the pressure drop in the length of the straight pipe as deduced from (4).†

The bend loss Δp is generally expressed in terms of a non-dimensional coefficient $\xi = \Delta p / \frac{1}{2}\varrho v^2$.

The bend resistance can also be expressed as a ratio ϕ of the pressure drop in the bend to the pressure drop in the same axial length of straight pipe, i.e.

$$\phi = \frac{\Delta p}{(\pi \gamma r/d)\varrho v^2}.$$

Thus ϕ represents the number of lengths of straight pipe, each equal to the axial length of the bend, that would have the same pressure drop as the bend.

Values of ξ and ϕ for smooth pipes deduced from Ward Smith's analysis[13] are plotted in Fig. 5.9 against r/d. They relate to bends with sufficiently long upstream and downstream tangents for the whole of the bend loss in fully developed pipe flow to be included. It will be seen that the bend resistance

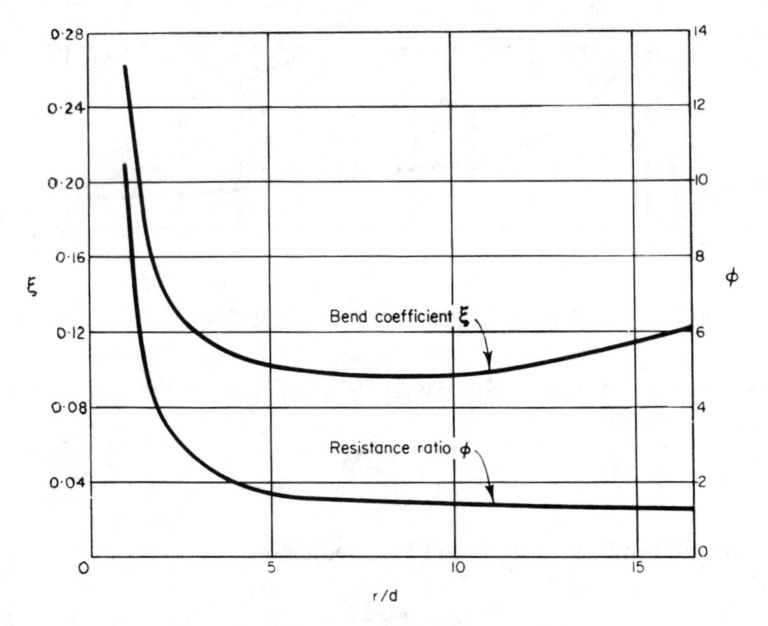

FIG. 5.9. Resistance of 90° bends ($r \neq 0$).

† It should be remembered that d is the diameter of the pipe, and is not equal to twice r, the radius of the bend.

increases rapidly with the sharpness of the bend. Thus, when the radius of the bend is equal to the pipe diameter, the total resistance of the bend is over 10 times the resistance of the same length of straight pipe. The curve of ξ against r/d shows that the excess loss due to the bend is a minimum when $r/d \simeq 8$. The reason for the existence of an optimum value of r/d is that, as r becomes larger, the total length, and so also the surface area of the bend, increases. Hence the wall friction increases, and there comes a stage at which the increase in friction balances the beneficial effect of increasing r; further increase in r leads to a rise in resistance. On the other hand, as Fig. 5.9 shows, the ratio of the resistance of the bend to that of the same length of straight pipe falls continuously as r is increased.

These results are valid for a Reynolds number of 10^5 based on pipe diameter d. To obtain approximate values of ξ for other Reynolds numbers the factor α shown in Fig. 5.10 may be used. This is the factor given by Ward Smith to be applied to a resistance coefficient, which he denotes by K and defines as the

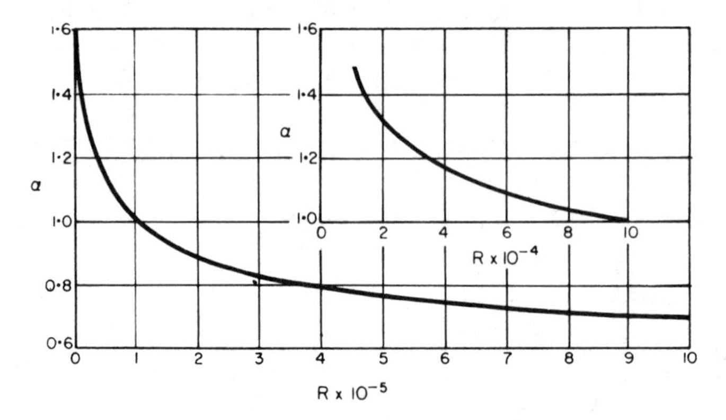

FIG. 5.10. Reynolds-number correction factor for the resistance of 90° bends.

pressure drop in a straight pipe having the same length as the centre line of the bend plus tangents.

Thus
$$K = \frac{\Delta p}{\frac{1}{2}\varrho v^2},$$

and, for 90° bends, K is related to ξ by the equation

$$\xi = K - \frac{2\pi\gamma r}{d}. \tag{25}$$

If we insert in (25) the appropriate value of γ for any Reynolds number, we find that Ward Smith's factor α obtained from Fig. 5.10 can be used to give a

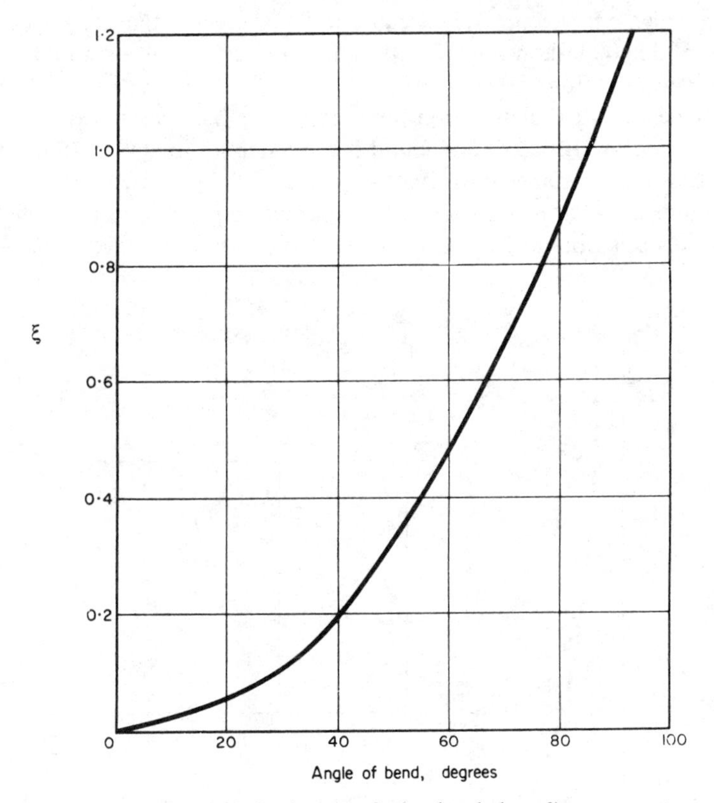

FIG. 5.11. Resistance of mitre bends ($r = 0$).

good approximation to the value of ξ as well as of K for Reynolds numbers within the range of Fig. 5.10.

When the radius of a 90° bend is zero, we have a 90° elbow or mitre bend, and the resistance is still higher. We may take as typical a result from some experiments with a 90° elbow which showed that the resistance was about 45 times that of an equivalent length of straight duct. It should be noted, however, that the measurements were taken fairly close to the elbow, so that the total loss of head due to the elbow may not have been measured: a still higher value for the resistance might therefore have been found from measurements made farther downstream.

For mitre bends with angles larger than 90° between their branches, the resistance coefficient ξ is less, as is shown by Fig. 5.11, which is based on results obtained in smooth pipes by Kirchbach(14) and Schubart.(15)

All the data so far quoted relate to smooth pipes; for rough pipes, including most of those used in practice, the losses in bends will probably be rather greater. Thus Hofmann,(16) in experiments with artificially sand-roughened pipes in which ε/d (see p. 84) was equal to 0·0058 found that, at a Reynolds number of 225,000, the value of ξ was about twice that for a smooth pipe in similar flow conditions.

So far bends in circular-section pipes only have been discussed. From the available data on ducts of square and rectangular section, Ward Smith concluded that the resistance coefficient K (see (25)) for a square duct of side h is not significantly different from that of a circular pipe of diameter h when the pressure drop is obtained from upstream and downstream measurements in

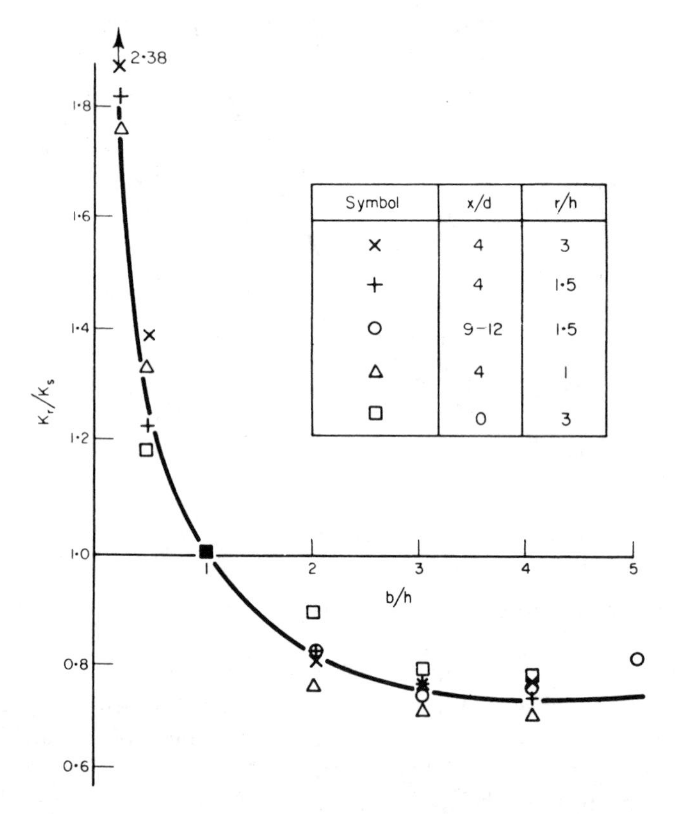

FIG. 5.12. Resistance of bends in pipes of rectangular cross-section.

fully developed flow. In a rectangular duct whose sides are h in the plane of the bend and b in the perpendicular direction, the effect of the ratio b/h on the ratio K_r/K_s can be obtained from Fig. 5.12, where K_r and K_s are Ward Smith's resistance coefficients for rectangular and square pipes respectively. This graph represents a mean curve through a number of points obtained from experiments in which the downstream pressure readings were taken at distances from the end of the bend ranging from 0 to 12 pipe diameters; but the departure from the curve of most of the points is so small that it seems reasonable to assume that the validity of the graph is not confined to these limits of downstream length of straight pipe.

Reduction of the Resistance of Mitre Bends

The sharp right-angled mitre bend (90° elbow) is much used in pipelines, particularly ventilating systems, because of its convenience for connecting lengths of piping running round sharp corners, e.g. at adjacent walls in buildings. It will be clear from what has already been said that a pipe system that includes a large number of mitre bends will suffer from serious energy losses; and that if the resistance of these bends could be appreciably reduced the running costs would be correspondingly lower, as would also the installation costs because a less powerful fan (or fans) would be needed.

Some improvement can be effected by rounding off the sharp corners of the elbow, particularly the inner corner. This is illustrated by the results of experiments by Frey[17] with the series of elbows for square-section ducts shown in

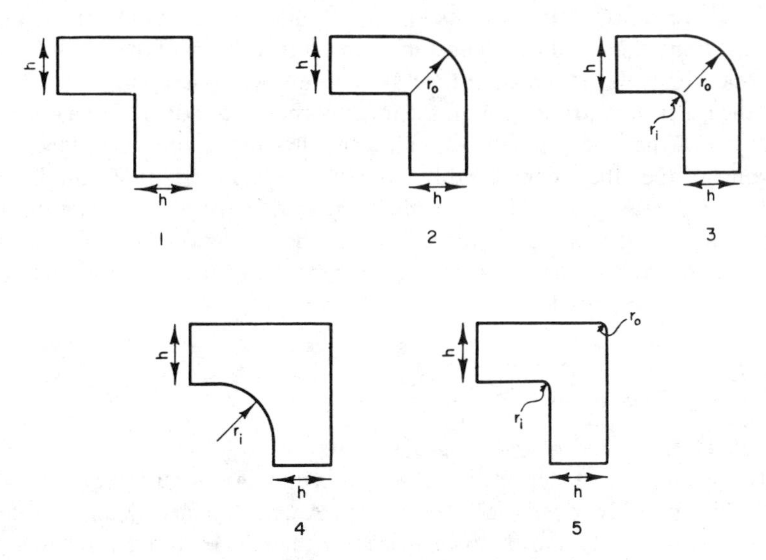

FIG. 5.13. Various designs of mitre bend.

Fig. 5.13. Particulars of these elbows are given in Table 5.6, together with the values of the resistance coefficient ξ measured by Frey, expressed in terms of ξ_1 for the plain elbow No. 1; r_o is the radius of the outer corner and r_i that of the inner.

TABLE 5.6. RESISTANCE OF 90° ELBOWS

Elbow no.	r_o/h	r_i/h	Relative resistance coefficient ξ/ξ_1†
1	0	0	1·00
2	1	0	1·64
3	1	0·25	0·84
4	0	0·25	0·60
5	0·083	0·083	0·90

† $\xi_1 = 1{\cdot}647$.

These results are interesting. In particular, they show that rounding off the outer corner only, so far from having any beneficial effect, actually increases the resistance coefficient if the radius of the corner is equal to the side of the duct. Thus, the resistance coefficient of No. 2 is 64 per cent higher than that of No. 1. This feature is again evident from a comparison of the results for elbows 3 and 4, where, even with the inner corner rounded, the sharp outer corner is better than the outer corner rounded to a radius $r_o/h = 1$. The explanation of this increase in resistance due to rounding off the outer corner while the inner one remains sharp is probably the reduction in the effective area at the bend† that this causes, and the resulting expansion of area downstream of the bend which, if too sharp, may be too much for the flow to follow without separation.

On the other hand, rounding off the inner corner to a radius of one quarter the length of the side, with the outer corner sharp or rounded, reduces the resistance of the elbow considerably (compare results for Nos. 1 and 4, and 2 and 3). It appears that radii much smaller than this, see No. 5, either on the inner or outer corners, have little effect. As the radius of the inner corner increases, however, so the constructional advantages of the 90° mitre joint are more and more impaired.

The Effect of Guide-vanes

An effective method of reducing the resistance of 90° elbows without adversely affecting their external shape is to insert a bank of guide-vanes which lead the air smoothly round the 90° corner.[18] A well-designed system of vanes reduces the resistance of an elbow to little more than the frictional resistance of the pipe and vane surfaces. Although the insertion of guide-vanes in all elbows increases the first cost of an installation, the running costs may be reduced sufficiently to enable the extra prime cost to be recovered in a reasonably short

† Although there is no reduction in the full cross-sectional area of the bend with a sharp inner corner and $r_o/h = 1$, there is an effective reduction in the area available for flow because the stream cannot follow the 90° turn at the sharp inner corner.

period. For the greatest reduction in resistance, the vanes should be of cambered aerofoil section, resembling aeroplane wings; but simpler vanes, made from thin sheet-metal bent into the form of circular arcs, can be designed to give results almost as good.[19, 20]

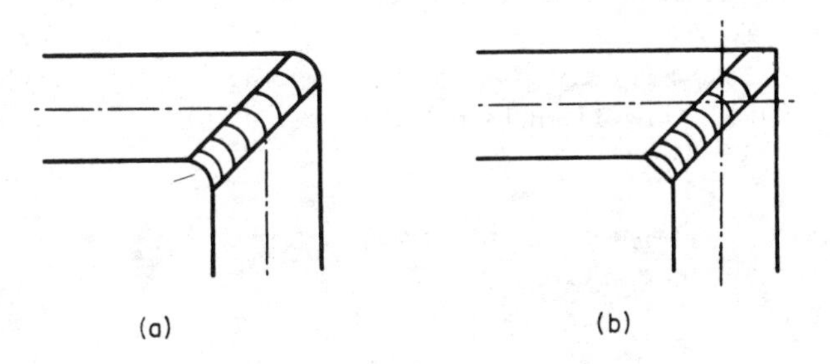

FIG. 5.14. Simple guide-vanes at 90° corners.

Most of the work on the effect of guide-vanes has been done on bends in ducts of square or rectangular section; and some Russian experiments[19] provide useful information on a comparatively simple sheet-metal guide-vane system for 90° bends in ducts of this type. The experiments show that the lowest resistance is obtained when the inner and outer corners of the elbow are rounded equally, with a radius about $b/4$ (Fig. 5.14(a)), where b is the length of the side of the duct in the plane of the bend. Six vanes are then sufficient, and they should be spaced so that the inner corner and the successive vanes are $0{\cdot}135b$, $0{\cdot}292b$, $0{\cdot}471b$, $0{\cdot}673b$, $0{\cdot}898b$, and $1{\cdot}145b$. With this arrangement, the vanes are spaced in an arithmetic progression, and the gap between the inner corner and the vane nearest to it is half that between the outer corner and the outer vane. If the inner and outer corners are not rounded, the arrangement shown in Fig. 5.14(b) is recommended. Here seven vanes are used, spaced at the following distances from the inner corner, which is chamfered off as shown $0{\cdot}118b$, $0{\cdot}253b$, $0{\cdot}404b$, $0{\cdot}573b$, $0{\cdot}759b$, $0{\cdot}960b$, and $1{\cdot}180b$. It should be noted that the chords of the circular-arc vanes should be inclined at an angle between 45° and 50° to the axis of the upstream arm of the bend, and the chord of the vanes should be one-quarter to one-third of the pipe diameter.

The Russian figures show that the resistance of an elbow fitted with guide-vanes in this way is from one-quarter to one-third of the resistance of the same bend without guide-vanes. Although this is still considerably above the purely frictional resistance, it does represent a valuable reduction. With about twice as many vanes, equally spaced, greater improvements can be obtained.

It is probable that the Russian system will also be effective in circular-section pipes, but positive information for such pipes is provided by experiments[21] with water flowing in a 15-cm pipe by Binnie and Harris, which led them to recommend two systems, one of which is shown in Fig. 5.15.† *ABCD* is a section of a ring fitting, containing the vanes, which was inserted into the pipe at the elbow. The blades were made from 20 s.w.g. brass,‡ bent into shape, the outer nine being circular arcs of the same radius, equally spaced. The backs of the leading edges were filed to an easy straight taper; and they were set, as shown in Fig. 5.15, at a positive incidence of about 5°. The trailing edges were sharpened in the same way. Another system shown in Fig. 5.16, which had eight vanes instead of ten, gave equally good results; but the vanes, which had to be irregularly spaced, were of rather more complicated form.

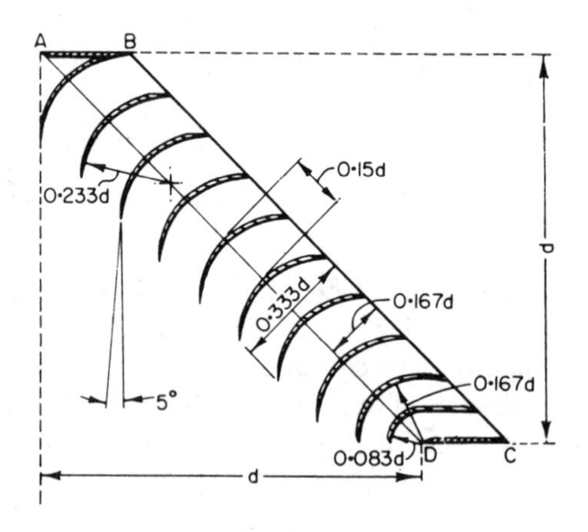

FIG. 5.15. Guide-vanes.

The resistance of a sharp right-angled bend fitted with systems of vanes to these designs was compared with that of a radiused bend with $r/d = 7$, i.e. about the optimum bend as shown in Fig. 5.9. Without guide-vanes, the resistance of the mitre bend between upstream and downstream sections at the positions occupied by the upstream and downstream ends of this radiused bend was about six times that of the radiused bend. Both systems of guide-vanes reduced this ratio to 1·5–1·6, i.e. by about 70 per cent. These figures, as indeed all others quoted in this section, should not be taken as general: they are true only for the particular Reynolds numbers of the various experiments. But there is no doubt about the general conclusion that simple guide-vanes reduce the

† Dimmock[22] also investigated thin guide-vanes at corners in circular pipes, but the type he recommended was rather more difficult to construct, being parabolic in section. One of his main objectives was to secure a good velocity distribution downstream of the corner. This is not as a rule of primary interest to engineers, who are concerned mostly with reducing power losses, but is important in special cases, such as the design of return-flow wind tunnels.

‡ It should be remarked that these vanes were designed for water flow. For air flow, thinner brass sheeting will probably serve in many cases provided that neither speed nor pipe diameter is so large that the vanes bend or vibrate appreciably.

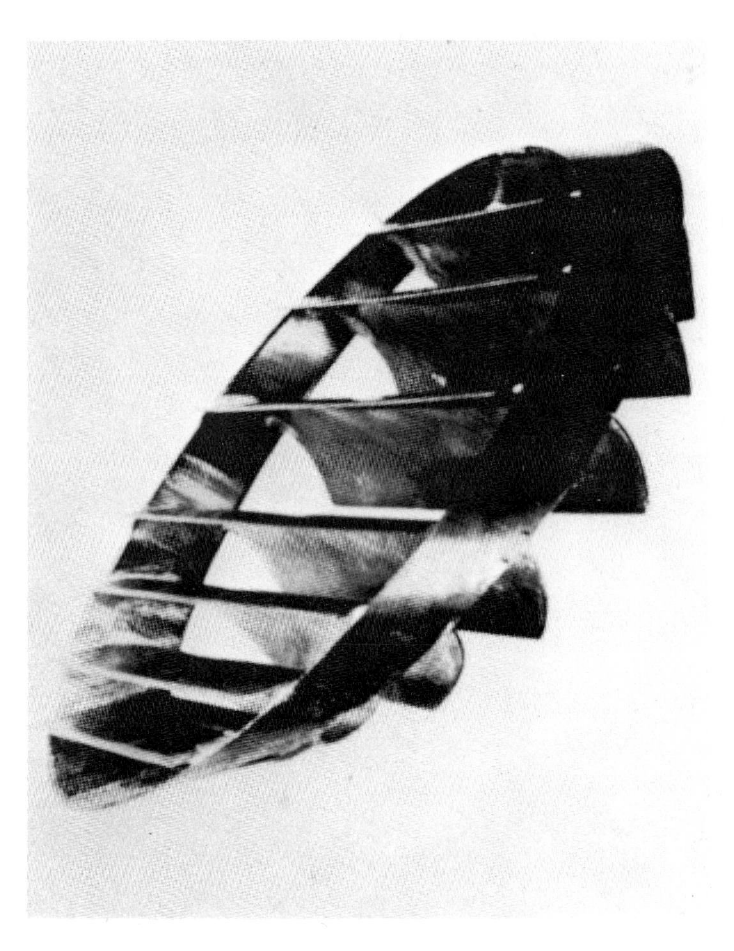

FIG. 5.16. Guide-vanes.

resistance of sharp bends very considerably; they are probably also beneficial in rounded corners of small radius.

A type of cascaded corner based on the results of Binnie and Harris has been designed by the National Engineering Laboratory. It is described in refs. 23 and 24. Air-flow tests, including some at a mean velocity of 38 m/sec in a 30-cm pipe, showed that they reduce the resistance of a 90° mitre bend by about 60 per cent and of a rounded corner (No. 3 in Fig. 5.13) by about 40 per cent.

Diffusers and Sudden Expansions

The steady flow of a perfect, frictionless fluid follows Bernoulli's law† that the total pressure is constant. Consider the incompressible flow of such a fluid through a diffuser (i.e. expanding duct) as shown in Fig. 5.17. The total pressure at the inlet *A* is equal to that at the outlet *B*. Thus:

$$p_A + \tfrac{1}{2}\varrho v_A^2 = p_B + \tfrac{1}{2}\varrho v_B^2,$$

and since v_B is less than v_A because the quantity flowing is the same at both sections, it follows that the static pressure p_B at *B* is greater than that at *A*. An expansion of area in incompressible flow is thus accompanied by an increase of static pressure. Air flowing into a region of higher pressure has to do work; it does this work at the expense of its kinetic energy, which is represented by the velocity pressure $\frac{1}{2}\varrho v^2$. In the absence of friction, the kinetic energy is just sufficient to overcome any increase in static pressure produced by the diffuser until the velocity is reduced to zero, i.e. until the outlet area of the diffuser becomes infinitely great.

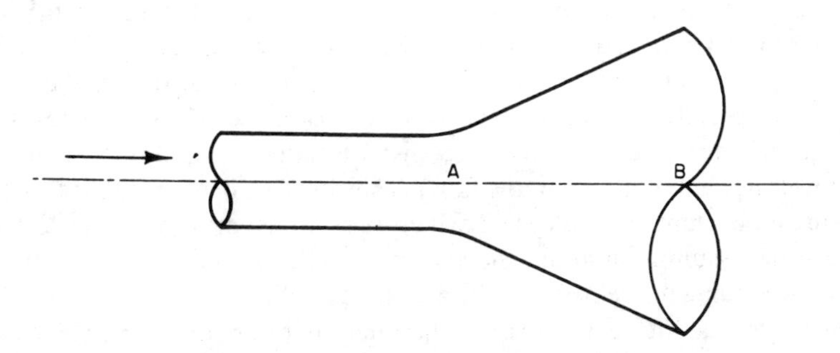

FIG. 5.17. Diffuser.

But friction introduces an additional resistance against which the air has to do work; and so, in the case of a real fluid, a stage may be reached, while the area is still finite, when the kinetic energy of the slowly moving particles near

† See Chapter II.

the walls is insufficient to overcome both the friction and the rising pressure. The velocity close to the walls then falls to zero, and the flow separates from the walls with the resulting increase of resistance previously noted. Usually the flow is turbulent, and the retarded fluid close to the walls receives appreciable energy from the faster-moving particles nearer the axis of the pipe. If the rate at which work has to be done against pressure and friction is less than the rate at which this additional energy is received, no separation takes place. Thus, it is found in practice that there is a limiting angle for a diffuser below which no separation occurs. Numerous investigations have shown that this angle is about 6° or 7° for a conical diffuser, the angle being defined as the total angle, i.e. twice the inclination of the side of the cone to the axis of the pipe. If the angle is greater than this, the diffuser becomes inefficient. The permissible limit also depends to some extent on the area ratio of the diffuser (ratio of outlet area to inlet area).

A diffuser may be used either as a transition piece joining a small pipe to a larger one; or else at the outlet from a pipe system where it discharges into atmosphere, in order to recover the velocity pressure, which would otherwise represent so much waste energy. It does this, as already explained, by converting the velocity pressure into static pressure. In either case, it is generally desirable to have as short a diffuser as possible in order to save space, but the small limiting angle of the diffuser means that a relatively long length will be necessary.

This conflict of requirements has led to numerous attempts to improve the efficiency of diffusers, and the mass of data that has been accumulated is so great that it is impossible to do more here than to summarize it briefly. Readers desiring more detailed information are referred to papers by Hofmann,[(25)] Patterson,[(26)] Cockrell and Markland,[(27)] and to recent data sheets.[(28, 29)] In using the numerical data, care is needed to distinguish between the various indexes of diffuser performance: some relate only to the recovery of static pressure (especially appropriate when the diffuser discharges to atmosphere) whilst others are based on considerations of total pressure (directly related to the power needed to maintain the flow). They nearly all assume that the static pressure is constant over entry and exit cross-sections; most of them effectively make some assumption about the velocity profile and flow direction at exit; and some assume the velocity profile at entry as well.

The angle of a diffuser in a rectangular duct can be about doubled if one pair of walls diverges and the other remains parallel. The efficiency of a diffuser in which the rate of divergence does not exceed the limiting angle depends also to some extent on the length of duct that precedes and follows it. Obviously, since the breakdown in flow in a diffuser is a boundary-layer phenomenon, the thinner the boundary layer at inlet the longer separation is delayed and the higher the efficiency. The inlet duct should therefore be as short as possible. On the other hand, it is found that a length of duct following the diffuser improves the efficiency, because the conversion of velocity pressure to static pressure is complete not at the outlet of the expansion but at some distance downstream, owing to changes in the velocity profile downstream of the diffuser outlet. Patterson states that if the inlet length is short the length of outlet duct should be about 4 times the maximum width or diameter of the diffuser, and with a long inlet the length of outlet should be about 50 per cent greater.

Experiments show that, in straight-sided diffusers whose divergence exceeds the limiting angle, separation of the flow always occurs near the inlet. This is because the pressure gradient (i.e. the rate of rise of pressure) along the diffuser is steepest there. Hence Gibson and others have investigated the effect of curving the walls so as to reduce the rate of expansion near the inlet, and have found that this flared type of expansion appreciably increases the efficiency. Gibson[30] concluded that the best results were given by the form of diffuser shown in Fig. 5.18, starting with a gentle curve and terminating in a straight-sided cone whose total angle should not exceed 35–40°. Patterson gives details of these and other experiments and quotes Gibson's equation for the curved portion of the wall.

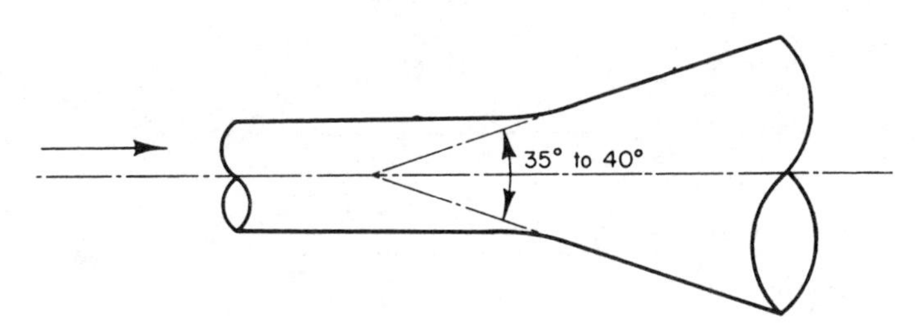

FIG. 5.18. Diffuser with curved inlet to expansion cone.

Other methods of improving the efficiency are: the insertion of guide-vanes (which take the form of annular rings in conical diffusers) near the inlet, as in Fig. 5.19; longitudinal splitter plates in the diffuser itself; boundary-layer control by blowing, suction, or vortex generators; the use of gauze screens in wide-angle expansions;[31] and the imposition of a certain type of rotation on the air before it reaches the diffuser.

Splitter plates are a helpful means of preventing incipient separation because the thickness of the boundary layer on each reduces the effective area ratio of the diffuser. Blowing air into the diffuser tangentially at the wall re-energizes the boundary layer and so is able to delay or prevent separation: this method may be convenient when a supply of compressed air is available. The method of suction is interesting but probably of little practical use except in special cases. There is, however, no doubt about its efficiency. One or two circumferential slits are cut in the walls of the diffuser about one-third of the length from the inlet, and the retarded air in the thickened boundary layer is removed by applying suction to these slits from a separate suction pump or fan. Alternatively, the slits may be replaced by porous walls. The air thus removed is replaced by air from the central part of the diffuser having higher kinetic energy, and separation of the flow is avoided. Large increases in the efficiency of the diffuser are thereby possible for angles of expansion as high as 50° or more, even when account is taken of the power required for suction.

The action of vortex generators, usually in the form of small vanes shaped like half of an aeroplane wing and projecting from the wall of the duct, depends on turbulent mixing: they shed vortices which bring fluid with higher kinetic energy into the more slowly moving surface air, thus re-energizing it and decreasing the tendency for the flow to separate. A similar effect can in some conditions be obtained by machining a series of circumferential grooves on the walls of the diffuser, starting at the inlet before any separation has occurred.

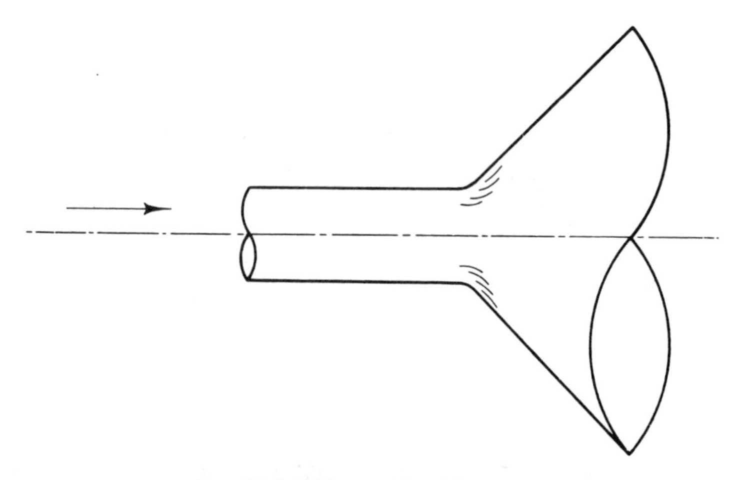

FIG. 5.19. Diffuser with guide-vanes.

Migay[32] showed that this increased the pressure-recovery characteristics of diffusers whose total included angle lay between 20° and about 50°, with a maximum improvement at 40–45°.

The method of imposing rotation is also interesting, and is quite simple to carry out. Fixed radial deflector vanes are inserted into the pipe ahead of the diffuser, so designed that the rotational velocity component they produce is proportional to the radius. After it leaves the deflectors, the air in the pipe therefore rotates as if it were a solid body, the tangential velocity being zero at the axis and a maximum at the walls. Each deflector is twisted along its length somewhat like an aeroplane propeller, the angle of twist increasing outwards from the centre. Patterson[26] gives full details for the design of a simple set of radial vanes, and the method is well worth applying where it is essential for diffuser losses to be reduced. It should be noted, however, that the rotation should be of the "rigid-body" type. A propeller-type fan in a duct, on the other hand, produces a type of rotation such that the rotational component at a given radius is approximately inversely proportional to the radius. This type of rotation is found to improve the efficiency of diffusers only when the rotation is small.

The Turbulent-velocity Components in Fully Developed Pipe Flow

The concept of turbulent flow as consisting of a steady mean flow on which are superimposed randomly fluctuating components was introduced in Chapter III. Experimental investigations[33, 34] have established the relationships that exist between these turbulent components and the mean axial velocity v in fully developed pipe flow at various Reynolds numbers. Typical results are shown in Table 5.7, calculated from the data given in ref. 34.

TABLE 5.7. TURBULENT-VELOCITY COMPONENTS IN FULLY DEVELOPED PIPE FLOW AT A REYNOLDS NUMBER BASED ON PIPE DIAMETER AND VELOCITY AT PIPE CENTRE OF 50,000

r/a	0	0·5	0·8	0·85	0·90	0·93	0·97
u'/v_r	0·031	0·057	0·082	0·091	0·104	0·120	0·187
v'/v_r	0·031	0·041	0·052	0·054	0·055	0·059	0·057
w'/v_r	0·031	0·048	0·066	0·071	0·076	0·081	0·093

N.B. u', v', and w' are the r.m.s. values of the velocity fluctuations in the axial, radial, and tangential directions respectively; v_r is the time-averaged axial velocity at radius r; and a is the pipe radius.

In turbulent flow the mean value of the static pressure varies with distance from the pipe wall: the radial variation is given by the equation[35]

$$r\frac{\mathrm{d}}{\mathrm{d}r}\left(\frac{\bar{p}}{\varrho}+v'^2\right)=(w'^2-v'^2),$$

where $\bar{p}$ is the mean (time-averaged) static pressure at radius r. Since v' and w' are zero at the wall, integration of the above equation gives

$$\bar{p}=p_w-\varrho v'^2-\int_r^a \varrho\frac{w'^2-v'^2}{r}\,\mathrm{d}r,$$

where p_w is the static pressure recorded by the hole in the wall. Numerical values of the corrections needed if $\bar{p}$ is identified with $\bar{p}_w$ in pipe-flow measurements with pitot and static tubes are given in Chapter VI.

References

1. O. Reynolds, An experimental investigation of the circumstances which determine whether the motion of water shall be direct or sinuous, and of the law of resistance in parallel channels, *Phil. Trans. Roy. Soc.* A **174** (1883) 935.
2. T. E. Stanton and J. R. Pannell, Similarity of motion in relation to the surface friction of fluids, *Phil. Trans. Roy. Soc.* A **214** (1914) 199.
3. J. Nikuradse, Gesetzmässigkeiten der turbulenten Strömung in glatten Rohren, *V.D.I. Forschungsheft* 356 (1932).
4. S. Gôldstein (Ed.), *Modern Developments in Fluid Dynamics* II, 331 *et seq.*, Clarendon Press, Oxford, 1938.
5. Engineering Sciences Data Unit, Friction losses for fully-developed flow in straight pipes, E.S.D.U. Data Sheet 66027 (1966).
6. Engineering Sciences Data Unit, Friction losses for fully-developed flow in straight pipes of constant cross section: subsonic compressible flow of gases, E.S.D.U. Data Sheet 74029 (1974).
7. A. J. Ward Smith, Subsonic adiabatic flow in a duct of constant cross-sectional area, *J. Roy. Aero. Soc.* **68** (1964) 17.
8. J. Nikuradse, Strömungsgesetze in Rauhen Rohren, *V.D.I. Forschungsheft* 361 (1963).
9. C. F. Colebrook and C. M. White, Experiments with fluid friction in roughened pipes, *Proc. Roy. Soc.* A **161** (1937) 367.
10. C. F. Colebrook, Turbulent flow in pipes with particular reference to the transition. region between the smooth and rough pipe laws, *J. Instn Civil Engrs* **11** (1938–39) 133.
11. T. E. Stanton, The mechanical viscosity of fluids, *Proc. Roy. Soc.* A **85** (1911) 366.
12. S. Goldstein (Ed.), *Modern Developments in Fluid Dynamics* II, 339–40, Clarendon Press, Oxford, 1938.
13. A. J. Ward Smith, The flow and pressure losses in smooth pipe bends of constant cross section, *J. Roy. Aero. Soc.* **67** (1963) 437; see also E.S.D.U. Data Sheet 67040 (1967).
14. H. Kirchbach, [Loss of energy in mitre bends], *Mitt. der Hydraulischen Inst. d. Tech. Hochschule, München*† **3** (1929) 68.
15. W. Schubart, [Energy losses in smooth- and rough-surfaced bends and curves in pipe lines], *ibid.*† **3** (1929) 121.
16. A. Hofmann, [Loss in 90-degree pipe bends of constant circular cross-section], *Mitt der Hydraulischen Inst. d. Tech. Hochschule, München* **3** (1929) 45.
17. K. Frey, Verminderung des Strömungswiderstandes von Körpern durch Leitflächen, *Forsch. IngWes* **5** (1934) 105.
18. A. R. Collar, Some experiments with cascades of aerofoils, *R. & M.* 1768 (1936).
19. G. N. Abramovitch, The aerodynamics of resistance, *Trans. Central Aerodynamic Inst. of Prof. Joukowski* 211 (1935).
20. C. Salter, Experiments on thin turning vanes, *R. & M.* 2469 (1946).
21. A. M. Binnie and D. P. Harris, The use of cascades at sharp corners in water pipes, *Engr* **190** (1950) 232.
22. N. A. Dimmock, Cascade corners for ducts of circular cross-section, *British Chem. Engng* **2** (1957) 302.
23. Royal Aeronautical Society, *Cascades of Guide Vanes for 90° Elbows*, Data sheet 00.02.09 (1964).
24. J. W. McHarg, *Air tests on a 12-inch Cascade Bend*, N.E.L. Report 243 (1966).
25. A. Hofmann, *Mitt der Hydraulischen Inst. d. Tech. Hochschule, München* **3** (1929) 160 and **4** (1931) 44.
26. G. N. Patterson, Modern diffuser design, *Aircr. Engng* **10** (1938) 267.
27. D. J. Cockrell and E. Markland, A review of incompressible diffuser flow, *Aircr. Engng* **35** (1963) 286.
28. Engineering Sciences Data Unit, Performance of conical diffusers in incompressible flow, E.S.D.U. Data Sheet 73024 (1973).

† Translated by the A.S.M.E. under title: *Transactions of the Hydraulic Institute of the Munich Technical University*, Bulletin 3.

29. ENGINEERING SCIENCES DATA UNIT, Performance in incompressible flow of plane-walled diffusers with single-plane expansion, E.S.D.U. Data Sheet 74015 (1974).
30. A. H. GIBSON, On the flow of water through pipes and passages having converging or diverging boundaries, *Proc. Roy. Soc.* A **83** (1910) 366; see also *Hydraulics and its Applications*, p. 92, Constable, London, 1945.
31. H. B. SQUIRE and K. G. WINTER, The R.A.E. 4 ft × 3 ft experimental low-turbulence wind tunnel. Part I. General flow characteristics, *R. & M.* 2690 (1948).
32. V. K. MIGAY, [Study of ribbed diffusers], *Teploenergetika*, No. 4 (1940) and No. 10 (1962).
33. A. FAGE, Turbulent flow in a circular pipe, *Phil. Mag.* (7) **21** (1936) 100.
34. J. LAUFER, The structure of turbulence in fully-developed pipe flow, N.A.C.A. Tech. Rep. 1174 (1954).
35. S. GOLDSTEIN (Ed.), *Modern Developments in Fluid Dynamics* I, 254 Clarendon Press, Oxford, 1938.

CHAPTER VI

THE MEASUREMENT OF INCOMPRESSIBLE FLOW IN PIPES BY PITOT AND STATIC TRAVERSE METHODS

THE pressure difference measured by a pitot–static tube is related to the local velocity and density of the fluid at the position of the instrument. The engineer concerned with pipe flow, however, is usually more interested in the rate of flow through the pipe, i.e. the quantity of fluid flowing per unit time. As noted in Chapter V, the velocity generally varies from point to point in the pipe cross-section. When the flowrate is determined from pitot–static explorations, therefore, enough readings have to be taken to enable the rate to be calculated by one or other of the integration methods to be described in this chapter.

The argument will be developed on the assumption that the flow is incompressible, so that the fluid density ϱ can be regarded as constant and the differential pitot–static pressure observed at each point can be converted into local velocity by a simple calculation. Integration of these local velocities over the pipe section gives the volumetric rate of flow and the mean flow velocity. Since the density is known, the mass flowrate follows directly from the volumetric flowrate.

At speeds at which compressibility effects assert themselves, i.e. when the variation of density with velocity is no longer negligible, additional measurements have to be made besides pitot and static readings in order to determine velocity (Chapter IV). Moreover, volume flowrate is then of less significance than mass flowrate, and, indeed, can only be defined in terms of some datum value of density. The integrations needed in order to obtain mass flowrate from measurements at various points across the pipe can still be effected by the same numerical methods as will be described in this chapter for incompressible flow, except those that depend on some assumption about the velocity profile across the pipe, which will not be the same as the mass-flow profile in compressible flow. Near and above the speed of sound, care is needed to avoid spurious effects on static-probe readings due to the proximity of a shock wave to the pressure holes at transonic speeds, or to the impingement of the bow wave at supersonic speeds after reflection from the pipe wall. Further, a sufficiently strong shock wave may upset the flow conditions in a duct by provoking separation from the wall.

In practice, pipe-flow measurements are seldom necessary or desirable in high-speed flow conditions: for reasons of power economy alone, the speed of flow is usually reduced at the earliest opportunity in the pipeline. The flow can then be measured in a low-speed section of the ducting.

Determination of the Mean Velocity

If we consider a small element of area s_1 of the pipe through which the mean velocity of flow is v_1 the volume flowing through this element is $s_1 v_1$, and

$$V = \sum s_1 v_1 = S v_m,$$

where V is the total volume flowing in unit time, S is the full cross-sectional area of the pipe, and v_m is the mean velocity over the cross-section.

Thus

$$v_m = \frac{1}{S} \sum s_1 v_1. \tag{1}$$

Suppose we divide the section of the pipe into n equal parts of area s, through which the mean velocities are v_1, v_2, v_3, etc., then

$$v_m = \frac{1}{S} \sum s v_1 = \frac{s}{S}(v_1 + v_2 + \ldots + v_n);$$

so that, since $S = ns$,

$$v_m = \frac{1}{n}(v_1 + v_2 + \ldots + v_n). \tag{2}$$

Thus the mean pipe velocity can be obtained by dividing the section into a number of parts of equal area and measuring the mean velocity through each. If the number of parts is sufficiently large, i.e. if each part is sufficiently small, the mean velocity through any part may be taken as the velocity at its centre. The number of velocity readings necessary will then be equal to the number of parts into which the section is divided. The more irregular the velocity distribution across the section, the greater will be the number of parts into which the section must be divided, if the velocity measured at the centre of area of each part is to approximate to the mean velocity across that part. Further, if we consider two pipes, similar in section but different in size, and assume that the velocity distributions across any pair of corresponding sections are similar, then, for the same percentage accuracy in the determination of v_m, the two sections must evidently be divided into the same number of parts. This fact is often overlooked; it must be remembered that if, for practical reasons, fewer measurements are taken in a small pipe, some accuracy will probably be sacrificed. The number of readings taken should depend primarily on the nature of the velocity distribution across the pipe, and not on the size of the pipe.

In considering how best to divide up a pipe section in practice, we shall discuss only pipes of circular and rectangular cross-section, since it is exceptional to use sections of any other form.

Pipes of Circular Cross-section

These pipes occur most often in practice, and most of the work that has been done on developing methods of determining the mean velocity from observations of local velocity relates to pipes of this kind. The most accurate method is that already indicated in Chapter V, pp. 92–95. At any radius r, enough observations are made on a number of different diameters – usually two will suffice, but more may be necessary if the readings differ widely, indicating a highly irregular velocity distribution – to define the mean velocity v_r at that radius. Then the mean pipe velocity v_m is obtained from the equation

$$V = \pi a^2 v_m = 2\pi \int_0^a v_r \, r \, \mathrm{d}r, \tag{3}$$

where V is the volume flowing in unit time, and a is the full pipe radius.

Hence

$$v_m = \frac{2}{a^2} \int_0^a v_r r \, \mathrm{d}r. \tag{3a}$$

This integration can be performed graphically as described in Chapter XIII, where (3a) is used in its equivalent non-dimensional forms

$$v_m = 2v_c \int_0^1 \frac{v_r}{v_c} \frac{r}{a} \, d\left(\frac{r}{a}\right) = v_c \int_0^1 \frac{v_r}{v_c} \, d\left(\frac{r}{a}\right)^2,$$

where v_c is the velocity at the centre of the pipe, i.e. when $r = 0$.

At the cost of a slight increase in the arithmetic, expressing all quantities non-dimensionally in this way has the advantages that the values are then independent of units systems, so that comparisons can at once be made with theoretical results, with results for standard forms of velocity profile, and with results from other investigations.

This method of graphical integration, using a planimeter or "counting squares" is, however, laborious, and methods of numerical integration are generally used. They are all of two types: either they determine the area under the curve of rv_r against r (or $r/a \; v_r/v_c$ against r/a) by some arithmetical rule, such as Simpson's, or they assume that the velocity distribution across the pipe section can be expressed in a mathematical form that allows the mean velocity to be determined from observations of velocity pressure made at a number of points at specified positions along a diameter.

A useful summary of the various methods that have been devised from time to time has been given by Kinghorn and McHugh.[1]

The tangential method. Formerly the most widely used of these numerical methods, but now almost entirely superseded by others found more accurate, applied the principle from which (2) was obtained. The pipe was considered to be divided into n zones of equal area consisting of one central circular zone and $(n - 1)$ annular zones by circles of radii $r_1, r_2, r_3 \ldots r_n$; and it was assumed that the mean velocity in each zone was obtained by measurement at the radius which divided the zone into two parts of equal area.

It is a simple matter to determine the radii of observation on this basis; and the pitot–static tube was located at these positions, generally along two mutually perpendicular diameters of the pipe. Thus, if n is the number of zones, the total number of observations is $4n$ ($2n$ along each diameter).

The accuracy of this method, which has been termed the tangential method, hinges on how nearly the mean velocity in each zone can be obtained from velocity measurements at a radius that divides the zone into two parts of equal area. Errors caused by this assumption will be less the greater the number of zones into which the section is divided, i.e. the narrower each zone is.

The log-linear method. An alternative method, recommended by the B.S.I.,[2] gives higher accuracy for the same number of measuring points. Termed the log-linear method, it was devised by Winternitz and Fischl,[3] who assumed – and quoted published supporting evidence – that the velocity distribution along a diameter can be represented by the equation

$$v_y = A + B \log \frac{y}{d} + C \frac{y}{d}, \tag{4}$$

where v_y is the point velocity at a distance y from the wall of the pipe of diameter d, and A, B, and C have the dimensions of velocity. The pipe is again divided into a number of annular zones and one central circular zone, all of equal area; but the measuring points, instead of being at the centre of area of each zone, are at positions at which the exact mean zone velocities occur if the velocity distribution can be adequately represented by (4). The calculation of the positions of these points is somewhat involved, and Table 6.1 quotes the positions for 4, 6, 8, and 10 measuring points per diameter.

Winternitz and Fischl made an extensive investigation of this method (and others discussed below) both by using Nikuradse's published velocity profiles

TABLE 6.1. LOCATION OF MEASURING POINTS FOR THE LOG-LINEAR METHOD

No. of measuring points per diameter	Distance from wall in pipe diameters
4	0·043, 0·290, 0·710, 0·957
6	0·032, 0·135, 0·321, 0·679, 0·865, 0·968
8	0·021, 0·117, 0·184, 0·345 0·655, 0·816, 0·883, 0·979
10	0·019, 0·077, 0·153, 0·217, 0·361 0·639, 0·783, 0·847, 0·923, 0·981

for fully developed flow in smooth and rough pipes† (see Chapter V) and by conducting a large number of pitot–static traverses, with up to 100 measuring points per diameter, in pipe flows that were not fully developed, including some with irregular velocity distributions. From this work they drew the following conclusions: For fully developed pipe flow, the four-point log-linear method gave an error of less than 0·5 per cent, whereas the ten-point tangential method overestimated the mean velocity by about 1 per cent. For pipe flow that was not fully developed and for irregular velocity distributions, including some that were markedly asymmetric, the four-point log-linear method was inferior, the six-point was about equivalent, and the eight-point and ten-point were superior, all by comparison with the ten-point tangential method. The ten-point log-linear traverse, which has been specified by the B.S.I. for Class A accuracy,[(2)] resulted in a mean-square error of about 0·5 per cent, or about half that of the ten-point tangential method.

It seems, therefore, from this investigation that, for the determination of the positions of the measuring points, the log-linear method is to be preferred to the tangential method; and that often equal accuracy can be achieved by using four or six points per diameter determined by the newer method as compared with ten points by the older. An additional reason for preferring the log-linear method is the apparent tendency (mentioned above) of the ten-point tangential method to overestimate slightly.

If high accuracy is not required, even fewer measuring points can be used. For example, Winternitz and Fischl give the following formula as an alternative to the log-linear formula for calculating the mean velocity v_m from three observations per diameter:

$$v_m = \tfrac{1}{5}[v_{0\cdot500} + 2(v_{0\cdot081} + v_{0\cdot919})], \qquad (5)$$

where $v_{0\cdot081}$, etc., represent the velocities measured at distances of 0·081, 0·500, and 0·919 pipe diameters from one wall.

† In the roughest pipes included in that investigation the average height of the sand roughness was one-thirtieth of the pipe diameter.

Aichelen's method. An even simpler method is due to Aichelen,[4] who, from an empirical analysis of Nikuradse's results, estimated that the mean velocity in fully developed pipe flow occurs at a radius of 0·762 times the full pipe radius. His method accordingly specifies that the velocity should be measured at 0·119 and 0·881 pipe diameters from one wall.† Winternitz and Fischl found that, in fully developed flow, in both smooth and rough pipes, this method gave a mean error of less than 1 per cent with differences comparable to those of the ten-point tangential and the four-point log-linear methods. For asymmetric velocity distributions, however, the errors were considerably higher: in two cases in which the four-point and six-point log-linear methods gave errors not exceeding 1 per cent, Aichelen's method overestimated the mean velocity by 6·8 and 4·3 per cent, and the three-point method ((5) above) by 3·6 and 1·3 per cent. The velocity distributions corresponding to these two sets of errors are

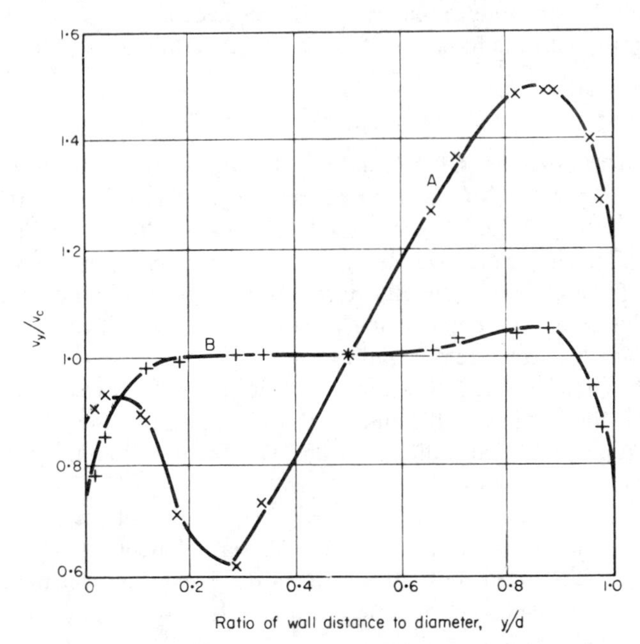

FIG. 6.1. Types of asymmetric velocity distribution in pipes.

shown by the curves *A* and *B* respectively in Fig. 6.1 in terms of the ratio v_y/v_c, where v_c is the velocity at the axis of the pipe and v_y is the velocity at a distance y from one wall.

It appears, therefore, that, if high accuracy is not required, measurements at three or even two points per diameter will suffice in many cases (see, for example, the three-quarter radius flowmeter; Chapter XI, p. 279). The number

† The two-point log-linear rule gives 0·112 and 0·888 pipe diameters.

of observation points needed will be greater the greater the asymmetry of the velocity profile; but even in cases such as that illustrated in Fig. 6.1, the six-point log-linear method will probably give results within 1 per cent.

The B.S.I.[2] specify ten log-linear points per diameter, located according to Table 6.1, on at least two diameters. In this connexion, it should be noted that for asymmetric velocity profiles, Salami has shown in an unpublished communication to the International Standards Organization that better results are obtained for a given total number of observation points by increasing the number of radii at the expense of the number of observations along each than by using more points on fewer radii. Thus, in a markedly asymmetric profile, much better accuracy was obtained by measuring at three points on each of eight radii than from six points on four radii or even eight points on three radii. Whenever accuracy of at least ·99 per cent is required, a preliminary survey traverse on three or four diameters should be made as a basis for the decision on the number of radii to be used for the final determination of the mean pipe velocity.

The method of cubics. Of the methods depending on the determination by numerical integration of the area of the curve of v_r/v_o on an $(r/a)^2$ base, Kinghorn and McHugh[1] recommend the "method of cubics" on which a series of cubic curves are fitted between adjacent pairs of measured velocities such that the gradients of successive curves are equal at their common boundaries. This leads to a complicated equation for v_m which can be integrated mathematically across the pipe radius.

Kinghorn *et al.*[5] have shown that this method gives results as accurate as those of the log-linear method, and has the advantage that measurements need not be taken at specified radial positions. On the other hand, it entails much more computational work.

The velocity-pressure observations required for any of the above methods can be made with pitot–static tubes of the standard diameter – about 8 mm – in pipes from about 30-cm diameter upwards. For measurements in smaller pipes, smaller instruments or different methods may have to be used. This subject is discussed on pp. 138–40.

Pipes of Rectangular Cross-section

So far it has not been possible to formulate rules whereby the mean velocity in a rectangular pipe can be determined with the accuracy possible in a circular pipe. Even a graphical integration method similar to that for which equation (3a) applies for circular pipes would need a very large number of observations, probably well over 50, to limit errors to 1 or 2 per cent. It is probably true to say that if better accuracy than this is required, an easy transition length followed by an adequate settling length, should be fitted to convert the rectangular section into a circular one in which to measure the mean velocity by one of the methods already described.

Failing this, the best method seems to be that developed at the N.E.L. by Myles, Whitaker, and Jones,[6] who devised an application to rectangular ducts of the log-linear method. This entails measurements of local velocity at 26 points in a cross-section along four traverse lines I, II, III, and IV in Fig. 6.2. Seven observations are made along each of I and IV, and six along each of II

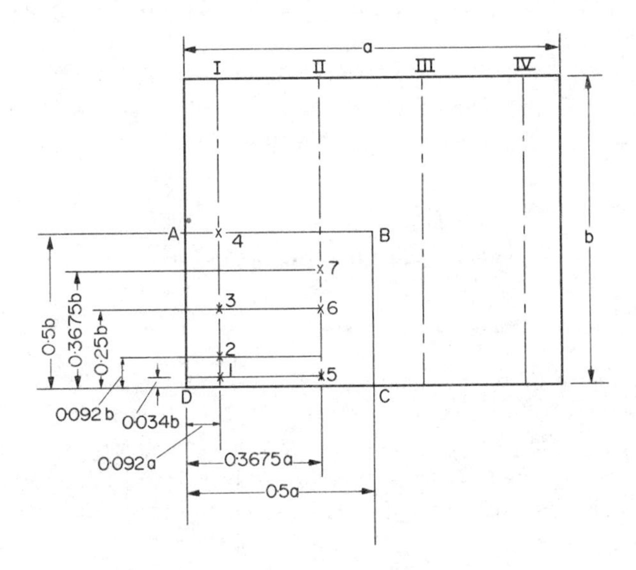

FIG. 6.2. Measurements in a rectangular pipe.

and III at the points marked in the quarter-duct $ABCD$. Denoting the velocities at points 1 . . . 7 by $v_1 \ldots v_7$, the formula for the mean velocity v_{AB} in $ABCD$ is

$$v_{AB} = \tfrac{1}{24}[2(v_1 + v_2) + 5v_3 + 3(v_4 + v_5 + v_6) + 6v_7].$$

The mean velocity v_m for the whole duct is one-quarter of the sum of the mean velocities in each quarter-duct thus obtained.

This method was checked[6] by measuring the mean velocity with a standard orifice plate (see Chapter VII) inserted in a circular section to which the rectangular section had been converted (see above). The log-linear results showed errors of $+0{\cdot}4$ to $+2{\cdot}0$ per cent, the latter in cases where considerable variations of velocity along the traverse lines, leading to marked asymmetry of the flow, had been purposely imposed.

Correcting for Variations in Mean Flowrate During a Traverse

All the methods discussed in this chapter are based on the assumption that no change occurs in the mean rate of flow while an exploration is being made. Measures should therefore be taken to ensure that this condition is fulfilled as far as possible. Even so, however, the mean rate of flow may vary by small but not negligible amounts during a complete exploration, so that all the individual traverse observations may not correspond to the same mean flowrate. To provide against this possibility, concurrently with each traverse reading a reference pressure differential should be taken from an independent source, such as a separate pitot tube and side hole in the pipe, sufficiently remote from the traverse plane not to affect conditions there. Each pressure differential of the traverse should then be reduced to the same standard reference pressure differential p_s by multiplying it by the ratio p_s/p'_s, where p'_s is the reference pressure differential observed when that particular traverse reading was obtained.

Evaluation of Pitot–Static Observations

Determination of Mean-velocity Head

Measurements of velocity may be made either by means of the combined pitot–static tube, or, as described later (see p. 126), by means of separate total-pressure and static tubes. In either case, the resultant reading at each point of measurement will be a differential pressure p which, on the assumption that the calibration factor of the instrument is 1, is related to the local speed v by the equation

$$p = \tfrac{1}{2}\varrho v^2, \tag{6}$$

where, if v is to be in metres per second, ϱ must be expressed in kilograms per cubic metre and p in pascals (N/m^2).

Now ϱ will depend upon the temperature and pressure of the air;† it is therefore necessary to measure the temperature in the pipe and the barometric pressure. Strictly, the air in the pipe will be at a pressure $b + p_s$, where p_s is the static pressure in the pipe measured above or below the barometric pressure b; but as a rule p_s can be neglected in relation to b. For the present we shall assume that this is permissible; the correction to be applied when p_s is too large to be neglected is discussed later (p. 123).

When the temperature is 15°C and the pressure 760 mm Hg, $\varrho = 1{\cdot}225$

† If necessary, the density of the air must also be corrected for humidity (see p. 337).

kg/m^3, so that, for any other temperature t°C and pressure b mm Hg, the value of the density in kilograms per cubic metre is

$$1{\cdot}225 \times \frac{288}{273+t} \times \frac{b}{760} = 0{\cdot}4642 \frac{b}{273+t}, \tag{7}$$

taking the absolute zero as −273°C (strictly −273·15°C).

The velocity pressure in (6) will as a rule be measured on some form of water manometer (see Chapter X), which will give it in terms of centimetres or millimetres of water-column. Now 1 cm of water-column at ordinary temperatures is equivalent very closely to a pressure of 98·1 Pa. Thus if h is the velocity head expressed in centimetres of water,

$$h = p/98{\cdot}1, \tag{8}$$

and, on substitution of (7) and (8), (6) becomes

$$h = 0{\cdot}002366v^2 \times \frac{b}{273+t}, \tag{9}$$

which may be written

$$v = k\sqrt{h}, \tag{10}$$

where

$$k = 20{\cdot}56\sqrt{\left(\frac{273+t}{b}\right)}, \tag{11}$$

and v is in metres per second.†

We can now apply these results to the calculation of the mean rate of flow through a pipe. The observations will consist of a number of measurements of velocity head h_1, h_2, etc., taken at points determined by one of the methods previously described. For simplicity we shall develop the argument on the assumption that, as is often sufficient in practice, observations are made along two mutually perpendicular diameters only. The general formulae for N diameters will be quoted afterwards.

Three cases have to be discussed.

(*a*) *Tangential and log-linear methods.* The pipe is divided into n zones, and a total of $4n$ observations of velocity head ($2n$ on each diameter) are made at the specified points.

† For a rapid estimation of the air speed corresponding to a velocity-head reading of h *millimetres* of water at ordinary temperatures and pressures, the formula

$$v = 4\sqrt{h}$$

will be found useful. It is accurate to better than 0·1 per cent for air at 15°C and 760 mm Hg barometric pressure.

Equation (2) applies to this case. For v_1, v_2, etc., substitute their equivalents (see (7)) $k\sqrt{h_1}$, $k\sqrt{h_2}$, etc. We thus obtain for the mean pipe velocity

$$v_m = \frac{k}{4n}(\sqrt{h_1} + \sqrt{h_2} + \sqrt{h_3} + \ldots). \tag{12}$$

If h_m is the velocity head corresponding to v_m, then

$$\sqrt{h_m} = \frac{v_m}{k} = \frac{1}{4n}(\sqrt{h_1} + \sqrt{h_2} + \sqrt{h_3} + \ldots). \tag{13}$$

(*b*) *Three-point method of Winternitz and Fischl.* Equation (5) gives the mean velocity as determined from observations along one diameter. In terms of the velocity head $\sqrt{h_m}$ this becomes

$$\sqrt{h_m} = \tfrac{1}{5}[\sqrt{h_{0\cdot500}} + 2(\sqrt{h_{0\cdot081}} + \sqrt{h_{0\cdot919}})]. \tag{14}$$

Since a distance of 0·919d from one end of a diameter is equal to 0·081d from the other end, we can write (14) as

$$\sqrt{h_m} = \tfrac{1}{5}[\sqrt{h_{0\cdot500}} + 2\sum\sqrt{h_{0\cdot081}}], \tag{15}$$

where $\sum\sqrt{h_{0\cdot081}}$ represents the sum of the square roots of the velocity heads measured at distances of 0·081d from each end of the diameter.

For two diameters, (15) becomes

$$\sqrt{h_m} = \tfrac{1}{5}[\sqrt{h_{0\cdot500}} + \sum\sqrt{h_{0\cdot081}}], \tag{16}$$

where $\sum\sqrt{h_{0\cdot081}}$ now represents the sum of the square roots of the four readings of velocity head that will be made at a distance of 0·081d from the pipe wall, two on each diameter.

(*c*) *Aichelen's two-point method.* In the notation used for the previous method we have for Aichelen's method (since $0\cdot119d = d(1 - 0\cdot881)$),

$$\sqrt{h_m} = \tfrac{1}{4}\sum\sqrt{h_{0\cdot119}}. \tag{17}$$

Equations (13), (16), and (17) show how $\sqrt{h_m}$ is obtained from measurements along two diameters. If the distribution is so irregular that more than two diameters have to be explored, the following equations for N diameters must be used:

Method (*a*):
$$\sqrt{h_m} = \frac{1}{2nN}[\sqrt{h_1} + \sqrt{h_2} + \sqrt{h_3} + \ldots]. \tag{13a}$$

Method (*b*):
$$\sqrt{h_m} = \frac{1}{5}\left[\sqrt{h_{0\cdot500}} + \frac{2}{N}\sum\sqrt{h_{0\cdot081}}\right]. \tag{15a}$$

Method (*c*):
$$\sqrt{h_m} = \frac{1}{2N}\sum\sqrt{h_{0\cdot119}}. \tag{17a}$$

Calculation of the Flowrate

Having determined $\sqrt{h_m}$ by whichever method we decide to use, we calculate the flowrate as follows:

The volume of air flowing per second will be $Sk\sqrt{h_m}$ and the mass $S\varrho k\sqrt{h_m}$, where S is the area of the pipe section.

If the pipe area is measured in square metres and h in centimetres of water column, we have

$$\left.\begin{aligned} V \text{ (cubic metres per minute)} &= C_1 S\sqrt{h_m}, \\ Q \text{ (kilograms per minute)} &= C_2 S\sqrt{h_m}, \end{aligned}\right\} \tag{18}$$

where C_1 and C_2 have the following numerical values when t is expressed in degrees Centigrade and b in mm Hg:

$$C_1 = 1234\sqrt{\left(\frac{273+t}{b}\right)}, \tag{19}$$

$$C_2 = 573\sqrt{\left(\frac{b}{273+t}\right)}. \tag{20}$$

We have already remarked that a correction may be necessary to allow for the fact that the pressure of the air in the pipe is actually not b, the barometric pressure, but $b + p_s$, where p_s is the static pressure in the pipe above or below atmospheric pressure. Since the value of b is nearly always about 760 mm Hg, i.e. about 1000 cm of water, and since only the square root of the pressure occurs in the equations (see (18)), it is clear that p_s can as a rule be as much as 20 cm of water (i.e. 2 per cent of b) without causing the values of V and Q as computed from (18), (19), and (20) to be more than 1 per cent in error. Hence, as remarked earlier, it is usually sufficiently accurate to assume that the air in the pipe is at atmospheric pressure. If the static pressure is more than 20 cm of water, the errors will exceed 1 per cent; and then, if higher accuracy is required, in place of b in the above formulae we must use $b + p_s$, with p_s expressed in mm Hg. If p_s is measured in centimetres of water, therefore, its numerical value must be multiplied by 10/13·6 before being added to or subtracted from b in the formulae, according to whether the static pressure is above or below the atmospheric pressure.

Temperature and pressure charts. If long series of quantity measurements have to be undertaken, and the variation of air temperature throughout the series is likely to be appreciable, it will save time to construct a chart like that in Fig. 6.3, from which the coefficients C_1 and C_2 can be read off directly.

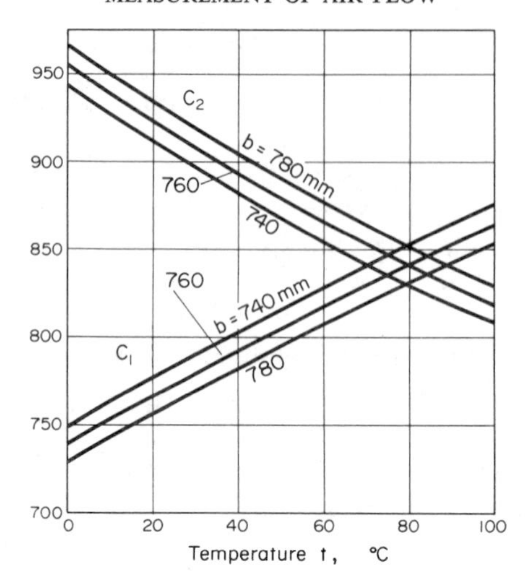

FIG. 6.3. Temperature and pressure charts.

Measurement of the Mean Total Pressure at a Section of a Pipe

For reasons explained in Chapter V, if we wish to determine the resistance of a length of pipe of varying cross-section, we may have to measure the mean total pressure at a section, which is not in general equal to the total pressure corresponding to the mean velocity and static pressure. This can be done accurately by graphical integration of point values of total pressure across the section as explained on pp. 92–5 and illustrated by a practical example on pp. 329–33. We can, however, use a quicker if possibly less accurate method of numerical integration which follows closely the lines of the numerical determinations of mean velocity set out in the early part of this chapter.

As before, imagine that the flow is incompressible and that the pipe section is divided into a number of small zones of area $s_1, s_2, s_3, \ldots$, in which the velocities and static pressures are $v_1, v_2, v_3, \ldots$, and $p_1, p_2, p_3, \ldots$ respectively. The mass of air passing through the element s_1 per second is $\varrho s_1 v_1$; the rate of passage of kinetic energy is $\frac{1}{2}\varrho s_1 v_1^3$, and of potential energy $p_1 s_1 v_1$. Hence the total energy per unit volume passing the section per second, which, as shown in Chapter V, is the mean total pressure T, is given by

$$T = \frac{1}{2}\varrho \frac{\sum s_1 v_1^3}{\sum s_1 v_1} + \frac{\sum p_1 s_1 v_1}{\sum s_1 v_1}.$$

If we make the zones all equal in area, this reduces to

$$T = \frac{1}{2}\varrho \frac{\sum v_1^3}{\sum v_1} + \frac{\sum p_1 v_1}{\sum v_1}, \tag{21}$$

where, in S.I. units, T is in pascals.

Observations taken for the determination of v_m by the use of (2) will suffice for the calculation of the first term of (21); for the second term we need also measurements of static pressure at the same positions in the pipe section.

The observations will generally consist of a number of velocity-pressure readings h_1, h_2, etc., for the velocities v_1, v_2, etc., together with static-pressure readings h_{s_1}, h_{s_2}, etc., for p_1, p_2, etc., all in centimetres of water column; and it will be convenient to express the mean total pressure in the same units. Let T_h be the mean total pressure so expressed. Then, using (6) and (8), we can make the following substitutions in (21): $98{\cdot}1\ T_h$ for T and $\sqrt{(196{\cdot}2\ h_1/\varrho)}$ for v_1, with similar substitutions for v_2, v_3, etc. We thus obtain

$$T_h = \frac{\sum \sqrt{(h_1)^3}}{\sum \sqrt{h_1}} + \frac{\sum h_{s_1}\sqrt{h_1}}{\sqrt{h_1}}. \tag{22}$$

As a rule, the static pressure will be fairly uniform across the section. Any significant departure from uniformity means that the flow is not parallel to the walls of the pipe, since the air will tend to flow towards the regions of low pressure. If the departure from axial flow is very pronounced, it is doubtful whether any method of measurement will give good accuracy, partly because both velocity-pressure and static-pressure readings at the various points of measurement may be in error on account of the inclination of the local direction of flow to the axis of the pitot–static probe. Measurements at sections where such conditions prevail should therefore always be avoided when possible. If, then, we can assume that the static pressure is uniform across the section, $h_{s_1} = h_{s_2} = \text{etc.} = h_s$, and (22) simplifies to

$$T_h = \frac{\sum \sqrt{(h_1)^3}}{\sum \sqrt{h_1}} + h_s. \tag{23}$$

Practical Notes on the Use of Pitot–Static Tubes in Pipes

To facilitate the location of the pitot–static head at the various measuring points necessary for a traverse it is convenient to provide a number of wire rings fitting the stem of the instrument fairly tightly. The positions of these rings can be adjusted before the tube is inserted into the pipe, so that when each ring in turn is up against the outside of the pipe the axis of the head is at one of the desired positions inside. A linear scale marked on the stem is also useful.

The hole cut in the pipe wall should not be larger than is necessary for the introduction of the tubes, and leakage through this hole must be prevented if accurate results are required.

The head of the instrument should, of course, point upstream, with its axis parallel to the direction of motion of the air, which is usually assumed to be parallel to the walls of the pipe. It is always advisable to check this assumption, however, by searching for an attitude in yaw corresponding to a minimum reading, as explained on p. 52. A useful method of ensuring that the head is correctly aligned is to have, permanently fixed to the stem outside the pipe, a straight-edge, about 15 cm long, parallel to the head, which can be sighted along the outside wall of the pipe.

Combined pitot–static tubes made to the N.P.L. ellipsoidal-nose pattern† are on the British market in a range of sizes suitable for pipes up to 2 or 2·5 m in diameter. The external diameter of the instrument in most common use is 8 mm, but this is not suitable for pipes of less than about 30-cm diameter. The reasons for this are explained on pp. 138–40.

Sometimes standard instruments are unsuitable or not available; and, if high accuracy is not essential, serviceable instruments can readily be made from material generally to be found in any factory. It obviously simplifies the construction of such instruments to have the total-head and static tubes separate, and to use them separately, or clamped together in a large pipe. Some care is necessary in shaping the heads of both tubes if it is intended to use them together. Although any open-ended tube facing the stream will indicate total pressure correctly, if the tubing is at all thick-walled the flow of air around the outside will be somewhat disturbed and may therefore lead to error in the measurement of the static pressure by means of another tube placed near the first. For this reason it is advisable to round off the bluff end of the total-pressure tube, or at least file the outside to a gradual taper so that it presents only a thin edge to the wind. Even so, there will be interference errors on the static-pressure readings if the tubes are too close together. Experiments by Ower‡ have shown that such errors will be negligible provided that the axes of the tubes are more than 5 diameters apart. All clamps and brackets should be confined to the stems of the tubes, well clear of the static orifices, and if possible, outside the pipe or duct.

For the static tube it will probably be found simplest to adopt the type having a hemispherical nose. Such a tube can easily be constructed by bending a straight piece of tubing to form the head and the stem, and closing the open end of the head by means of a hemispherical plug having a short shank that forms a push fit or screws into the tube, the joint being made airtight by finishing it off with solder. Care should be taken to leave no gap between the plug and the end of the tube: the outline of the head should be smooth and continuous. The static holes should be 6 tube diameters behind the junction of the plug and the head, and the stem 7 diameters behind the holes. The tube will then indicate the static pressure closely (see p. 33); but if it is more convenient to have the stem at some other distance behind the static holes, the correction to be applied to the observed pressure can be obtained from Fig. 3.9. The size of the static holes is not of great importance, provided that they are not too

† The tapered and hemispherical-nose types are now virtually obsolete.
‡ Results not published.

large; a diameter of about 1 mm in 8 mm diameter tubing is suitable. It is important that in drilling the holes the outside of the tube should be left entirely free from burrs.

Observations of total pressure and static pressure by separate tubes need not be made simultaneously, provided that the conditions of flow are reasonably steady and constant. It is more often convenient to read first the static pressure and then the total pressure at the point (or vice versa), the appropriate tubes being successively introduced into the pipe. This necessitates the cutting of smaller holes in the piping than when the tubes are inserted clamped together. If necessary, the procedure described on p. 120 can be used to allow for any small differences in the mean rate of flow there may be while all the observations are being made.

A further simplification is possible if the static pressure is sensibly uniform across the pipe section. In that case only the total pressure need be measured at the various points of observation; and the static pressure can be measured by cutting a small hole in the pipe wall and connecting it to the manometer via a suitable fitting on the outside of the pipe. The edges of the hole must be flush with the inner pipe surface. This method is discussed more fully on pp. 136–8.

The Reduction of Unsteadiness Due to Pressure Fluctuations

Unsteady readings in pitot–static measurements may be caused by the random pressure fluctuations that sometimes occur, particularly when one end of the pipe system is open to atmosphere and therefore influenced by draughts; remarkably large disturbances can then occur on a windy day. An effective method of reducing the resulting unsteadiness of the manometer readings in such conditions depends in principle on the fact that a tube of given dimensions exerts a certain amount of damping on the motion of a pressure pulse along its length. The damping due to the total-pressure tube will obviously be less than that of the static tube on account of the larger resistance of the small static orifices. Hence, even if the same pulse acts simultaneously both at the mouth of the total-pressure tube and at the static holes, the fluctuations will not be damped out to the same extent by the time the pressures reach the manometer unless extra resistance is introduced in the total-pressure connecting tube. Such extra resistance may be provided by using a longer length of connecting tubing

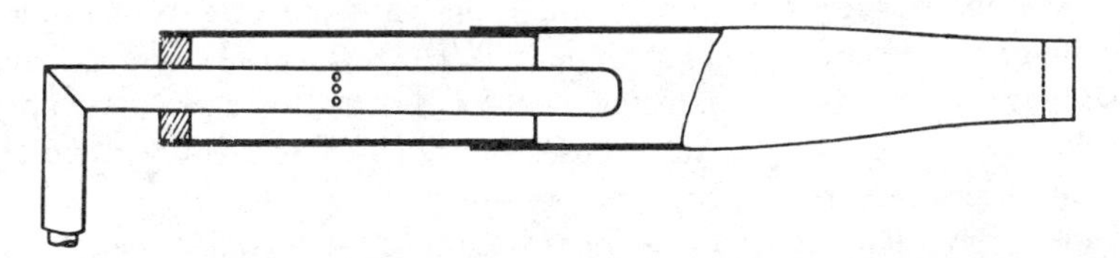

FIG. 6.4. Equalization of damping on both sides of a pitot–static tube.

on the total-pressure side, but a more convenient method is to insert a piece of capillary tube into the total-pressure connecting tube. It will generally be found that a 3 or 4 cm of capillary is enough.

The necessary lengths of tubing can readily be determined by the method illustrated in Fig. 6.4. The head of the pitot–static tube is inserted through a cork into a length of metal tubing of larger diameter, the other end of which is connected, as shown, to a convenient length of rubber tubing closed by a second cork. By squeezing this rubber tubing in the hand, fluctuating pressures can be applied simultaneously to the mouth of the total-pressure tube and the orifices in the static tube. If the other ends of these tubes are connected to the manometer in the usual manner, the lengths of the connecting tubes, or the amount of capillary inserted, can be altered until the oscillations of the manometric liquid become small even for large applied pressure fluctuations. When the connexions have been thus adjusted by trial and error, it will be found that, even with unsteady conditions of measurement, sufficiently steady readings will generally be obtained on the manometer.

A system balanced in this manner for pressure variations will not give entirely steady readings when used to measure flow in a pipe where the velocity and static pressure are fluctuating in the same way; but it will not usually result in greatly improved steadiness.

Effects of Turbulence in Pipe-Flow Measurements

As indicated in Chapter III, pitot–static observations are affected by turbulent-velocity components, which are present in most practical cases in which measurements are made. In general these effects are probably small, but they should be taken into account when the highest accuracy is demanded. The turbulent components are definite and known for fully developed flow in smooth, straight pipes (see Chapter V, p. 109); and, as shown in Chapter III, their effect on the measurements of total pressure can be allowed for by applying Goldstein's correction to the reading of the total-pressure side of the pitot–static combination. Thus the total pressure can be obtained with reasonable confidence. On the other hand, because of the uncertainty about the value of the turbulence correction to be applied to the readings of the static tube, it seems desirable, when high accuracy in pipe-flow measurement is required, to traverse only the pitot tube and to obtain the static pressure from a hole or a number of holes (see below) in the pipe wall. The turbulence causes a variation in static pressure along a diameter, which is known for a circular pipe (see p. 109); hence the static pressure at any point in a section can be calculated if the wall static pressure is measured.

Thus, from a knowledge of the distribution of the turbulent-velocity components along a diameter, corrections for turbulence effects can be applied to velocity-pressure readings obtained from a manometer of which one side is connected to a pitot tube (which can be traversed across the pipe) and the other to a suitably designed hole in the pipe wall.† These corrections for fully developed flow, based on the turbulent-intensity data then available,‡ have been given by Fage[7] for use when this method is employed to measure the volume rate of flow; and Ower[8] has extended the analysis to determine the corrections to individual readings of velocity at different radii. The results are set out in Table 6.2 in the form of percentage corrections to be applied to the results of calculations of velocity and quantity carried out in the ordinary way from observations of total pressure and wall static pressure.

TABLE 6.2. PERCENTAGE CORRECTIONS TO BE APPLIED TO ALLOW FOR THE EFFECT OF TURBULENCE IN FULLY DEVELOPED PIPE FLOW

(*a*) Corrections on local velocity

$r/a = \dfrac{\text{radius of observation}}{\text{radius of pipe}}$	Percentage correction
0	+0·1
0·5	0
0·8	−0·7
0·85	−1·1
0·9	−1·8
0·93	−2·5

Note: These corrections are calculated for a pipe Reynolds number of 18,340, for which $\gamma = 0{\cdot}0066$. For other Reynolds numbers the tabulated corrections must be multiplied by $\gamma/0{\cdot}0066$. Values of γ can be obtained from Fig. 5.3. or Table 5.2 Since γ decreases as the Reynolds number increases, the corrections at Reynolds numbers greater than 18,340 are smaller than those tabulated above.

(*b*) Corrections on total quantity obtained by traverse and numerical or graphical integration

Pipe Reynolds number	γ	Percentage correction
3,000	0·0109	−1·0
25,000	0·0062	−0·6
100,000	0·0045	−0·4
200,000	0·0039	−0·4
400,000	0·0034	−0·3

† Practical aspects of this method are discussed on pp. 136–8.

‡ More refined measurements at pipe Reynolds numbers of 50,000 and 500,000 have since been reported by Laufer;[9] corrections based on these measurements are very close to those calculated by Fage.

Corrections on total quantity have also since been published (see fig. 11 of ref. 2); the numerical values are quite close to those given in Table 6.2(b). It should be noted, however, that the Reynolds numbers quoted in ref. 2 are based on the axial pipe velocity, and are therefore some 15–20 per cent higher than those of Table 6.2(b).

The figures in Table 6.2(a) show that the corrections on local velocity increase in magnitude towards the pipe wall. They must, however, vanish actually at the wall; and it is probable that the maximum corrections are not much different from the values given at $r/a = 0{\cdot}93$. At this radius, turbulence causes appreciable errors because the r.m.s. velocity components are appreciable fractions of the mean axial velocity (see p. 109). The allowance for the radial variation in static pressure is of the same sign as the correction to be applied to the pitot readings, and thus reduces the correction to be applied to the differential pressure from which the velocity is calculated.

It should be remarked that the above corrections are based on the assumption of fully developed turbulent pipe flow, for which the turbulent-velocity components are known. But since the corrections are small, the values thus calculated can be assumed to be accurate enough for most of the cases in which the turbulence is not fully developed.

Influence of Displacement of Effective Centre of Pitot-tube in Pipe-flow Measurements

As was explained in Chapter III, the pressure at the geometrical centre of a pitot tube in a transverse velocity gradient, such as occurs in pipe flow, is higher than the local total pressure. Unless appropriate corrections are applied, therefore, the local velocities and the integrated flowrate will be overestimated.

The results of investigations of the characteristics of pitot tubes in transverse gradients are usually presented in the form of an effective displacement z of the centre of the probe in the direction in which the velocity is increasing (see p. 43). In pipe flow, therefore, the velocity measured at radius r is to be interpreted as relating to the radius $(r - z)$. Alternatively, the velocity at the measurement position may be taken to be approximately $(v + z\,\mathrm{d}v/\mathrm{d}r)$.

The value of z would be expected to depend on the magnitude of the velocity gradient, and hence to vary across the pipe; but available data provide no reliable information about the way in which it might do so. Reference 7 postulated that z increases linearly from zero at the pipe axis $(r = 0)$ to a maximum value at $r/a = 0{\cdot}85$, thereafter falling linearly to zero at the wall; but the derivation of the resulting correction to flowrate appears to have omitted a term in $\mathrm{d}z/\mathrm{d}r$ which would have considerably reduced the magnitude of the flowrate corrections. The British Standard on flowrate measurement by pitot traverse[2] does not accept the variation of z with r postulated in ref. 7; instead, it elects to treat z as constant (equal to 0·16 δ for a square-ended pitot-probe† whose inner diameter is 0·6 times its outer diameter δ, and 0·08 δ for an ellipsoidal head); the probe displacement was then converted to equivalent velocity change at each of the ten-point log-linear stations r/a = 0·277, 0·566, 0·695, 0·847, and 0·962 (see Table 6.1) by assuming power-law approximations to the velocity profile (equation 10),‡ and the mean velocity ($Q/\pi a^2$) was obtained by summation. The results are given in Table 6.3§; the value to be used for the exponent m can either be determined from the measured velocity profile itself or assumed to be that corresponding to the Reynolds number of the experiment by using Table 5.4.

TABLE 6.3. CORRECTIONS TO BE APPLIED TO ALLOW FOR DISPLACEMENT OF EFFECTIVE CENTRE OF SQUARE-ENDED PITOT PROBE (WITH INTERNAL DIAMETER 0·6 TIMES EXTERNAL DIAMETER) IN FULLY DEVELOPED PIPE FLOW. (VALUES CALCULATED BY THE METHOD OF REF. 2)

(*a*) Percentage corrections on flowrate

	$m = 1/5$	$m = 1/7$	$m = 1/9$
$\delta/d = 0{\cdot}01$	−0·4	−0·3	−0·2
$\delta/d = 0{\cdot}02$	−0·8	−0·6	−0·5
$\delta/d = 0{\cdot}03$	−1·2	−0·9	−0·7

(*b*) Data for correcting local velocity: values of percentage correction divided by δ/d

	$m = 1/5$	$m = 1/7$	$m = 1/9$
r/a = 0·277	−9	−6	−5
0·3	−9	−7	−5
0·4	−11	−8	−6
0·5	−13	−9	−7
0·566	−15	−11	−8
0·6	−16	−11	−9
0·695	−21	−15	−12
0·7	−21	−15	−12
0·8	−32	−23	−18
0·847	−42	−30	−25
0·9	−64	−46	−36
0·95	−128	−91	−71
0·962	−171	−122	−94

Note: The values of r/a given to three decimal places in the first column of Table 6.3(b) denote the radial positions specified for the ten-point log-linear traverse (see Table 6.1).

† The Young and Maas value is 0·18 δ (see p. 43).

‡ This information was provided by Mr. R. W. F. Gould of the National Physical Laboratory.

§ These have been calculated afresh and are somewhat less in magnitude than their corresponding values deduced from ref. 2.

The corrections to local velocities corresponding to the bi-linear variation of z with r assumed in ref. 7 were given in ref. 8 and correspond to a velocity-profile power-law index of about one-fifth. Values corresponding to the constant-displacement assumption of ref. 2 are given in Table 6.3 over the same range of radius as is covered by the specified ten-point log-linear radial stations. (The values of z cannot remain constant right to the pipe axis, since the displacement here must be zero if the velocity profile is symmetrical; nor can z remain constant right to $r = a$, for the corrected velocity profile would then fall to zero before reaching the wall.)†

The displacement-effect corrections obtained by the method of ref. 2 (and given in Table 6.3(a) and (b)) are larger than those obtained by applying correctly the variation of z with r of ref. 7. Regarding the larger values as possibly conservative, we may conclude that the displacement-effect correction to flowrate is hardly likely to exceed 1 per cent for the probe sizes normally used in practice, and that it should not exceed 0·5 per cent if δ/d (where d is the pipe diameter) is less than 0·015 when $m = 1/5$, or less than 0·02 when $m < 1/7$. The numerical values of the corrections should, however, be regarded as highly tentative if higher precision in flowrate measurement is aimed at: for this purpose, definitive data concerning the variation of z with r are required.

Stem-retraction Effect

The design of a static-pressure probe (whether or not incorporated into a pitot–static instrument) is based on the principle of locating the pressure orifices where the pressure fall due to flow acceleration over the forward part of the probe is balanced by the pressure rise due to flow deceleration caused by the stem upstream of itself (see p. 26). When the instrument is being used in a pipe, only part of the stem protrudes into the flow (the rest of the stem being "retracted", i.e. lying outside the pipe), and, indeed, the extent of the exposed part varies as the probe is traversed across the pipe. Because not all of the stem is exposed, the flow approaching the instrument is not decelerated as much as it is in the calibration conditions when the whole stem is exposed: there is a short-fall in the pressure rise shown by the upper curve of Fig. 3.4. The static-pressure reading will therefore be too low, and the observed pitot–static pressure difference too high. This is known as the *stem-retraction effect*.

Instead of correcting the pressure readings, in practice it is convenient instead to "correct" the tube coefficient K (defined as the ratio of the pitot–static

† The log-linear method used for determining flowrate does not require values of local velocity closer to the pipe axis than $r/a = 0{\cdot}277$ nor nearer the wall than $r/a = 0{\cdot}962$. Note also that the British Standard combines the displacement effect correction with the additional correction due to wall proximity effect: this, too, causes a displacement of effective centre, and reduces the values of z when the axis of the probe is nearer the wall than $2{\cdot}5\,\delta$. Further, the Standard stipulates that the probe is not to be used within a distance of $0.75\,\delta$ of the pipe wall.

pressure difference to $\frac{1}{2}\varrho v^2$: see p. 24). The "correction" to K is essentially positive; it increases as the stem retraction is increased (i.e. as the extent of the exposed stem is decreased). It only assumes practical significance when the probe head lies close to the wall through which the stem passes; this peripheral zone does, however, contribute substantially to the total flow-rate, and the stem retraction correction therefore cannot be ignored when the highest measurement accuracy is sought. Numerical values are given in ref. 2; typical curves are shown in Fig. 6.5, in which ϕ is the amount to be added to K and y denotes the length of the exposed stem divided by the stem diameter. The curves relate to a pitot–static probe whose static-pressure holes are at a distance of 10 probe diameters upstream of the stem.

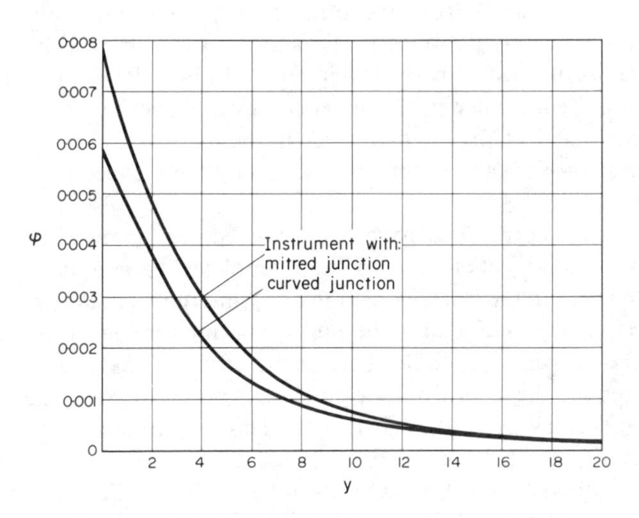

FIG. 6.5. Typical values of stem-retraction correction.

Stem-blockage Effect

We now consider the effect of the partial obstruction presented to the flow by the presence of the pitot or pitot–static probe in the pipe. By far the greater part of the obstruction comes from the probe stem, and for this reason the effect is usually referred to as *stem blockage*. The reduced cross-sectional area at the position of the stem forces the fluid velocity here to increase correspondingly; and since the acceleration starts upstream of the stem plane, the associated fall in static pressure is already manifest at the static-pressure orifices of the pitot–static probe or at the wall pressure tappings when a pitot probe is being used in conjunction with hole-in-the-wall measurement of static pressure. Unless a correction is applied, therefore, velocities and flowrate will be over-estimated. The magnitude of the correction increases the closer the static pressure holes are to the stem axis, and the greater the stem diameter. It also increases with the distance to which the stem protrudes into the flow: in the course of a traverse, therefore, the blockage varies from one reading to another. (The flowrate itself is kept the same throughout, by maintaining constant the reading of a reference meter or some reference differential pressure that is unaffected by the presence of the probe, such as that across a contraction well upstream or downstream.) Hence it is necessary to deduce from each pitot-probe reading what the velocity would have been, at the radial position concerned, in the absence of blockage effect altogether. The required correction is evaluated by considering the blockage-induced velocity rise and pressure fall separately from the other flow perturbations on which it is superimposed, namely the velocity rise and pressure fall over the nose of the instrument (as illustrated by the lower curve of Fig. 3.4), and the velocity fall and pressure rise indicated by the upper curve but reduced in magnitude by the stem retraction effect already discussed. In the reduction of the observations, therefore, stem-blockage effect and stem-retraction effect are allowed for separately.

The blockage corrections have been derived semi-empirically in ref. 10, using considerations analogous to those employed in considerations of the increase in speed in the working section of a wind tunnel in the presence of the model as compared with that in the empty working section when the flowrate through the tunnel is the same. The results of ref. 10 have been adopted in ref. 2 and require the recorded static pressure to be increased, or the pitot–static differential decreased, by $(0{\cdot}7\ kS/A)\ \Delta p_c$, where k is the stem-blockage factor, S the frontal projected area of the exposed portion of the stem, A the pipe cross-sectional area, and Δp_c the pitot–static pressure differential on the pipe axis; the value of k was found to vary little across the pipe. Alternatively, a single correction can be applied to flowrate instead of to local velocities. Numerical values of the corrections are illustrated in Figs. 6.6 and 6.7, in which x denotes the distance of the static-pressure holes upstream of the stem axis. These values are taken to apply equally whether the measurements are made using a pitot–static probe or a pitot probe in conjunction with a static-pressure tapping in the pipe wall. Full lines in Fig. 6.7 relate to wall-to-wall traverse; broken lines to wall-to-axis traverses. The curves for $x = 0$ correspond to a cantilever pitot cylinder.

The effects that occur when the probe head is carried by a cylindrical support extending right across the pipe have been reported in ref. 11. There, however, the comparisons with probe-free conditions are made on the basis of the same values of total pressure far upstream and the same values of static pressure far downstream instead of on the basis of the same flowrate.

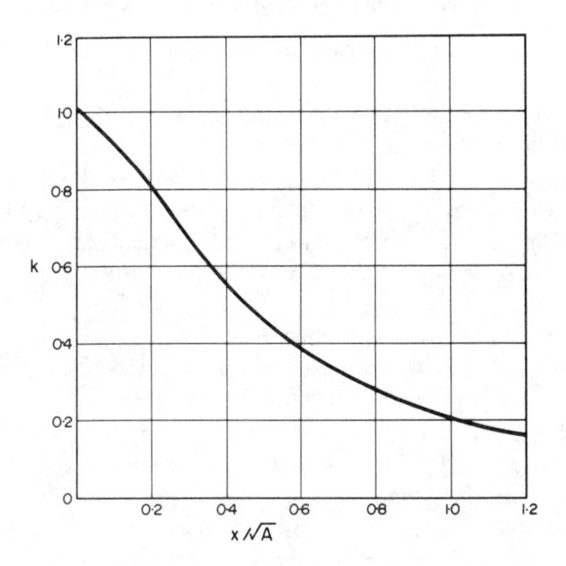

FIG. 6.6. Average value of blockage factor k over traverse planes at distances x upstream of axis of stem.

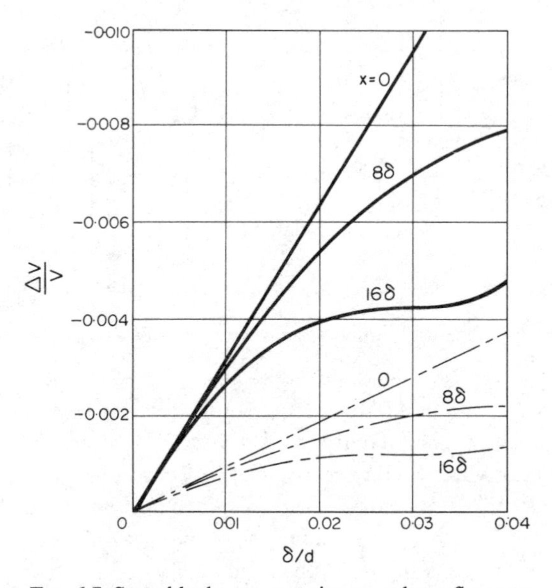

FIG. 6.7. Stem-blockage correction to volume flowrate.

Measurement of Wall Static Pressure

Some care is necessary in inserting pressure tappings into a pipe to measure wall static pressure. It is not sufficient just to cut a small hole in the pipe wall and to braze or solder a length of tubing to it to convey the pressure to a manometer. The primary requirement is that the mouth of the hole on the inside of the pipe must be smooth and truly flush with the internal pipe surface. It is essential that there shall be no burrs or other surface irregularities in the vicinity of the hole; and the hole should be so located that any errors due to the upstream pressure field of the stem of the pitot probe are negligibly small (see p. 27).

The size of the hole also has some effect because the absence of a solid boundary over the area of the hole changes the local flow conditions. As shown theoretically by Thom and Apelt[12] and observed experimentally by Ray,[13] fluid is deflected into the hole; and this in general causes the pressure acting in the hole to differ from the static pressure which would be measured by a hole of infinitely small diameter. The size of the error depends not only on the hole diameter d' but also on its depth l', and on the geometry of the internal arrangements of the pressure connexion. Shaw[14] has established values of the static-pressure error for sharp-edged holes of the type shown in Fig. 6.8(a) in

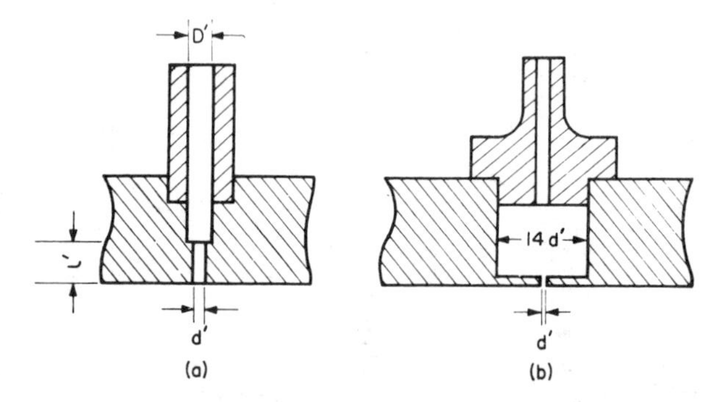

FIG. 6.8. Pipe-wall fittings for measurement of static pressure.

which the internal diameter of the pressure connexion is twice the hole diameter. Shaw found that the error increases with the ratio l'/d' up to a value of this ratio of 1·5, beyond which further increase of l'/d' ceases to have any effect. The value of the static-pressure error Δp for $l'/d' \geqslant 1{\cdot}5$ is given in non-dimensional form in Fig. 6.9 in which τ_0 represents the local shear stress at the wall (see p. 80). We may note in passing that the parameter

$$\frac{d'}{\nu}\sqrt{\left(\frac{\tau_0}{\varrho}\right)}$$

against which Δp is plotted is the equivalent of a Reynolds number since

$$\sqrt{\left(\frac{\tau_0}{\varrho}\right)}$$

has the dimensions of velocity. In fact, Shaw calls this parameter the hole Reynolds number.

As shown in Chapter V, for fully developed flow in smooth pipes $\tau_0 = \frac{1}{2}\varrho v^2 \gamma$, where γ is found from the data given in that chapter in diagrammatic, tabular, and algebraic form. If we use this expression for τ_0, we see from Fig. 6.9 that the maximum static-pressure error in a smooth pipe for fully developed turbulent flow is about $3\gamma \frac{1}{2}\varrho v^2$ for holes of the type investigated by Shaw. We see also from Fig. 5.3 that the maximum value that γ can have in these conditions is about 0·01, from which it follows that the maximum static-pressure error represents only 3 per cent of the velocity pressure $\frac{1}{2}\varrho v^2$ or 1·5 per cent of v. And this is for holes in which $l'/d' \gtrless 1{\cdot}5$. For smaller values of l'/d', Δp is less: results given by Shaw in the same paper show that when l'/d' has the values 1 and 0·5, Δp over the whole range of hole parameter is about half and one-third respectively of the values given in Fig. 6.9.

In a large number of practical cases conditions will be such that the errors will be much less even when $l'/d' \gtrless 1{\cdot}5$. Consider, for example, air flowing at 15 m/sec in a 30-cm pipe with a static side-hole 5 mm in diameter. Assuming fully developed smooth-pipe turbulent flow, we find that $\gamma \simeq 0{\cdot}004$, and the hole parameter

$$\frac{d'}{\nu}\sqrt{\left(\frac{\tau_0}{\varrho}\right)}, \quad \text{i.e.} \quad \frac{vd'}{\nu}\sqrt{\left(\frac{\gamma}{2}\right)},$$

is about 230. Figure 6.9 shows that Δp is then only about one-third of its maximum value. As a rule it will be possible to use side-tappings in which l'/d' is considerably less than 1·5, so that the errors on static pressure will be much less than this, particularly if the hole diameter does not exceed 3–4 mm. It is therefore justifiable to conclude that the hole-size error will generally be negligibly small; burrs or discontinuity of surface near the hole are far more likely sources of error.

Later work by Livesey, Jackson, and Southern[(15)] substantially confirmed Shaw's conclusions that Δp can be made negligibly small in most cases. This work was done mainly with holes with large cavities between them and the pressure connexion. For the arrangement shown in Fig. 6.8(b), it was found that when $l'/d' = 2{\cdot}5$ the pressure registered was very close to the true static pressure for all values of the non-dimensional hole parameter up to about 150.

Franklin and Wallace,[(16)] by using flush-fitting pressure transducers, subsequently established datum values by direct measurement instead of deducing them by extrapolation to zero hole size. The hole geometry was similar to that

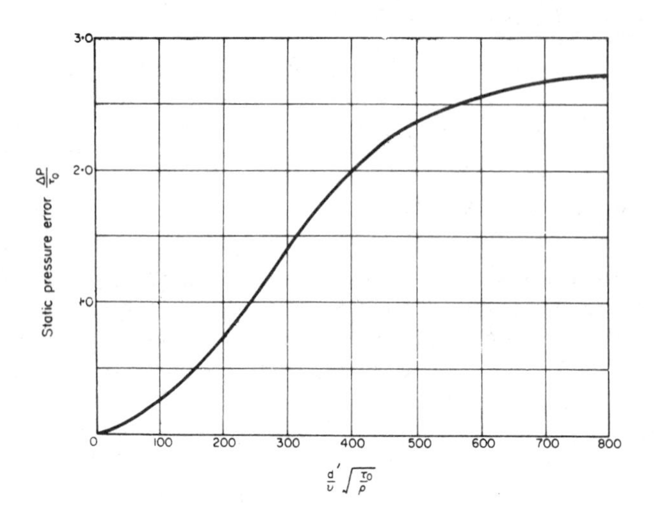

FIG. 6.9. Static-pressure error due to size of hole in wall.

of Fig. 6.8(a), with l'/d' equal to 15 and d'/D' to 0·45. The measurements were made on a wall of the working section of a wind tunnel at Mach numbers below 0·5 and with boundary layers that were fully turbulent at the measurement section. The static-pressure error curve obtained differed little from Shaw's (shown in Fig. 6.9); hence the conclusions drawn in the above discussion remain valid. Data obtained by Rainbird[17] in supersonic flow at Mach numbers up to 3·6 fall close to the pressure-error curve of Franklin and Wallace for values of $(d'/v)\sqrt{(\tau_0/\varrho)}$ up to 400 for all Mach numbers, but lie above that curve at higher hole Reynolds numbers than 400 when the Mach number exceeds 0·6.

Minimum Pipe Diameter in Which the Standard Pitot or Pitot–Static Tube can be Used

As has already been shown, the flowrate in a pipe can be determined from measurements of velocity made at a number of points in a cross-section. This can be done by locating a pitot–static combination at the appropriate positions, or by placing a pitot only at these positions and obtaining the static pressure from a hole in the pipe wall. The maximum size of probe that can be used in a pipe of given diameter will be influenced by the nearest approach to the pipe wall required by the numerical-integration method adopted (log-linear, cubics, etc.), by the possible wall-proximity effects on both components of the pitot–static combination, and by possible errors due to the displacement of the effective centre of a pitot tube in a transverse total-pressure gradient.

Probe Size

From the preceding discussion of the various traverse procedures, it seems probable that accuracy better than that given by the ten-point log-linear rule will seldom be required, i.e. that the axis of the probe head will not have to approach more closely than $0{\cdot}019d$ to the pipe wall. The smallest pipe in which this is possible is one in which $0{\cdot}019d = 0{\cdot}5\delta$ (where δ is the outside diameter of the probe), i.e. a pipe whose diameter is 26 times that of the probe tubing: the probe will then be in contact with the pipe wall at the innermost measuring point.

We must consider now whether the two effects already mentioned, namely those of wall proximity and transverse velocity gradient, are likely to cause significant errors with this relative size of probe and pipe diameter.

Wall-proximity Effects

Figure 3.16 shows that with the tube in contact with the pipe wall, i.e. when its axis is distant $0{\cdot}5\delta$ from the wall, a correction of 1·5 per cent has to be added to the observed velocity to allow for the wall-proximity effect on a square-ended pitot tube whose internal diameter is $0{\cdot}6\delta$. There is no published information about similar effects for tubes with the N.P.L. standard ellipsoidal head; but, since Livesey[18] found the effect to be negligible for heads with sharp-edged conical noses, it appears likely that the corrections will be smaller for hemispherical heads than for those to which Fig. 3.6 relates. This was, in fact, assumed in the preparation of the B.S.I. Standard.[2] However, even if the corrections are about the same, the error on mean flowrate must be less than 1·5 per cent because the correction applies only to the velocity measured nearest to the wall, which is as a rule considerably less than the mean; moreover, since for the ten-point log-linear method the pipe is divided into five equal-area zones, the wall-proximity correction will most probably only apply to 20 per cent of the total flow at most. At the next measuring point, the axis of a pitot of diameter one-twenty-sixth of the pipe diameter is distant about 2δ from the pipe wall, where the wall-proximity error shown by Fig. 3.16 is zero.

Some wall-proximity effect might be expected on the static side of a pitot–static combination, particularly if, as in all standard types except the Prandtl tube with its annular slit, the static orifices are arranged all round the periphery. The only experimental evidence on this point seems to be that provided by some unpublished results obtained by N. E. Sweeting in 1945 with a small static tube 1·9 mm in diameter. These showed a complete absence of any systematic variations in static-pressure indications (as compared with the static pressure recorded at a hole in the surface of a flat plate) when observations were made with the tube at a number of distances from the plate, in one of which plate and tube were in contact.

Effects Due to Transverse Velocity Gradient

Only the pitot reading will be affected by the transverse velocity gradient, and Table 6.3(a) shows that to limit the error on flowrate due to this cause to 1 per cent, the diameter of a square-ended pitot tube should not exceed one-fiftieth of the pipe diameter. However, the corrections for the N.P.L. standard ellipsoidal-nosed pitot are, according to ref. 2, half those for the square-ended pitot. Moreover, as already noted, the correction for wall proximity is of opposite sign to that for velocity gradient, and so will offset part of the gradient correction. It therefore appears a reasonable assumption that to limit the sum of the errors due to the gradient and wall-proximity effects to 0·5 per cent for the ellipsoidal-nosed pitot, the value of 50 for the ratio d/δ can be relaxed to, say, 35–40.

The Relative Merits of Measuring Static Pressure at a Side Hole or by Static-tube Traverse

For the highest accuracy in the measurement of the flowrate in pipes, the balance of existing knowledge appears to be slightly in favour of measuring the static pressure at a hole in the side of the pipe rather than by traversing a static tube either by itself or as part of a pitot–static combination. The corrections to be applied for the errors incurred by the assumption that the static pressure at all points is equal to the wall static are incorporated in Table 6.2. In addition further corrections can be applied for the effect of hole size, if necessary (see pp. 136–8).

For the successful use of the method, however, certain precautions are essential. Considerable care must be taken in making and installing the side-hole fittings in the pipe wall, and in ensuring that there are no burrs or interruptions of the interior pipe surfaces near the fittings. Even slight irregularities of this kind, or a small amount of dirt in or near the hole, can cause errors much greater than those due to hole size. In addition, unless there is no possibility of the existence of any swirl or asymmetry of the flow which might cause differences in wall static pressure around the circumference, there should be a number of holes spaced around the section of the pipe at which measurements are being made and communicating with an annular chamber, to give an average value of the wall static pressure. Even then it is not certain that the corrections appropriate to swirl-free fully developed flow can be applied; and steps should be taken to ensure that such conditions substantially exist, e.g. by arranging for a suitably long, straight, inlet length of pipe upstream of the measurement plane, including, if necessary, some form of flow straightener (see pp. 189–93).

The axial location of the hole is also of some importance. The stem-retraction effect discussed on pp. 132–3 arises because the influence of the stem of a pitot-static probe, when inserted into a pipe, is less than in the probe design and calibration conditions. The upstream influence that remains occurs also with a pitot probe, and can markedly affect the reading of a wall static-pressure tapping located upstream of the stem and in line with it. Since the influence of a stem is appreciable for more than 10 diameters upstream (see Fig. 3.4), it is safer to locate the wall pressure tappings at a distance of one pipe diameter upstream of the mouth of the pitot probe (see Fig. 6.11) rather than opposite the mouth as is commonly recommended (e.g. in ref. 2), and to allow for the pressure drop along the pipe (see p. 144).

It seems, therefore, that provided that proper care is taken, we can measure static pressure with good accuracy by the hole-in-side method. On the other hand, there must be some uncertainty about results obtained from a static-tube traverse, firstly because the experimental evidence that there is no wall-proximity effect is too scanty – it comprises only a few results obtained in a single experimental investigation – and, secondly, because of the uncertainty about the turbulence effects on the static-tube readings.

There is, however, some reason to believe that errors due to this cause will also be small. As stated in Chapter III, the differential pressure indicated by the pitot–static combination in isotropic turbulence is $\frac{1}{2}\varrho v^2(1 + \alpha\overline{v^2}/v^2)$, where α can vary between 1 and 5 according to the scale of the turbulence. Before the latest work on which this conclusion is based was available, it was thought that Goldstein's correction (see p. 41) could be applied, in which case α would have the value 2. Using Goldstein's correction, Ower[(8)] showed that the corrections to be applied to the total quantity or the mean velocity in a pipe, as deduced from a pitot–static traverse, ranged from $-1{\cdot}6$ per cent at a Reynolds number of 3000 to $-0{\cdot}5$ per cent at 400,000. This was for a value of α of 2, i.e. about midway in the theoretical range of 2–5 for isotropic turbulence. Although the turbulence in fully developed pipe flow is not isotropic (see p. 109), it is not thought – but here again further experimental evidence is highly desirable – that the departure from this condition would have serious effects. Therefore, in view of the comparative smallness of the corrections calculated by Ower, it seems likely that the rate of flow as deduced from a careful pitot–static traverse of a pipe will not as a rule be more than ± 1 per cent in error, provided that the tube diameter is not greater than one-fortieth of the pipe diameter.

This conclusion is consistent with the results of an investigation made at the N.E.L.[(19)] in which the mean velocities obtained by ten-point log-linear pitot–static traverses at various flowrates in circular pipes were compared with those determined from standard plate orifices (see Chapter VII) installed in the same pipeline. The orifices had been previously calibrated by tests in water in separate pipelines, in which the mean rates of flow were accurately determined by direct weighing.

Measurement of Flow in Small Pipes

(a) By Pitot–Static Tube

As shown on p. 140, the pitot–static tube of the usual external diameter of 8 mm should not be used in pipes of diameter less than 30 cm. Smaller pitot–static combinations conforming to the standard geometry can be obtained: British instrument-makers' catalogues include one with an external diameter of 2·3 mm (0·09 in.). On the same assumption, as on p. 140, that the ratio of pipe diameter to tube diameter should not be less than 40, we find that the diameter of the smallest pipes in which this instrument can be traversed is about 10 cm for the ten-point log-linear integration. This minimum pipe diameter will be greater if it is decided that more points of measurement per diameter are needed. For smaller pipes, still smaller pitot–static tubes could no doubt be specially constructed. Measurements have been made with an instrument of the type shown in Fig. 6.10, in which some of the difficulty of concentric construction

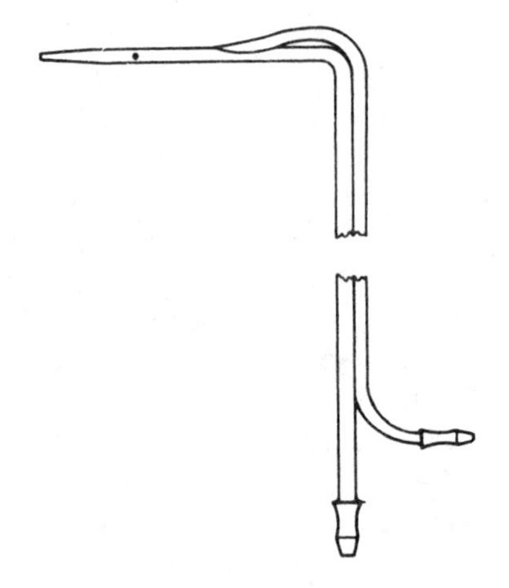

FIG. 6.10. Small pitot–static tube.

with tubing of such small diameter has been overcome by arranging the tubes side by side except over a short length of the head itself. In Fig. 6.10 the static tube is shown emerging above the total-pressure tube; actually it should emerge from one side so that the static holes themselves can approach nearer to the pipe wall and thus give the maximum traversing range.

An instrument of this kind needs calibration before use. Even with small instruments conforming to the geometry of the standard concentric type, there is a risk of error if they are used to measure speeds not high enough for the Reynolds number to exceed about 700: at lower Reynolds numbers the pitot–static calibration factor may vary appreciably (see Table 3.1). For the small instrument of 2·3 mm diameter mentioned above, this value of the Reynolds number corresponds to an air speed of 4·5 m/sec at normal temperature and pressure.

(b) *Pitot Tube and Hole in Side*

The general principles of this method have already been described; provided that the necessary precautions are taken, particularly in the manufacture and fitting of the wall pressure tapping and in avoiding swirl, it offers a better solution than the pitot–static traverse of the problem of measuring the flow in small pipes.

A sketch of the general arrangement is given in Fig. 6.11. The pitot should be traversed, along at least 2 diameters, in a plane about 1 pipe diameter downstream of the side hole so as not to disturb the static-pressure readings. The difference between the static pressures at the two pipe sections one pipe diameter apart will not, as a rule, be large enough in reasonably smooth pipes to lead to errors of more than 2 or 3 per cent in the measurement of velocity. From equation (4) of Chapter V the pressure drop per one-diameter length of pipe is $4\gamma\,\frac{1}{2}\varrho v^2$; and from Fig. 5.3 the maximum value of γ for fully developed turbulent flow in smooth pipes is about 0·01, so that the pressure drop per

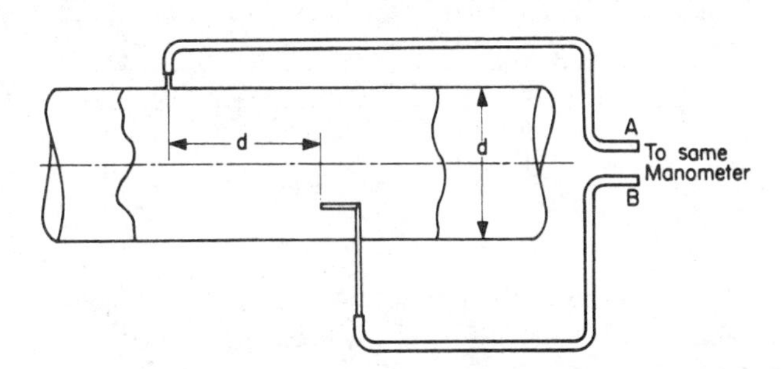

FIG. 6.11. Measurement of flowrate by total-head traverse and side hole.

diameter length for these flows cannot exceed 4 per cent of the velocity pressure. In any event, an allowance for the pressure drop can be made, either on the assumption of smooth-pipe conditions or by preliminary measurements of static pressure at side holes in the traverse plane as well as the upstream plane. The principles that govern the design and size of the side hole have been explained on pp. 136–8.

The external diameter of the pitot tube which, it may be useful to repeat, can be any open-ended tube facing the stream, should preferably not exceed one-fortieth of the pipe diameter. If necessary, hypodermic tubing can be used, and the correction due to Young and Maas (see Table 6.3) should be applied as appropriate.

In using this method, swirl must be eliminated as far as possible by means of flow straighteners (see pp. 189–93). Even so, it is desirable for the highest accuracy to have a number of side holes in the static-pressure plane, all connected to a common airtight reservoir from which another lead to the manometer is taken. The reservoir may consist of a glass bottle with a neck sufficiently wide to take a large cork through which all the pressure connexions pass. A method of making the cork airtight is described on pp. 272–3.

(c) *The Smooth Calibrating Pipe and Micrometer Pitot*

The most accurate method of measuring the flow of air in small pipes is an elaboration of that described in the preceding section. It requires the insertion into the pipeline of a length of smooth-bore pipe, at a suitable section of which a small pitot tube can be traversed by means of a micrometer screw. The smooth pipe is most conveniently attached to an open end of the pipeline if this is available; otherwise it may be inserted at some suitable place in the line. A sketch of the calibrating pipe is given in Fig. 6.12. The traversing pitot is placed

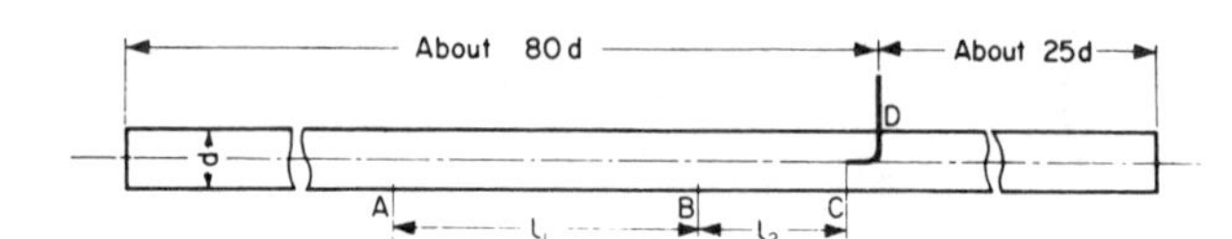

FIG. 6.12. Calibrating pipe.

at *D*, and *A*, *B*, *C* are three static-pressure side holes; *C* is opposite the mouth of the pitot, and *A* and *B* are upstream, *B* being some 4 or 5 pipe diameters forward of the pitot to be quite clear of interference. The velocity pressure at the mouth of the pitot is obtained by connecting the pitot and *B* to opposite sides of a differential manometer, and by adding to the reading obtained the drop in static pressure between *B* and *C*. In most cases, the pressure drop along the pipe will be linear, so that this correction will be simply $(l_2/l_1)(p_A - p_B)$, where l_1 is the length *AB*, l_2 is the length *BC*, and p_A and p_B are the static pressures at *A* and *B*. It is advisable, however, always to make a few preliminary measurements of the static pressures at *A*, *B*, and *C*, with the pitot removed or screwed back against the far pipe wall, to check this linearity of the pressure drop and, if it is found not to be true owing to special circumstances, to determine the relation between the static pressure at *C* and the pressure drop between *A* and *B*.

When the correction to the pitot pressure at C minus the static at B has been established, the quantity of air flowing is measured by taking readings with the pitot at various points in the section, whose radii can be determined from the micrometer readings (see below). Readings are not now taken at specified radii, which is, after all, only a convenient practical procedure, but must be sufficient in number to define the curve of velocity across the diameter. Then the volumetric rate of flow or the mean velocity is calculated from (3) or (3a); the integral can be evaluated graphically by measuring the area under the curve of vr plotted against r. The results have then to be corrected for the effect of turbulence and the size of the pitot as described on pp. 138–40. A numerical example is worked out in Chapter XIII. It should be noted that if the calibrating pipe is not shorter than that shown in Fig. 6.12, the pitot need not as a rule be traversed along more than 1 diameter; but one or two check readings along a second diameter should be made initially to verify that this is unnecessary.

This method is usually used for the measurement of flow in small pipes, but there is no reason why it should not be adapted to larger pipes if desired. A drawing of a form of traversing pitot suitable for small pipes is given in Fig. 6.13. The pitot itself is made of hypodermic tubing, which can be obtained in very small diameters. For 12·5-mm bore calibrating pipe, the outside diameter

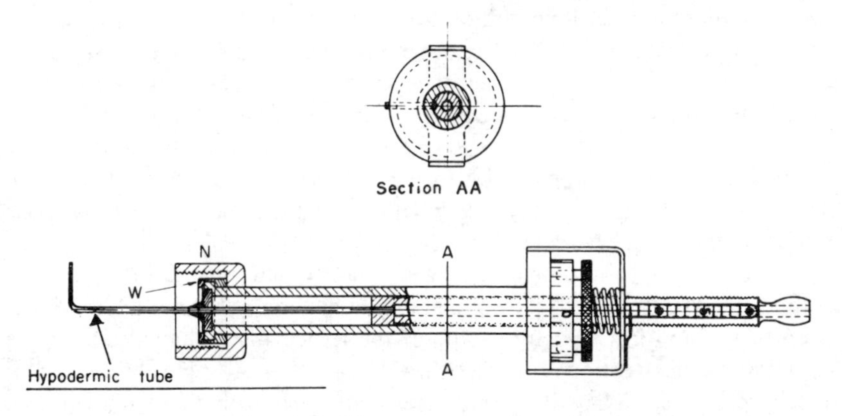

FIG. 6.13. Micrometer traversing pitot.

of the hypodermic tube should not exceed 0·5 mm. The head of the pitot is bent at right angles to face the stream in the usual way; and this portion should be flattened to reduce the correction due to the relative sizes of pipe and pitot (see p. 139). By measuring the pipe diameter and the thickness of the mouth of the pitot, and observing the micrometer reading when the pitot head is just in contact with the far wall of the pipe, we can deduce the radial position of the head for any micrometer reading. Contact with the far wall of the pipe can be simply indicated by the use of an electric-torch bulb and battery if the pitot tube itself, when mounted in position in the pipe, is electrically insulated from it. Insulation is effected by means of the ebonite nut *N* and washer *W* which secure the pitot to the boss fixed to the outside of the pipe. The hole in the wall of the pipe through which the pitot passes should not be larger than necessary to allow the pitot to be introduced without straining it. Leakage through the passages between parts of the pitot gear having relative movement can be reduced to a negligible amount by the use of vaseline.

Calibrating pipes of this kind, of smooth internal bore and equipped with static side holes and traversing pitot, are extremely useful for many kinds of air-flow measurements and calibrations, not only for small pipes. If there is room available to make the pipe of such length that all inlet disturbances are damped out before reaching the section of measurement, the flowrate can be deduced from a single velocity measurement at the axis of the pipe, once the relation between the axial velocity and the mean velocity has been established experimentally (see p. 280 and the example worked out in Chapter XIII). In this way, the amount of experimental work is greatly shortened.

(*d*) *The Shaped Nozzle*

A convenient practical method of measuring the mean flow in a small pipe is to use a standard orifice, nozzle, or venturi (see Chapter VII), or any other constriction for which the coefficient is known. This method is particularly suitable when there is a sufficient length of parallel pipe upstream of the section at which the nozzle or orifice is inserted for the flow at that section to have assumed the fully developed turbulent-velocity distribution, and when, in all other respects, the conditions comply with those specified for the application of the standard discharge coefficients.

Very often it is not possible to meet these requirements, but the same method may still be used with a slight modification, whereby advantage is taken of the uniform velocity profile of the jet leaving a shaped nozzle – a feature to which attention is drawn on pp. 175 and 179. A single observation of velocity pressure, taken with a pitot–static tube arranged co-axially with the nozzle in the outlet plane, will give a good approximation to the mean velocity of the jet leaving the nozzle; and this velocity, multiplied by the area of the nozzle throat, will give the volume flowing. When the highest accuracy is required, a preliminary exploration of the jet by means of the pitot–static tube may be desirable in order to verify the constancy of velocity across the jet. A uniform distribution of this nature has only been observed when there has been fully developed smooth-pipe flow upstream of the nozzle; and it is conceivable that if conditions here are very widely different from this, a lack of uniformity may occur in the jet. For most purposes, however, traversing the jet will be found to be unnecessary. If the nozzle can be conveniently placed at the outlet end of the pipe, a further simplification is possible, since the static pressure in the jet will now be equal to that of the surrounding atmosphere, i.e. zero if we adopt the convention by which atmospheric pressure is taken as the datum. The total pressure at a point in the jet is therefore equal simply to the velocity pressure; so that all that is required in these circumstances is a single observation of total pressure at the axis of the jet in the outlet plane of the nozzle.

References

1. F. C. KINGHORN and A. MCHUGH, An international comparison of integration techniques for traverse methods of flow measurement, *N.E.L. Rep.* (in preparation); International Standards Organization Document I.S.O./TC30/SC3 (U.K. 19) 85.
2. B.S.1042: Part 2A: 1973, *Methods for the measurement of fluid flow in pipes.* Part 2. *Pitot tubes. 2A. Class A accuracy*, British Standards Institution, London, 1973.
3. F. A. L. WINTERNITZ and C. F. FISCHL, A simplified integration technique for pipe-flow measurement, *Water Power* **9** (1957) 225.
4. W. AICHELEN, Der geometrische Ort für die mittlere Geschwindigkeit bei turbulenter Strömung in glatten und rauhen Rohren, *Z. f. Naturforschung* **2** (1947) 108.
5. F. C. KINGHORN, A. MCHUGH, and W. DUNCAN, An experimental comparison of two velocity-area numerical integration techniques, *Water Power* **25** (1973) 330.
6. D. J. MYLES, J. WHITAKER, and M. R. JONES, A simplified technique for measuring volume flow in rectangular ducts, *N.E.L. Rep.* 251, 1966.
7. A. FAGE, The estimation of pipe delivery from pitot-tube measurements, *Engng* **145** (1938) 616.
8. E. OWER, Measurement of the flow of liquids and gases, *Trans. Inst. Chem. Engrs* **18** (1940) 87.
9. J. LAUFER, The structure of turbulence in fully-developed pipe flow, N.A.C.A. Tech. Rep. 1174 (1954).
10. R. W. F. GOULD, *Pitot stem blockage corrections in uniform and non-uniform flow*, British Aero. Res. Council Current Papers 1175, 1971.
11. L. N. KRAUSE and C. C. GETTELMAN, Effect of interaction among probes, supports, duct walls, and jet boundaries on pressure measurements in ducts and jets, I.S.A. Jl **9** (1953), 95 (*Instruments* **26** (1953) 1381).
12. A. THOM and C. J. APELT, The pressure in a two-dimensional static hole at low Reynolds numbers, *R. & M.* 3090 (1958).
13. A. K. RAY, On the effect of orifice size on static-pressure reading at different Reynolds numbers, *Ingenieur-Archiv* **24** (3) (1965) 171.
14. R. SHAW, The measurement of static pressure, *J. Fluid Mechs* **7** (1960) 550.
15. J. L. LIVESEY, J. D. JACKSON, and C. J. SOUTHERN, The static hole error problem, *Aircr. Engng* **34** (1962) 43.
16. R. E. FRANKLIN and J. M. WALLACE, Absolute measurements of static-hole error using flush transducers, *J. Fluid Mech.* **42** (1970) 33.
17. W. J. RAINBIRD, Errors in measurements of mean static pressure of a moving fluid due to pressure holes, *Quart. Bull. Div. Mech. Engng, Nat. Aero. Est., Nat. Res. Council Canada*, Rep. DME/NAE, 1967 (3), 1967.
18. J. A. LIVESEY, The behaviour of transverse cylindrical and forward-facing total-head probes in transverse total-pressure gradients, *J. Aero. Sci.* **23** (1956) 949.
19. F. C. KINGHORN, A. MCHUGH, and W. DUNCAN, Accuracy of air flowrate measurements using pitot traverses and orifice plates, *N.E.L. Rep.* 518, 1972.

CHAPTER VII

THE PLATE ORIFICE, SHAPED NOZZLE, AND VENTURI TUBE

THE pitot tube, in conjunction with a properly designed static tube or a side hole, has the important advantage that its calibration has been established by careful experiment, so that it constitutes a fundamental standard for the measurement of air speed. In the absence of any other standard of equal precision, it must play an important part in the calibration of all other instruments for air-speed measurements. It is, however, subject to certain shortcomings for practical work. In the first place, it is not well suited to a rapid determination of quantity, since the pipe section has to be traversed and a certain amount of arithmetical work has to be performed. Secondly, the pressures to be observed are very often small, so that sensitive manometers are required to give the necessary accuracy; and it is difficult to use these sensitive instruments under ordinary industrial conditions, except in the laboratory.

For these reasons, methods have been sought in which the pressures to be measured are automatically augmented, and in which single readings enable the mean velocity in a pipe to be determined. Although such methods lack the certainty of the pitot tube when used with a manometer of the requisite precision – in fact, it is often safest to calibrate the apparatus employed by means of the pitot tube if the highest accuracy is required – they are undeniably superior to that instrument as regards the ease and rapidity with which the necessary observations can be performed. Briefly, these methods require the insertion into the stream of some device, such as a nozzle or a plate with a circular hole cut in it, which constricts the free area through which the air can pass and so effects a local increase in velocity. It is then found that a comparatively large drop of static pressure is experienced by the air in passing from the full diameter of the pipe to the section at which the constriction, and hence also the velocity, is a maximum, and that this pressure drop can be used as a measure of the air speed and flowrate.

Equations for the Theoretical Flow Through a Constriction

Suppose that in a given pipe the walls are shaped as shown in Fig. 7.1, so that the area at section BB is less than at AA. Let p_1, v_1, ϱ_1, a_1 be respectively the absolute static pressure (referred to vacuum), the mean velocity, the density of the air, and the area of the section at AA; and p_2, v_2, ϱ_2, a_2 be the corresponding quantities at BB. Also imagine that the pipe walls at AA and BB and the directions of flow at these sections are parallel to the axis of the pipe.

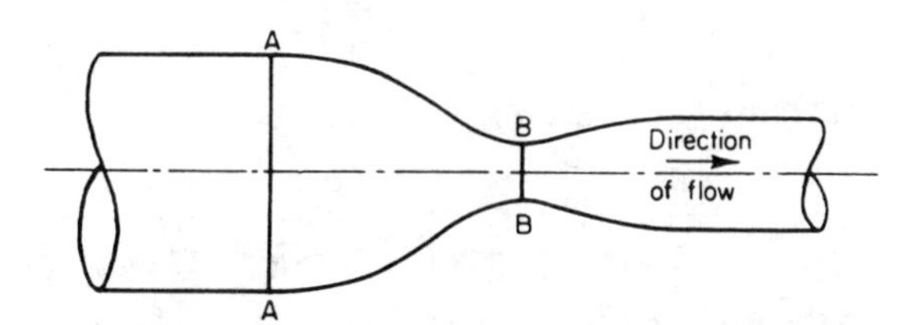

FIG. 7.1. Flow through a constriction.

We shall assume the flow to be frictionless, and at a later stage introduce numerical coefficients to cover the errors caused by this and other assumptions. Such coefficients will have to be determined experimentally.

The difference between the static pressures p_1 and p_2 will depend mainly on the ratio of a_1 to a_2. Frequently, the ratio p_2/p_1 does not differ much from unity,† and in these circumstances ϱ_1 and ϱ_2, the densities of the air at the two sections, may be considered as effectively equal.

We start with the general Bernoulli equation derived in Chapter II, viz.,

$$\frac{v^2}{2} + \int \frac{dp}{\varrho} = \text{constant}$$

and apply it to the air at sections AA and BB. We then have

$$\frac{v_2^2 - v_1^2}{2} = \int_{p_2}^{p_1} \frac{dp}{\varrho}. \tag{1}$$

Case (a): p_2/p_1 nearly equal to unity. In this case, ϱ_2 may be taken as equal to ϱ_1, which is equivalent to regarding the air as incompressible. Equation (1) becomes

$$\frac{v_2 - v_1^2}{2} = \frac{p_1 - p_2}{\varrho}. \tag{2}$$

Also, the mass of air flowing per unit time across AA is equal to that across BB, i.e.

$$\varrho a_1 v_1 = \varrho a_2 v_2,$$

† Since the pressures p_1 and p_2 are referred to vacuum, their mean value in most cases will be approximately atmospheric pressure, i.e. about 10 m of water. Hence in such cases they may differ by about 10 cm of water — an easily measurable amount — before the ratio of one to the other differs from unity by more than 1 per cent.

or
$$v_1 = \frac{a_2 v_2}{a_1}. \tag{3}$$

Substituting (3) in (2), we have

$$v_2^2 = \frac{2(p_1 - p_2)}{\varrho[1 - (a_2^2/a_1^2)]}, \tag{3a}$$

and the theoretical mass of air flowing per second is given by

$$q_t = \varrho a_2 v_2 = a_2 \sqrt{\left[\frac{2\varrho(p_1 - p_2)}{1 - (a_2^2/a_1^2)}\right]}; \tag{4}$$

or, writing m for the ratio a_2/a_1, we have finally

$$q_t = a_2 \sqrt{\left[\frac{2\varrho(p_1 - p_2)}{1 - m^2}\right]} \tag{5}$$

as the equation for the flow of an incompressible fluid through a constriction.

Case (b): p_2/p_1 appreciably less than unity. Compressibility can no longer be neglected, so that ϱ_2 cannot now be taken as equal to ϱ_1: the density will change with p according to the usual laws for gases. In practice the flow along the pipe is almost always sufficiently rapid for the expansion (p_2 is less than p_1) in the constriction to be taken as adiabatic. We now use the general Bernoulli equation (1) and apply it, as in the previous case, to sections *AA* and *BB*. Using the adiabatic relations between the pressure and density ratios set out in Chapter II (eqns. (7)), we thus obtain

$$\frac{v_2^2 - v_1^2}{2} = \frac{\gamma}{\gamma - 1}\frac{p_1}{\varrho_1}\left[1 - \left(\frac{p_2}{p_1}\right)^{(\gamma-1)/\gamma}\right]. \tag{6}$$

The equation for continuity of flow, i.e. equal masses of air passing sections *AA* and *BB* in unit time, is, as before,

$$\varrho_1 a_1 v_1 = \varrho_2 a_2 v_2,$$

or, since
$$\frac{\varrho_2}{\varrho_1} = \left(\frac{p_2}{p_1}\right)^{1/\gamma},$$

$$v_1 = \left(\frac{p_2}{p_1}\right)^{1/\gamma} \frac{a_2}{a_1} v_2.$$

Putting this value of v_1 in (6), and reducing, we have

$$v_2 = \sqrt{\left\{\frac{2\dfrac{\gamma}{\gamma - 1}\dfrac{p_1}{\varrho_1}\left[1 - \left(\dfrac{p_2}{p_1}\right)^{(\gamma-1)/\gamma}\right]}{1 - \dfrac{a_2^2}{a_1^2}\left(\dfrac{p_2}{p_1}\right)^{2/\gamma}}\right\}}, \tag{7}$$

so that, if we write, as before, $m = a_2/a_1$, the theoretical mass of air flowing per second becomes

$$q_t = a_2 \varrho_2 v_2 = a_2 \left(\frac{p_2}{p_1}\right)^{1/\gamma} \varrho_1 v_2$$

$$= a_2 \sqrt{\left\{ \frac{2 \frac{\gamma}{\gamma - 1} \varrho_1 p_2^{2/\gamma} p_1^{(\gamma-2)/\gamma} \left[1 - \left(\frac{p_2}{p_1}\right)^{(\gamma-1)/\gamma}\right]}{1 - m^2 \left(\frac{p_2}{p_1}\right)^{2/\gamma}} \right\}};$$

i.e.

$$q_t = a_2 \sqrt{\left\{ \frac{2 p_1 \varrho_1 \frac{\gamma}{\gamma - 1} \left[1 - \left(\frac{p_2}{p_1}\right)^{(\gamma-1)/\gamma}\right] \left(\frac{p_2}{p_1}\right)^{2/\gamma}}{1 - m^2 \left(\frac{p_2}{p_1}\right)^{2/\gamma}} \right\}}. \quad (8)$$

This equation can be thrown into a form more convenient for practical purposes, containing the pressure drop $p_1 - p_2$ across the constriction, which is easily measurable. Moreover, the modified equation is more readily comparable with (5), its counterpart for the flow of an incompressible fluid.

Let us write ϕ for the expansion ratio p_2/p_1 wherever it occurs in (8); then we may make the following substitution for p_1 when it occurs by itself (not as part of the ratio p_2/p_1):

$$p_1 = \frac{p_1}{(p_1 - p_2)} \times (p_1 - p_2) = (p_1 - p_2) \times \frac{1}{1 - \phi}.$$

Equation (8) may now be written:

$$q_t = a_2 \sqrt{\left[\frac{2\varrho_1(p_1 - p_2)}{1 - m^2}\right]} \sqrt{\left[\frac{(1 - m^2) \frac{\gamma}{\gamma - 1} \frac{1}{1 - \phi} (1 - \phi^{(\gamma-1)/\gamma}) \phi^{2/\gamma}}{1 - m^2 \phi^{2/\gamma}}\right]}. \quad (9)$$

Comparing this equation with (5), we see that it is the equation for the discharge of an incompressible fluid, modified by the term under the second radical, which allows for the effect of compressibility. Equation (9) is the general equation for the flow of a fluid through a constriction.

The Practical Use of Constrictions

In practice, the constriction in a pipeline consists of some device such as an orifice plate, a nozzle, or a venturi tube, each of which is considered in detail later in this chapter. When these appliances are suitably proportioned to the flow, a pressure difference $(p_2 - p_1)$ in (9) is set up of such magnitude that it can usually easily be measured to the necessary accuracy (better than 99 per cent), without a manometer of high sensitivity.

It will be seen that this pressure difference depends upon the ratio of the area of the constriction to that of the pipe, a decrease in this ratio producing a corresponding increase in the pressure difference to be measured. Practical considerations may set a lower limit to the minimum area of the constriction in a pipe of given size, but even so the pressure difference can be made much larger than the velocity pressure in the open pipe. As an example of the advantage to be derived from this feature of a constriction, we may take the case of a 15-cm diameter pipe in which the mean velocity is 6 m/sec. The corresponding velocity pressure is about 2·25 mm of water; whereas if a 7·5-cm diameter plate orifice is inserted in the pipe, the pressure drop across it will be about 9 cm of water, i.e. about 40 times the velocity pressure.

A further merit of the method of measuring the flow in a pipe by means of a constriction is that, since no traverse of the pipe is necessary, as it is for measurements by pitot tube, it lends itself readily to the taking of continuous records of the quantity of air flowing. For this purpose, one has merely to transmit the upstream and downstream pressures to a recording differential pressure gauge, the charts for which may be graduated to read in centimetres of water, or, if great accuracy is not of importance, directly in kilograms of air per minute. In the latter case, of course, the corrections for compressibility and the variations of density of the air with temperature and pressure have to be neglected; and, to reduce errors as far as possible, it is advisable to adjust the gauge to read correctly at the estimated mean conditions of temperature, pressure, and pressure drop across the constriction to which it is likely to be subjected when in use.

The chief difficulty in the practical application of this method of measurement arises from the fact that the actual rate of flow q is always less than the theoretical q_t. The ratio q/q_t is termed the discharge coefficient of the constriction, and its value naturally depends upon the type of the constriction. In practice, two main classes of constriction are employed, of which the venturi tube and the shaped nozzle, on the one hand, and the thin plate orifice, on the other, are typical. The characteristic of the former class is that the stream of fluid is confined within solid boundaries, shaped to avoid abrupt flow disturbances, at least as far as its minimum area. In fact the position of the minimum area, or "throat", of the constriction coincides with that of the stream, and the direction of flow at the throat is parallel to the axis of the pipeline. Thus the conditions postulated in the derivation of the theoretical equations of flow are reproduced except those relating to the uniformity of approach velocity over the entire upstream cross-section of the pipe and to the absence of friction and turbulence. And as losses of energy due to the latter cause upstream of the throat are generally relatively small, we find that the coefficients of discharge of such constrictions are usually nearly unity; and the differential pressures they produce are correspondingly close to those calculated from the theoretical equations (5) or (9) according to whether the flow is incompressible or not.

Conditions are different for the plate orifice, which consists essentially of a thin plate inserted into the pipe on a transverse section, and having cut in it a hole, usually sharp-edged, circular, and co-axial with the pipe. In this case, the

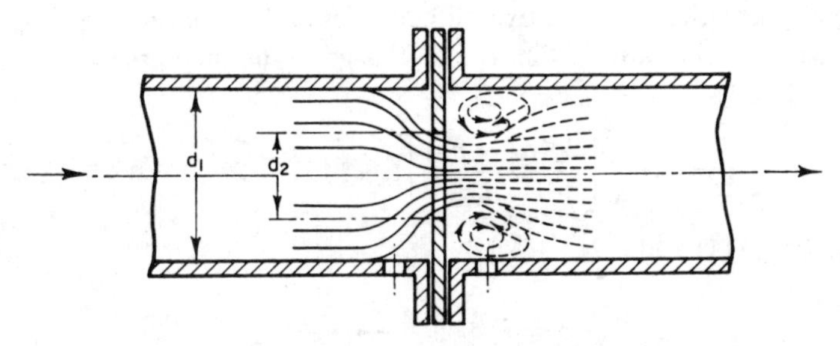

FIG. 7.2. Flow through a plate orifice.

stream issuing from the orifice does not do so as a parallel jet, but continues to contract for some distance downstream of the plate.† The section of minimum area of the stream itself is reached at a distance of 1 pipe diameter or so from the plane of the orifice, and is of the order of 0·6 or 0·7 times the area of the orifice. After passing this minimum section – sometimes called the "vena contracta" – the jet begins to expand, and the pipe again runs full some few diameters downstream.

Equations for the Flow Through a Thin Plate Orifice

The conditions of flow are illustrated in Fig. 7.2. The suffix 2 should now refer to the vena contracta, but for convenience we still take a_2 as the smallest area of the constriction itself, in this case the area of the orifice, and call the area of the vena contracta χa_2. Then, for the theoretical discharge when the flow can be treated as incompressible, we derive the following equation in precisely the same manner as obtained in (5):

$$q_t = \chi a_2 \sqrt{\left[\frac{2\varrho(p_1 - p_2)}{1 - \chi^2 m^2}\right]}, \tag{10}$$

and, if λ is the ratio of the true discharge q to the theoretical,

$$q = \lambda \chi a_2 \sqrt{\left[\frac{2\varrho(p_1 - p_2)}{1 - \chi^2 m^2}\right]}. \tag{11}$$

† Except at very low speeds with which the engineer is not usually concerned (see ref. 1).

Now we have no means of calculating λ and χ from theoretical considerations, and their separate determination by experiment is difficult. In fact it is doubtful whether we could make satisfactory measurements of λ, which depends on the frictional and turbulent losses of energy. The practical expedient that is therefore always adopted is to neglect the contraction of the stream after it passes the orifice, and to call the discharge calculated from (5) the theoretical discharge. An overall coefficient of discharge α† is then introduced, which is the ratio of the true discharge to the theoretical as calculated from (5). Thus,

$$q = \alpha a_2 \sqrt{\left[\frac{2\varrho(p_1 - p_2)}{1 - m^2}\right]}, \tag{12}$$

so that, from (11) and (12),

$$\alpha = \lambda\chi \sqrt{\left(\frac{1 - m^2}{1 - \chi^2 m^2}\right)}. \tag{13}$$

Although we can conveniently determine neither λ nor χ, the overall coefficient α, which is a function of both χ and λ, in accordance with (13), can be measured experimentally; for if, for a given orifice installation, q is measured by one of the other methods available, α can be calculated from (12). (See numerical example in Chapter XIII.)

It may be objected that the introduction of the quantity χ was an unnecessary complication since no use has been made of it. But it is helpful in giving us a more complete physical picture of the nature of the flow through an orifice; and we shall see immediately that it has a useful practical application when we come to consider the flow of compressible fluids.

The flow must be treated as compressible when the ratio p_2/p_1 is no longer nearly unity. We then proceed as follows: the minimum area of the stream is now somewhat larger than its former value of χa_2, for the gas expands in volume on account of the slight drop in pressure downstream of the orifice. We therefore take a new coefficient of contraction χ_c, and call the minimum area of the stream $\chi_c a_2$. It can then be verified that the equation corresponding to (10) is now

$$q_t = \chi_c a_2 \sqrt{\left[\frac{2\varrho_1(p_1 - p_2)}{1 - \chi_c^2 m^2}\right]} \sqrt{\left[\frac{(1 - \chi_c^2 m^2)\ \dfrac{\gamma}{\gamma - 1}\dfrac{1}{1 - \phi}(1 - \phi^{(\gamma-1)/\gamma})\phi^{2/\gamma}}{1 - \chi_c^2 m^2 \phi^{2/\gamma}}\right]}, \tag{14}$$

and, as before,

$$q = \lambda q_t. \tag{15}$$

† See note on definition of discharge coefficient, p. 156.

Combining (14) and (15) and rearranging slightly, we have

$$q = \lambda\chi a_2 \sqrt{\left[\frac{2\varrho_1(p_1 - p_2)}{1-\chi^2 m^2}\frac{\chi_c^2(1-\chi^2 m^2)}{\chi^2(1-\chi_c^2 m^2)}\right]} \sqrt{\left[\frac{(1-\chi_c^2 m^2)\dfrac{\gamma}{\gamma-1}\dfrac{1}{1-\phi}(1-\phi^{(\gamma-1)/\gamma})\phi^{2/\gamma}}{1-\chi_c^2 m^2 \phi^{2/\gamma}}\right]}. \tag{16}$$

We now introduce the *discharge coefficient* α *for incompressible flow*, by substituting in (16) the value of $\lambda\chi$ given by (13), and thus obtain

$$q = \varepsilon\alpha a_2 \sqrt{\left[\frac{2\varrho_1(p_1 - p_2)}{1-m^2}\right]}, \tag{17}$$

where

$$\varepsilon = \frac{\chi_c}{\chi}\sqrt{\left[\frac{(1-\chi^2 m^2)\dfrac{\gamma}{\gamma-1}\dfrac{1}{1-\phi}(1-\phi^{(\gamma-1)/\gamma})\phi^{2/\gamma}}{1-\chi_c^2 m^2 \phi^{2/\gamma}}\right]}. \tag{18}$$

As before, we have thus obtained the equation for compressible flow in a form in which it is exactly similar to that for incompressible flow, with the modification for the effect of compressibility confined to the single term ε.† It is important to notice that the coefficient of discharge α recurring in this equation is the *coefficient for incompressible flow*. In the sequel we shall have to consider various factors that can affect the value of α, but we need not include compressibility since its effect is confined to the term ε.

It should be remarked that the orifice equations (12) and (17) have been derived on the assumption that the upstream and downstream pressures p_1 and p_2 are measured at sections of the pipe where the flow is parallel to the axis. For reasons explained on p. 159 it is generally not convenient in practice to measure the pressures in regions of parallel flow; as a rule, we measure two pressures, which we may denote by p_1' and p_2', which differ somewhat from the values of p_1 and p_2 respectively appearing in the equations. This need not, however, cause us any concern: for example, taking the case of incompressible flow and introducing the *measured* pressure drop $p_1' - p_2'$ into (12), we have

$$q = \alpha a_2 \sqrt{\left[\frac{2\varrho(p_1' - p_2')}{1-m^2}\right]}, \tag{12a}$$

where α is the overall coefficient of discharge which, as before, represents the ratio of the true to the theoretical value of q, the latter being again obtained from (5). The true discharge is still given by (11), so that, from (11) and (12a), we have

$$\alpha = \lambda\chi \sqrt{\left(\frac{1-m^2}{1-\chi^2 m^2}\right)}\sqrt{\left(\frac{p_1 - p_2}{p_1' - p_2'}\right)}. \tag{13a}$$

† ε is termed the "expansibility" factor in British Standard 1042 (ref. 2).

Comparing this with (13), we see that the value of α differs from that previously derived by the factor represented by the second radical in (13a). All that this implies in practice is that, if we measure the upstream and downstream pressures at sections that do not comply with the conditions of parallel flow postulated in deriving the equations, we shall have to use a different value of the discharge coefficient α, as we should have expected on general grounds. But since α has in any case to be determined in the first instance by experiment, the only precaution we need subsequently take is to use the value of α determined for the appropriate disposition of the pressure taps at which the upstream and downstream pressures were observed in the initial calibration to determine α.

Note on Definition Discharge Coefficient

The International Organization for Standardization (I.S.O.) define α as the "flow" coefficient and $\alpha(1 - m^2)^{-1/2}$ as the "discharge" coefficient. Some authorities, for example the V.D.I.[(3)], publish values of this flow coefficient; and it should be remembered that these values must be multiplied by $(1 - m^2)^{1/2}$ to convert them to values of α as defined in this book, which follows the general practice in the United Kingdom as adopted by the British Standards Institution in ref. 2, where α is termed the "basic" discharge coefficient. This practice seems the more logical, since α so defined is the direct ratio of the actual to the theoretical flow, and is the only term in the flow equations that has to be determined experimentally and does not include any of the geometrical parameters of the constriction.

A further argument in support of U.K. practice is that, if high accuracy is not required, α for a sharp-edged plate orifice can be assumed to equal 0·6 over a range of m from 0·05 to 0·55 with an error of less than 1 per cent, whereas the I.S.O. values increase from about 0·6 to about 0·72.

The term $(1 - m^2)^{-1/2}$ is sometimes called the "velocity of approach" factor.

Variables Affecting the Discharge Coefficient

Before we consider the numerical values that have been established experimentally for the discharge coefficients of the different standard types of constriction, it will be useful to discuss from a general standpoint the variables that can influence these values. These variables are:

(a) The density ϱ of the fluid.
(b) The viscosity μ of the fluid.
(c) The compressibility of the fluid.
(d) A speed v characteristic of the flow.
(e) The diameter d_2 of the constriction at the throat.

(f) The diameter d_1 of the pipe.
(g) The roughness of the pipe, expressed in terms of the average height t of the excrescences.
(h) The velocity distribution in the pipe upstream of the constriction.

For the moment we neglect (h), and apply the principle of dimensional homogeneity (see Chapter II, pp. 17–21) to the other variables in this list. In this way, we obtain the following functional relation for α:

$$\alpha = f\left(R_2, M_2, \frac{d_2}{d_1}, \frac{t}{d_1}\right), \tag{19}$$

where R_2† is the throat Reynolds number, and M_2 is the throat Mach number v_2/c (c = speed of sound).

Our working equation for compressible flow (17) uses the value of α for incompressible flow, the effects of compressibility being allowed for by the term ε. Hence we can neglect the variable (c) in the above list, i.e. delete M_2 from (19), which then reduces to

$$\alpha = f\left(R_2, m, \frac{t}{d_1}\right), \tag{20}$$

where m is the ratio a_2/a_1, i.e., $(d_2/d_1)^2$. It is legitimate to write m in place of d_2/d_1 because (19) is a functional relation only.

Equation (20) takes into account all the variables listed except (h). In view of the unlimited number of possible distributions of velocity ahead of the constriction it is impossible to allow for this variable quantitatively. Instead, all published values of α for standard constrictions relate to an upstream velocity distribution for fully developed turbulent pipe flow. This important fact is sometimes overlooked in the practical use of the standard values of α; strictly, these values can be used only when the condition of fully developed flow exists ahead of the constriction. Any departure from this condition must affect the value of α; on general grounds one would expect the effect to be greater the larger the throat diameter, i.e. the larger the value of m. In view of the type of velocity distribution for fully developed flow, however (mean velocity generally over 80 per cent of the maximum – see Fig. 5.6), we should expect this effect to be small for most practical cases provided that the velocity profile is not irregular or markedly more peaked than the fully developed type, or the value of m too large.

The effect of pipe roughness is also usually small in practice. It had to be taken into account in the derivation of (20) because, for geometrical similarity to be completely defined for two similar types of constriction, equality of both m and t/d_1 is necessary; but t will usually be small in relation to d_1, since most

† We could equally well use the Reynolds number of the main flow based on the diameter of the pipe d_1; some of the standard codes do in fact base their data on this Reynolds number. In Ref. 2 R_2 is designated R_t.

piping systems will have reasonable internal smoothness. Hence the larger the pipe the smaller the relative roughness, that is the more nearly will it approach the ideal pipe in which $t = 0$, i.e. for which the term t/d_1 in (20) vanishes, and the functional variation for α can be expressed by

$$\alpha = f(R_2, m) \quad \text{or} \quad \alpha = f(R_1, m) \tag{21}$$

according to whether the Reynolds number is based on d_2 or d_1.

Equation (21) is in fact the basis of all the numerous investigations from which the published values of the discharge coefficient α have been derived for the various types of constriction standardized in different countries; and the data are always given as functions of the two main variables, the area ratio m and the Reynolds number

$$R_2 = \left(\frac{v_2 d_2}{\nu_2}\right) \quad \text{or} \quad R_1 = \left(\frac{v_1 d_1}{\nu_1}\right).$$

Any effect of roughness is allowed for as a small correction based either on the pipe diameter d_1 (as in the British Standard Code[(2)]), or (as in the German Standard[(3)]) on a value of the roughness ratio d_1/t (see below).

As was explained in connexion with the calibration of the standard pitot–static tube, a valuable consequence of the form of variation of α with Reynolds number and m as expressed by (21) is that the values of α determined with one liquid or gas can be applied to others; and calibrations can be carried out in any convenient fluid. Some of the published standard data have been based on or checked by tests in water; indeed, water calibrations are potentially among the most accurate available since the rate of flow can be determined by direct weighing.

The Plate Orifice

Mechanically, the plate orifice is the simplest of the devices under discussion in this chapter, and it is, moreover, the one most easily inserted into an existing pipeline with a minimum of alteration to the layout. Consequently, we find that it is the device for which discharge-coefficient data have been provided by most of the investigators of the flow through constrictions. As a result of their work, and particularly that carried out in Germany, the values of the discharge coefficient can be used with confidence provided that the conditions under which these values were measured are sufficiently closely reproduced.

As usually employed, the plate orifice consists simply of a thin plate or diaphragm clamped between two flanges of the pipe, and having a circular hole, co-axial with the pipe, cut in it, as in Fig. 7.2. Two pressure taps, one upstream and one downstream of the plate, serve as a means of measuring the static-pressure drop $(p_1 - p_2)$ from which the delivery is calculated according to (17).

Equation (17) can be thrown into a form which is more convenient for practical use. As it stands, it will give the quantity of air flowing in kilograms per second if the area a_2 is measured in square metres, ϱ_1 in kilograms per cubic metre, and p_1 and p_2 in pascals. In practice, engineers usually prefer to calculate the quantity Q in kilograms per minute, and to measure the area a_2 of the orifice in square centimetres, and the pressure difference $p_1 - p_2$ in centimetres of water. It will also be convenient to insert the value for ϱ under standard conditions of temperature and pressure, and to apply a correction for the conditions prevailing at the time of the observations. The procedure in modifying equation (17) is similar to that performed in detail in obtaining equation (18) of Chapter VI from (6) of the same chapter, and need not be repeated here. The final form of the modified equation as recommended for practical use is

$$Q = K\varepsilon\alpha a_2 \sqrt{\left[\frac{1}{1-m^2}(h_1 - h_2)\frac{b}{K_1 + t}\right]} \text{ kilograms per minute,} \qquad (22)$$

where a_2 is the area of the orifice in square centimetres, $(h_1 - h_2)$ is the pressure drop in centimetres of water, t is the temperature of the air in the pipe upstream of the orifice, and b is the barometric height. If the temperature and barometric height are measured in degrees Centigrade and millimetre of mercury respectively, $K = 0{\cdot}0573$ and $K_1 = 273$. If h_1, the static-pressure head in the pipe upstream of the orifice, is more than 20 cm of water above or below atmospheric pressure, b must be increased or diminished by the equivalent of h_1 in millimetres of mercury, according to the note on p. 123.†

We have now to consider the value of the discharge coefficient α and of the compressibility factor ε. It will be remembered that the equations of flow were derived on the assumption that the pressure p_2 is measured at the section of minimum area of the stream, where the flow is parallel to the axis of the pipe. To measure p_2 at this section in practice, it would be necessary to find the position of the vena contracta of a given orifice before the downstream pressure tap could be located. Apart from the fact that this position is not fixed, but depends to some extent on the rate of flow, it is obviously undesirable to have to make the necessary preliminary measurements. It has therefore become common practice to measure p_2 either at a fixed distance downstream of the orifice plate – 1 pipe diameter or so – or actually at the downstream face. The former method has the disadvantage that, for every size of orifice in a given pipe, the pressure is in effect measured at a different place with respect to the constriction, so that it is not possible to compare the results of different series of experiments on a geometrical-similarity basis. The majority of workers have therefore adopted the second course, which is not open to this objection. Thus, p_2 in the equations is taken to be the pressure at the downstream plane of the orifice plate instead of at the vena contracta. Since an empirical value of

† If necessary, the density of the air must also be corrected for humidity (see p. 337).

α has in any case to be used, there is no practical objection to our measuring p_2 at any convenient place. We may note here that the drop of pressure between the downstream plane of the plate and the vena contracta is normally only a few per cent of the pressure drop across the orifice. Similarly, p_1 is preferably measured at the upstream plane of the plate, where the pressure is slightly higher than it is a short distance upstream: there is a small rise of pressure as the stream approaches the orifice plate.†

The shape of the edge of the orifice is a variable that has been shown to have an appreciable influence on the discharge coefficient, which increases as the edge is more rounded. It is therefore necessary to standardize some form of edge if we wish to make use of values of the discharge coefficient established once for all by careful calibration tests, and so avoid the necessity of having to calibrate orifices whenever they are installed to measure flow (but see pp. 173–4). A slightly rounded edge has certain practical advantages, the chief of which is that it is less liable to damage or erosion than a sharp edge. On the other hand, a different radius of edge has to be used for each orifice size if geometrical similarity is to be preserved. This complication renders the rounded-edge type unsuitable for standardization, and we therefore find that only sharp-edged orifices have been standardized.

Standard Orifices

One of the earliest investigations of the orifice plate as a device suitable for the commercial metering of gases and liquids was that described by J. L. Hodgson in his classic paper(4) delivered in 1916. Intensive investigations were also undertaken in Germany and America.‡ The German work was by far the more complete and resulted in the specification of a standard sharp-edged orifice which has not only remained substantially unchanged since 1930 (ref. 3, 1932 edition) but has also been adopted, with minor differences of detail, by the International Organization for Standardization and by the British Standards Institution.(2) The type with corner tappings seems to be generally preferred in the German and British specifications. The main advantage of this type is that the positions of the pressure tappings are not variables that affect the conditions of geometric similarity if the size of the orifice or pipe is changed. The same is theoretically true also if the low-pressure tapping is sited at the vena contracta, but in practice this means that the position of the low-pressure tapping has to be changed for each different-sized orifice in a given pipe. Corner tappings are the only ones that allow data for different-sized

† Pressure tappings at the upstream and downstream faces are usually called "corner" taps.

‡ References will be found on pp. 194–6; owing to their large number and the similarity of scope of many of them, it has not been feasible in the summarized account here given always to identify any particular result quoted with the publication from which it came.

orifices to be compared on a Reynolds-number basis from experiments with a single set of pressure tappings in any given pipe. Nevertheless, the American codes[5, 6] do not favour corner tappings. For further discussion of the relative merits of the various pressure-tap positions the reader should consult ref. 7, but he will find that the protagonists of the different types failed to agree.

The essential features of the sharp-edged orifice as adopted by the B.S.I. with corner tappings are shown in Fig. 7.3. It will be seen that the passage through which the air flows has a short cylindrical entry followed by a conical portion with a half-angle of not less than 30 degrees. The pressures are tapped

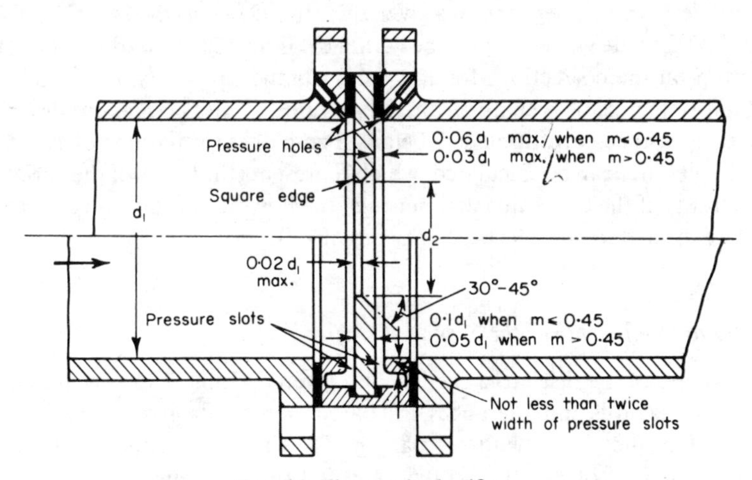

FIG. 7.3. The standard orifice.

off either through two single pressure holes, as shown in the upper portion of Fig. 7.3, or, as shown below, from two annular chambers which are in communication with the pipe pressure over the whole circumference. The latter arrangement is to be preferred when there is any possibility that the flow is not quite symmetrical, e.g. because of bends or obstructions in the pipe ahead of the orifice or downstream (see p. 167). It should be mentioned, however, that serious asymmetry would in any case render the standard coefficients invalid, so that the use of annular chambers must not be taken as a safeguard against anything more than a slight lack of symmetry of flow. Further details of the orifices will be found in refs. 2 and 3. Anyone contemplating the use of the device is strongly recommended to study these documents, which are admirably complete but far too long to be more than briefly quoted here.

The data that these publications give for the sharp-edged orifice take the form of datum or "basic" values of α and corrections to allow for variations of Reynolds number and roughness of surface. In the British code the roughness correction is given as a pipe-size factor for reasons explained on pp. 157–8;

the German corrections are based on tabulated values of the correction for different d_1/t values and Reynolds numbers, the appropriate selection being made either from roughness data specified for various internal pipe surfaces or by calculation from actual measurements of the resistance coefficient γ of the pipe. With a knowledge of γ, equation (7) of Chapter V can be used to calculate d_1/t.

In addition, both codes specify "tolerances" on these corrections and on the basic values of α. These are not tolerances as generally understood in engineering, but are percentages that indicate possible margins of error even when all specified conditions have, as far as is known, been fulfilled. In B.S. 1042 a tolerance of x per cent on a given quantity Q means that on 95 occasions out of 100 the true value of Q will be within the limits $Q(1 \pm 0{\cdot}01x)$. Tolerances are given on the corrections for Reynolds number, pipe diameter, and compressibility factor, as well as on the basic value of α. The magnitudes of the various tolerances are discussed below. From these separate tolerances an overall tolerance can be calculated, which represents the limits of the variations on each side of the true value within which the flowrate as calculated from (22) may be expected to lie on 95 occasions out of 100.

Discharge Coefficients of the Standard Orifice

The values of the discharge coefficient α given in Table 7.1, as well as of the various corrections and tolerances, are based on the data given in B.S. 1042.[(2)] For $m \not> 0{\cdot}5$, they agree within 1 part in 600 with the German data given in ref. 3; for values of m between 0·55 and 0·7 the difference does not exceed $1\frac{1}{2}$ parts in 600. The B.S.I. values are consistently higher than the German by these small amounts; there are also some minor differences in the values of the correction factors and tolerances. The data of Table 7.1 relate to corner tappings; B.S. 1042 gives also data for tappings 1 pipe diameter upstream and

TABLE 7.1. BASIC DISCHARGE COEFFICIENTS α FOR THE STANDARD ORIFICE WITH CORNER TAPPINGS

m	α	Minimum value of R_2 for which α can be used	m	α	Minimum value of R_2 for which α can be used
0	0·596	10^5	0·40	0·605	$3{\cdot}2 \times 10^5$
0·05	0·598	10^5	0·45	0·604	3·8
0·10	0·599	10^5	0·50	0·602	4·6
0·15	0·601	$1{\cdot}3 \times 10^5$	0·55	0·598	5·8
0·20	0·603	1·7	0·60	0·592	7·3
0·25	0·604	2·2	0·65	0·584	$9{\cdot}4 \times 10^5$
0·30	0·605	2·4	0·70	0·573	10^6
0·35	0·605	2·8			

0·5 diameter downstream; and also for flange tappings about 2·5 cm from each face of the orifice plate. The corner tappings have the additional advantage, which we have not yet mentioned, that the data may be used for the two following cases in addition to ordinary pipe flow:

(a) To measure the flow from a pipe into a large space; the maximum permissible value of m for this purpose is 0·15.
(b) To measure the flow from a large space into a pipe or into another large space separated from it by a partition; for such measurements the value of α for $m = 0$ (i.e. $\alpha = 0 \cdot 596$) is used.

Table 7.1 gives the values of α for the standard orifice for values of m ranging from 0 to 0·7. It is found that α is constant at all values of the throat Reynolds number R_2 above a certain minimum for each value of m. Values of these minimum Reynolds numbers are included in the table. For lower Reynolds numbers, corrections (see below) must be applied.

The I.S.O. have recently proposed revised values of α which have been obtained by fitting more precise curves to the experimental data from which the B.S.I. values are deduced. These new values are consistently lower than the basic values given in Table 7.1 (i.e. the values at Reynolds numbers higher than the minima there quoted); but the differences do not exceed 0·5 per cent over the entire range of m, and they are mostly about 0·25 per cent. If the exact new basic I.S.O. values are required, they can be calculated by the I.S.O. formula

$$\alpha = 0 \cdot 5959 + 0 \cdot 0312 m^{1 \cdot 05} - 0 \cdot 184 m^4,$$

and the effect of Reynolds number is represented by an additional term equal to

$$0 \cdot 0029 m^{1 \cdot 25} \left(\frac{10^6 \sqrt{m}}{R_2} \right)^{0 \cdot 75} .$$

If we assume that the minimum value of R_2 for which the basic values of α can be used without Reynolds-number correction is that for which the effect of R_2 on α is limited to 1 in the third decimal place, we can use the above term to derive the values of minimum R_2 given in Table 7.2.

TABLE 7.2. I.S.O. VALUES OF MINIMUM R_2

m	0·05	0·10	0·15	0·20	0·25	0·30	0·35	0·40 and above
R_2 min.	3×10^4	8×10^4	2×10^5	4×10^5	5×10^5	6×10^5	7×10^5	10^6

These values differ somewhat from those given in Table 7.1. But these differences are more or less of academic interest only. No one measuring air flow by means of an orifice should assume that the values of α he uses are accurate to 1 in 600 without having the orifice carefully calibrated (preferably by water-flow weighing) *in situ*.

Subject to the following restrictions, the basic values of α given in Table 7.1 can be used in all pipes of internal diameter not less than 2·5 cm, and for flow into and out of large spaces, as mentioned above:

(a) The orifice diameter must not be less than 6·5 mm.
(b) If the pipe diameter does not exceed 5 cm the area ratio m must not exceed 0·5.
(c) For pipes of above-average roughness, the minimum pipe diameter must be more than 2·5 cm. Minima of 5 cm are specified for rusty steel pipes and (not rusty) cast-iron pipes, and of 20 cm for slightly encrusted steel and cast-iron pipes. For all other pipes in common use – brass, copper, lead, plastic, steel, and bituminized cast iron – a minimum diameter of 2·5 cm is allowable.
(d) The ratio of the downstream pressure p_2 to the upstream pressure p_1 must not be less than 0·8 (see below).

It will be seen from Table 7.1 that as long as m does not exceed 0·55, the value of α is approximately constant; and the assumption of a mean value of 0·603 will lead to an error of only 1 per cent at $m = 0{\cdot}55$, whilst over most of the range of m up to 0·55 the error will be of the order of 0·5 per cent. Similar conclusions regarding the small variation in α for m less than 0·5 were drawn by Hodgson,[(4)] who quoted a mean value of 0·608 for square-edged orifices very similar to the standard: Hodgson stated that this value could be used provided that d_2/d_1 did not exceed 0·7 ($m = 0{\cdot}49$).

At Reynolds numbers lower than those given in Table 7.1, the value of α rises and then falls rapidly as the Reynolds number is further decreased. This is shown in Fig. 7.4, which is based on results obtained by Johansen[(1)] for sharp-edged orifices (not identical with the standard). A useful practical feature to be observed from these results is that the variation of α with Reynolds number is much less for small area ratios than for large. Hence, when doubt is entertained whether the flow in a given case is likely to be fast enough for the critical Reynolds number to be exceeded, it is preferable to use an orifice with as small a contraction ratio m as possible. We must bear in mind, however, that if m is too small the fan may not be able to maintain the desired rate of flow against the increased loss of pressure (see p. 170). Apart from this limitation, the use of small values of m is preferable also for other reasons, which are mentioned in the two following sections of this chapter.

Variation of Orifice Discharge Coefficient with Pipe Diameter

We have already seen that the discharge coefficient of a constriction depends to some extent on the pipe diameter, unless special care is taken to ensure a

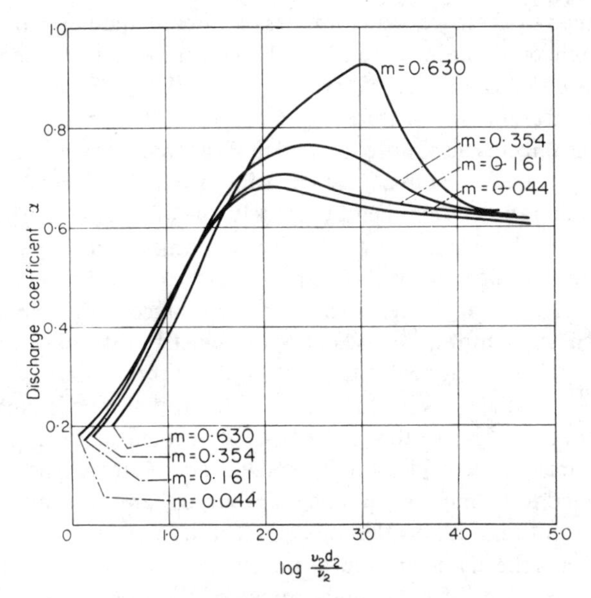

FIG. 7.4. Orifice coefficients at low Reynolds numbers.

smooth interior to the pipe. This effect, which is really a roughness effect, has been referred to by various investigators in relation to orifice discharge coefficients, and a useful analysis of existing data for pipes of normal surface finish has been made by Johansen.[8] He dealt with results obtained by Spitglass,[9] Witte,[10] Jakob,[11] and Kretschmer,[12] and expressed the discharge coef-

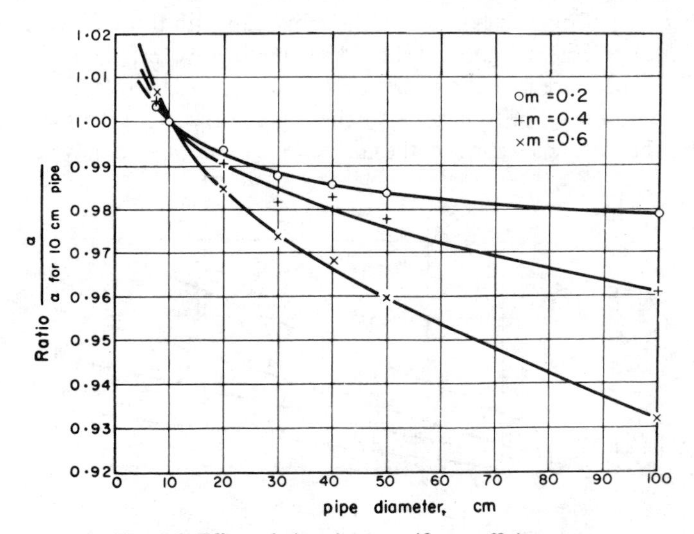

FIG. 7.5. Effect of pipe size on orifice coefficient.

ficients obtained by these workers for pipes of different diameter in terms of the discharge coefficients obtained in a 10-cm diameter pipe. The averaged results of the analysis are shown in Fig. 7.5. It will be seen that α decreases as the pipe diameter increases, the decrease being more rapid for small than for large diameters. The effect is also more marked at large than at small values of m – a further instance of the advantage of using a small value of m. Spencer, Calame, and Singer[13] recommend that the value of m should, if possible, be within the range 0·25 to 0·36, for which the sum of the effects of pipe roughness and possible loss of edge sharpness (see p. 167) are least. The effect of roughness in increasing the discharge coefficient is in accordance with expectation; for, as first shown by Stanton,[14] the curve of velocity distribution in a rough pipe is more peaked in the centre than in a smooth pipe. That is, for the same mean rate of flow, the speed at the centre of the rough pipe is higher than in the smooth pipe; and hence the orifice discharge coefficient is higher – more fluid passes through the orifice for the same pressure drop. The effect would also be expected to be more pronounced for large than for small values of m, and the curves of Fig. 7.5 show that this is in fact the case.

The decrease in the diameter effect as the diameter increases has already been explained: for a series of commercial pipes of the same finish, the relative roughness will decrease with increase of diameter, until the larger pipes finally approximate to perfectly smooth pipes. Witte (ref. 10; 1931) showed that, at pipe Reynolds numbers up to 700,000, with pipes turned and polished internally, there is no measurable change in α for diameters of 5–100 cm; and also that the coefficients are in very close agreement with those obtained in 100-cm pipes of ordinary roughness.

Corrections to α for pipe size, based on B.S.I. values,[2] are shown in Fig. 7.6. For the range of pipe sizes in which these overlap with those to which Fig. 7.5 relates, the agreement is good. It should be noted, however, that the corrections given in Fig. 7.6 are maxima to be applied to small pipes in a normal state of surface smoothness, or to larger, rougher pipes. According to the size of the pipe and the state of its surface, the corrections to be applied may be half those

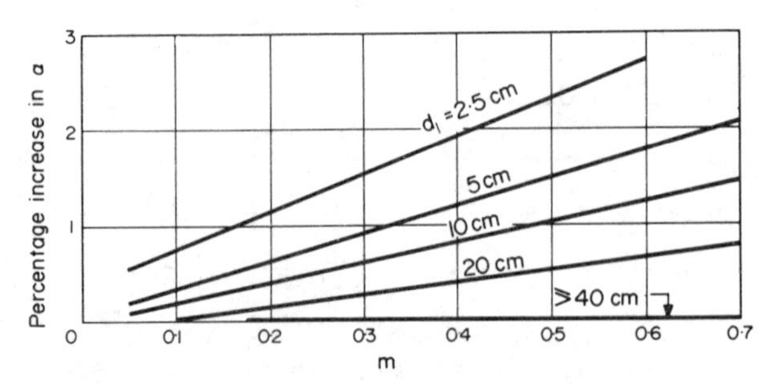

FIG. 7.6. Pipe-size correction for standard orifice.

shown in Fig. 7.6, or even zero. A table is given in B.S. 1042 from which the appropriate correction can be deduced.

Other Factors Affecting the Discharge Coefficient

Great care must be taken to make the upstream edge of the orifice very sharp if the standard values of α are to be used.[13] Even slight rounding increases α, and it is important that, besides being sharp, the edge should be true and free from dents. The extent to which the edge needs to approach perfection increases as the size of the orifice decreases. According to B.S. 1042, the edge may be considered sharp if the radius of curvature nowhere exceeds $0{\cdot}0004d_2$. The V.D.I. rules state that this condition implies that the edge of an orifice of 125-mm diameter is just able to reflect a visible ray of incident light.

Another important consideration is the presence of bends or changes in the pipe section, upstream or downstream of the orifice: if there is an insufficient straight length of pipe between them and the orifice, the discharge coefficient will be affected. The requisite free lengths of pipe between various fittings and the orifice, in order that α shall not be affected, are given in B.S. 1042. A selection of these figures is reproduced in Table 7.3; these lengths can be considerably reduced if an effective flow straightener (see p. 189) is fitted upstream of the orifice.[15]

TABLE 7.3. LENGTH (NUMBER OF DIAMETERS) OF STRAIGHT PIPE REQUIRED UPSTREAM OF VARIOUS FITTINGS FOR ORIFICES WITH ANNULAR PRESSURE TAPPINGS (SEE LOWER PORTION OF FIG. 7.3)

Type of obstruction	Value of m not greater than				
	0·05	0·1	0·3	0·5	0·7
Contraction (not more than $0{\cdot}5d_1$ in a length $3d_1$) or Expansion (not more than $2d_1$ over a length $1{\cdot}5d_1$)	16	16	20	26	33
Gate valve, fully open	12	12	13	20	38
Globe valve, fully open	18	18	23	32	49
Contraction, any reduction	25	25	25	26	33
Single bend, up to 90°	10	10	16	29	56
Two or more bends in same plane	14	15	22	36	57
Two or more bends in different planes	34	35	44	63	89

If, instead of annular tappings, there are single upstream and downstream pressure tappings as in the upper portion of Fig. 7.3, the lengths specified in Table 7.3 must be at least doubled; or, if they are the same, the basic tolerance on α (see below) must be increased by adding 0·5 per cent. Partly-open gate or globe valves are liable to cause swirl, and the lengths given for fully open valves in this table may have to be increased twofold or more unless a flow straightener is fitted.

From Table 7.3 we note one more advantage of using as low a value of m as the fan can accommodate – the smaller m, the shorter is the requisite upstream length of pipe, i.e. the less susceptible the value of α to upstream disturbances.

As regards the length of pipe downstream of the orifice, the minimum distance is given as 3 diameters, but if it is less than 5 diameters when $m \leq 0{\cdot}4$ or less than 7 diameters when $m > 0{\cdot}4$, the basic tolerance on α is to be increased by adding 0·5 per cent.

Corrections and Tolerances for the Standard Discharge Coefficients

B.S. 1042 gives corrections Z_R to be applied to α for Reynolds numbers (see Fig. 7.7) when these are below the minima for constant α listed in Table 7.1, and, as already noted, for pipe diameter (see Fig. 7.6). Tolerances are also given for basic α, which are equal to 0·5 per cent up to $m = 0{\cdot}5$, 1 per cent at $m = 0{\cdot}6$, and 1·5 per cent at $m = 0{\cdot}7$. This increased tolerance (i.e. uncertainty – see p. 162) at higher m is no doubt partly due to the fact already noted on p. 157 that departures from the fully developed turbulent velocity distribution ahead of the constriction will have greater effects on α the higher the value of m.

B.S. 1042 also gives tolerances for pipe size (see Fig. 7.8), for Reynolds-number correction (which is equal to 0·33 Z_R, Z_R being taken from Fig. 7.7), and for the compressibility factor ε (which may amount to 1 per cent). The tolerance on the final value of Q calculated from (22) is the r.m.s. of the individual tolerances for these three factors and on basic α, and can, in a combination of unfavourable conditions, amount to rather more than 3 per cent. Usually, however, it will be possible from the data given in B.S. 1042 to match the selected orifice with the flow conditions in such a way that the overall tolerance is less than this. On the other hand, additional tolerances are specified in B.S. 1042 to allow for possible errors in the measurement of pipe and orifice diameters, of the air density, and of the pressure differential across the orifice.

The V.D.I. rules[3] give correction factors (greater than 1) to be used if the radius of curvature of the edge of the orifice exceeds $0{\cdot}0004d_2$. No similar corrections or tolerances are given in the B.S.I. publication which does, however, emphasize the importance of a truly sharp edge. Corrections have also to be made if the temperature of the air stream is appreciably different from that at which the orifice diameter was measured, to allow for the change in diameter with temperature. This change is calculated in the normal way, using the coefficient of linear expansion of the material of which the orifice is made. A similar correction should be applied to the diameter of the pipe; the value of m will be unchanged if the orifice plate and the pipe are made of the same material.

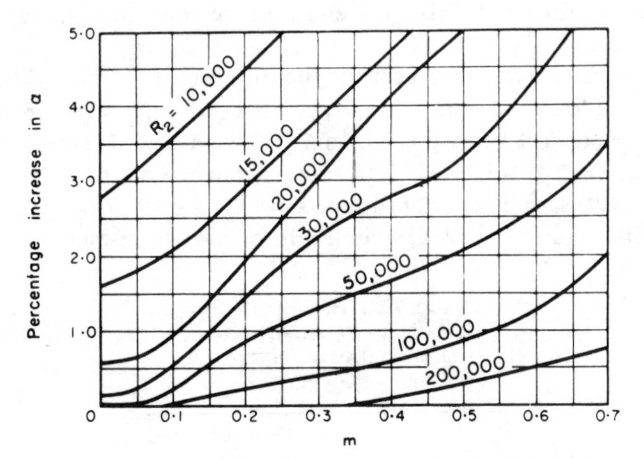

FIG. 7.7. Reynolds-number correction for standard orifice.†

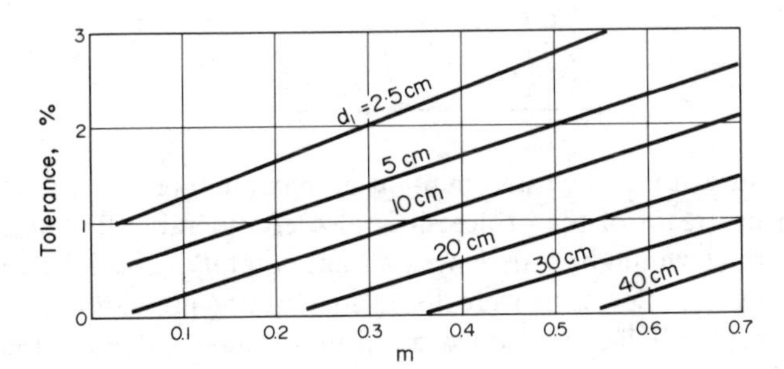

FIG. 7.8. Pipe-size tolerance for standard orifice.

For full details of the various corrections and tolerances the user should refer to B.S. 1042.

Loss in Pressure Due to the Passage of Air Through an Orifice Plate

As we have seen, the static pressure at the downstream face of an orifice plate is considerably less than that upstream of the plate. There is a further slight decrease as the air leaves the orifice, a pressure minimum being reached at the minimum section of the jet about 0·5 to 1 pipe diameter from the orifice. This pressure minimum is followed by a recovery of pressure, rapid at first and then more gradual, which persists up to a distance downstream depending on the

† The curves for R_2 = 10,000 and 15,000 are applicable only for pipes of less than 5-cm internal diameter.

value of m as shown in Table 7.4, and probably also to some extent on the Reynolds number.

At the point of maximum recovery, the jet, which at its throat had a section considerably less than that of the pipe, has expanded once more to the full pipe diameter. After this pressure maximum has been reached, there is a gradual pressure drop due to wall friction, but not at the normal rate (see Fig. 5.3) because fully developed pipe flow, even if it exists upstream of the orifice, will require a considerable length of pipe to re-establish itself. The

TABLE 7.4. DISTANCE OF POINT OF MAXIMUM PRESSURE RECOVERY DOWNSTREAM OF SHARP-EDGED ORIFICE†

m	Distance (pipe diameters)
0·7	2·6
0·5	3·5
0·3	4·7
0·1	7·6
0·04	10·2

static pressure at the downstream point of maximum recovery is considerably less than upstream of the orifice; it is also considerably less than the static pressure which would be experienced at that point if the orifice were removed, and the only loss of pressure were the normal friction pressure drop in a parallel pipe. The orifice, in fact, introduces a resistance equal to that of a long length of pipe, and causes a corresponding loss of energy in the flow. In other words, a fan delivering a given quantity of air per unit time along a pipeline would take more power if an orifice were introduced into the line and the quantity of air flowing were maintained at its previous value.

The last-mentioned point has an important practical significance; for before a decision is made to install an orifice in a given pipe system to measure the flowrate, it is necessary to make sure than the fan available will deliver the requisite quantity against the increased pressure difference due to the loss of pressure through the orifice. Experiments show that a sufficiently close estimate can be made of the pressure loss from the assumption that, expressed as a fraction of the difference between the pressures at the upstream and downstream faces, it is numerically equal to $(1 - m)$, i.e.

$$\frac{p_D}{p_1 - p_2} = 1 - m,$$

† Table 7.4 is based on data given in ref. 4 for a sharp-edged orifice similar to but not identical with the standard of B.S. 1042.

where P_D is the unrecoverable loss of pressure through the orifice. This formula gives results that agree closely with those plotted in Fig. 7.12.

The Contraction of the Jet from an Orifice

Witte (ref. 10; 1931) used the published values of α for the German standard orifice to make approximate estimates of the contraction of the jet of an incompressible fluid after flowing through that orifice. That is to say, he used (13) to calculate χ, the ratio of the area of the minimum section of the jet to the area of the orifice. For this purpose, he had to assume that λ was equal to unity, i.e. that losses of energy due to friction and turbulence upstream of the orifice are small enough to be negligible. Although this assumption is near the truth for constrictions such as shaped nozzles and venturis, as is clear from the fact that their discharge coefficients are very near unity (except at low Reynolds numbers when frictional effects become relatively more important), it cannot be so nearly true for an orifice whose sharp edge must cause eddying and consequent loss of energy. However, with this limitation in mind, we shall find it of some interest to see the results of Witte's calculations, as they will provide us with an indication of the order of size of the minimum section.

Taking λ as unity, we can rewrite (13) thus:

$$\chi = \frac{\alpha}{\sqrt{[1 - m^2(1 - \alpha^2)]}},$$

and, inserting the appropriate value of α from Table 7.1, Witte obtained the values of χ for different values of m shown in Table 7.5. According to these assumptions, therefore, the minimum section of the jet has an area between 60 and 70 per cent of that of the orifice itself.

TABLE 7.5. APPROXIMATE VALUES OF χ FOR INCOMPRESSIBLE FLOW

m	0.1	0·2	0·3	0·4	0·5	0·6	0·7
χ	0·60	0·61	0·62	0·64	0·66	0·68	0·70

The Compressibility Factor ε for Orifices

A number of investigators have discussed the compressibility factor ε for orifices, and formulae for calculating it have been given by Witte (ref. 10; 1930, 1931), Ruppel,[16] Busemann,[17] and Buckingham.[18] These formulae are, however, all somewhat complicated and based on various assumptions which may or may not represent real conditions. It is now standard practice to use values of ε based on experimental investigations by Witte (ref. 10; 1930, 1931) with air and superheated steam flowing through the German standard orifice and the German standard nozzle. It will be seen from (9) and (18) that for a given gas, i.e. for a given value of γ, ε is a function of the pressure ratio $\phi(=p_2/p_1)$† and the area ratio m. In practice, p_2 is not usually measured directly: the technique most commonly employed is to measure the pressure difference $p_1 - p_2$ across the orifice. The British Standard[(2)] code accordingly presents the values of ε as functions of m and $(p_1 - p_2)/p_1$ for different values of γ, the results being given in the form of graphs for a range of values of $(p_1 - p_2)/p_1$ that correspond to a range of p_2/p_1 between 1 and 0·8. The German code tabulates the values of ε for a range of p_2/p_1 of 1 to 0·75, i.e. $(p_1 - p_2)/p_1$ of 0 to 0·25. Both codes specify that pressure ratios smaller than these limits (0·8 and 0·75 respectively) should not be used;‡ this restriction can usually be satisfied by suitable choice of the area ratio m. Table 7.6 gives values of ε for air ($\gamma = 1{\cdot}4$) compiled from the German data. Values of ε for intermediate values of $(p_1 - p_2)p_1$ can be obtained by interpolation; curves of ε plotted on a base of p_2/p_1 for constant m will be found to be linear. Both codes give values of ε for other gases, but only figures for air are reproduced here.

TABLE 7.6. VALUES OF THE COMPRESSIBILITY FACTOR ε FOR STANDARD ORIFICES ($\gamma = 1{\cdot}4$)

	$(p_1 - p_2)/p_1$								
m	0·25	0·20	0·15	0·10	0·08	0·06	0·04	0·02	0
0	0·923	0·938	0·953	0·968	0·974	0·980	0·987	0·993	
0·1	0·922	0·938	0·953	0·968	0·974	0·980	0·987	0·993	
0·2	0·920	0·936	0·952	0·967	0·973	0·980	0·986	0·993	
0·3	0·917	0·933	0·950	0·966	0·972	0·979	0·986	0·992	1·000
0·4	0·913	0·930	0·947	0·964	0·971	0·978	0·985	0·992	
0·5	0·907	0·925	0·943	0·961	0·969	0·976	0·984	0·991	
0·6	0·899	0·919	0·939	0·958	0·966	0·974	0·983	0·991	
0·7†	0·890	0·911	0·933	0·955	0·963	0·972	0·981	0·990	

† Values obtained by extrapolation: the German data are not given for $m > 0{\cdot}64$.

The formula $\varepsilon = 1 - \beta(p_1 - p_2)/p_1$ gives results that agree within 1 or 2 parts in 1000 with the figures given in Table 7.6, where β has the values shown in Table 7.7 for different values of m.

TABLE 7.7

m	0·05	0·1	0·2	0·3	0·4	0·5	0·6	0·7
β	0·31	0·31	0·32	0·33	0·35	0·37	0·40	0·45

† The pressure p_1 and p_2 must be expressed as absolute pressures (referred to vacuum) in using them to determine ε.

‡ The reason for this, not stated in either code, is that the I.S.O. consider that the available experimental data are insufficient to justify the international standardization of values of ε for lower pressure ratios.

Practical Notes on the Use of Orifices

The literature about orifice plates is very extensive.[19] Some writers have tended perhaps to overstate the arguments for adopting them rather than other methods of flow measurement, and to underestimate the care that must be taken if the basic values of the discharge coefficient α published by the various standardizing authorities are to be used without verification. Strictly, these basic values of α are valid only for smooth pipes and fully developed turbulent flow upstream of the constriction; but the various correction factors and tolerances allow for certain departures from these conditions. Even so, however, quite appreciable errors may arise: as already noted on p. 168, in adverse circumstances in which all the tolerances given in ref. 2 have high values simultaneously. For example, if $m = 0{\cdot}5$, $d_1 = 2{\cdot}5$ cm, and $R_2 = 20{,}000$ the overall tolerance is 3·3 per cent; i.e. there is a 5 per cent chance that in any particularly case the calculated value of α may be 3·3 per cent in error.

This is an extreme case: as a rule orifice installations meeting the B.S.I. requirements will give considerably better accuracy.[20] Indeed, if great care is taken, the errors will probably be less than 0·5 per cent. This was shown by some experiments[21] made at the National Engineering Laboratory with water flowing through a carefully made and installed orifice, with a length of over 30 diameters of straight pipe ahead of the orifice so that a velocity distribution closely approaching that for fully developed turbulent flow must have existed. Moreover, the pipe was very smooth. Calculations of the mean rate of flow using B.S.I. values for α agreed within 0·25 per cent with the rate of flow as obtained by direct weighing of the water flowing in a measured time.

It can thus be accepted that, provided that the B.S.I. requirements are fully met, there will be no need to calibrate the orifice: the values of the discharge coefficient can be taken from Table 7.1 and corrected as necessary in accordance with B.S. 1042. The most important conditions to observe are those relating to the sharpness of the edge of the orifice, the smoothness of the pipe, particularly for a pipe of small diameter, the location of the pressure taps close to the orifice plate, and the length of unobstructed straight pipe upstream and downstream of the orifice (see Table 7.3). If for some reason it is impossible or inconvenient to reproduce the standard conditions sufficiently closely, a calibration of the orifice should be performed *in situ* over the range of flow at which it will be used in service.† The calibration is not a matter of great difficulty; an example of one method of carrying it out is given in Chapter XIII.‡ If the pipe is sufficiently large for the flowrate to be measured by means of the standard pitot–static tube, the work is somewhat simplified. It is only necessary to take measurements with this instrument at a convenient section of the pipe and

† Some information about the order of errors caused by certain departures from standard conditions for both orifices and nozzles is given in B.S. 1042: Part 3: 1965, *Guide to the Effects of Departures from the Methods in Part 1*.

‡ Another possible method employs a choked nozzle as mentioned on pp. 189–90.

simultaneous readings of the pressure drop across the orifice, the air temperature in the pipe, and the barometric pressure. Equation (12) or (17) (according to the value of p_2/p_1) for the discharge through the orifice can then be solved for α by inserting the measured value of q. If an accuracy of 1 per cent is good enough, the compressibility factor can be neglected, provided that p_2/p_1 is greater than 0·98, for the data given on p. 172 show that ε is greater than 0·99 for all values of p_2/p_1 between 0·98 and 1·00.

If it is decided that a calibration is advisable, there will probably not be much object in incurring the trouble and expense of making or buying a standard orifice. A simpler type may be installed, with, perhaps, slightly rounded edges, which are easier to make and less likely to suffer damage than sharp edges; and the pressure taps can be located at any convenient points near to the orifice plate, but not necessarily at the upstream and downstream faces.

Attention has already been drawn to the need for bearing in mind the increased pressure against which the fan will have to work to deliver the same quantity of air as before if an orifice meter is inserted in the pipeline.

The Shaped Nozzle

Another form of constriction, which has found favour in many quarters, is the shaped nozzle. In this device, the air is guided smoothly to the section of minimum area of the stream, which coincides with the minimum area of the constriction itself from which the stream issues as a parallel jet that does not contract further. In consequence, the nozzle has a considerably higher discharge coefficient than the plate orifice; as we shall see in the next section, α for the nozzle is about 0·95 on the average as compared with 0·6 for the orifice. It follows from (22) that, for the same rate of flow and the same value of m, the differential head with the nozzle is about $(0{\cdot}6/0{\cdot}95)^2$, i.e. 1/2·5, that of the orifice; and, since the irrecoverable pressure loss is roughly the same percentage of the differential head as it is for the orifice, the fan will have to overcome a considerably smaller resistance for the same rate of flow if a nozzle is fitted instead of an orifice.

But it may not always be possible to make full use of this advantage that the nozzle has over the orifice, because the lower differential pressure from which the flow is calculated may not be high enough to be measured with the requisite accuracy by the manometer or pressure recorder available: a nozzle with a smaller value of m may then have to be used with a consequent increase in the net pressure loss. Whatever the constriction adopted, the value of m should be chosen – within any limits prescribed by the standard code – such that the differential pressure is no greater than can be measured to the desired degree of accuracy. For example, if the manometer can be read to 1 mm of water, the value of m should as a rule be such that the differential pressure across the constriction at the lowest rate of flow that it is required to measure does not exceed 10 cm of water; the error in $(h_1 - h_2)$ (see eqn. (22)) will then be 1 per cent and the error in Q will be 0·5 per cent, which will usually be sufficient. Whether we use an orifice or a nozzle, therefore, we shall need a pressure drop of 10 cm of water; but to get it we can use a smaller value of m with the nozzle. This is a distinct advantage because, as we have already noted in connexion with the orifice (and it is true also for the nozzle), the smaller the value of m the less the probable errors due to roughness and the presence of obstructions in the pipe ahead or downstream.

Thus, the higher value of α for the nozzle gives the latter an indirect advantage over the orifice; but the main superiority of the nozzle lies in the absence of any mechanical feature comparable with the sharp edge of the orifice, to the condition of which the discharge coefficient is so sensitive. This is of special advantage in the metering of streams of gas or air laden with dust or other material which might erode or tend to round off the sharp edge: in such conditions, the use of the nozzle is strongly to be recommended.

Another characteristic of the nozzle, of which use can often be made, is the uniform velocity distribution that exists over the greater part of the cross-section of the jet issuing from the throat of a well-designed nozzle. This is discussed in greater detail on pp. 179–80.

The Standard Nozzle

The British Standards Institution, the Verein deutscher Ingenieure, and the International Organization for Standardization have adopted as a standard nozzle a type that is in all essentials the same as that known as the German standard nozzle (see ref. 3, 1932 edition). It was also adopted by the International Federation of National Standardizing Associations† (I.S.A.) in 1932, and is sometimes called the I.S.A. 1932 nozzle. It is illustrated in Fig. 7.9 from which it will be seen that there are two types, one for use when $m < 0{\cdot}45$ and one when $m > 0{\cdot}45$. The two radii of curvature of the inner profile, namely $0{\cdot}333d_2$ and $0{\cdot}2d_2$, and their centres, are so arranged that when $m \leq 0{\cdot}45$ the profile runs tangentially into the flat upstream face of the nozzle and into the cylindrical throat portion; but when m exceeds 0·45, the curved part of the nozzle extends beyond the diameter d_1 of the pipe; a flat surface is then to be turned on the upstream side as shown in Fig. 7.9(b).

The arrangements for tapping off the pressures are the same as for the standard orifice (see Fig. 7.3); either single tappings or annular rings may be used. The mouth of the nozzle has a small lip, as shown, to protect it from mechanical damage; its use is optional: if there is no likelihood of damage, it may be omitted.

† The title of the International Organization that existed before the war of 1939–45.

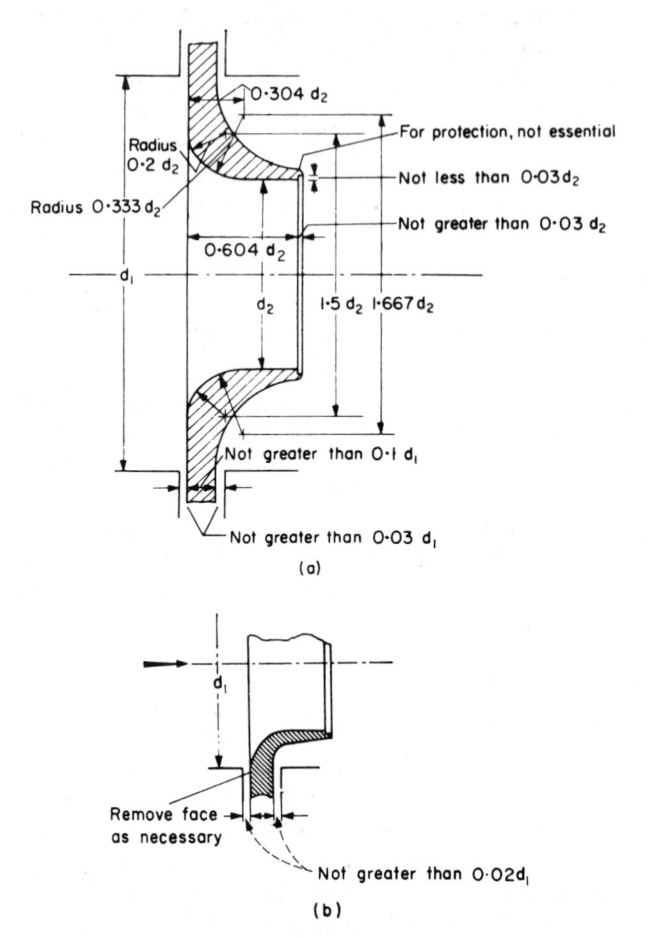

FIG. 7.9. The standard nozzle.

Discharge Coefficients of the Standard Nozzle

As a result of a large amount of work, mainly by the same investigators quoted in connexion with the standard orifice, values of α for the standard nozzle have been established (refs. 10, 11, 22, 23, 24), and accepted by the B.S.I. and the V.D.I. They are as in Table 7.8.

According to B.S. 1042, these basic values of α vary more with Reynolds number than those for the standard orifice. Variations are liable to occur at all values of m at Reynolds numbers below 300,000; the corrections are shown in Fig. 7.10. Corrections for pipe-size roughness effect, based on throat Reynolds number, are shown in Fig. 7.11; these corrections need not be applied if the diameter of the pipe exceeds the figures shown in Table 7.9.

TABLE 7.8. DISCHARGE COEFFICIENTS FOR STANDARD NOZZLE

m	α	m	α
0	0·987	0·30	0·969
0·05	0·986	0·35	0·963
0·10	0·985	0·40	0·955
0·15	0·982	0·45	0·946
0·20	0·979	0·50	0·935
0·25	0·975	0·55	0·924

The basic values of α, corrected as necessary for Reynolds number and pipe size, can be used at Reynolds numbers of 20,000 and above for nozzles of throat diameter not less than 1·2 cm for values of m up to 0·55 in pipes of 5 cm internal diameter and upwards. The standard nozzle may also be used, as may the orifice, at the end of a pipe delivering into or receiving from a large space, or between two large spaces separated by a partition. It may also be used for critical-flow metering (see below). The requisite upstream and downstream

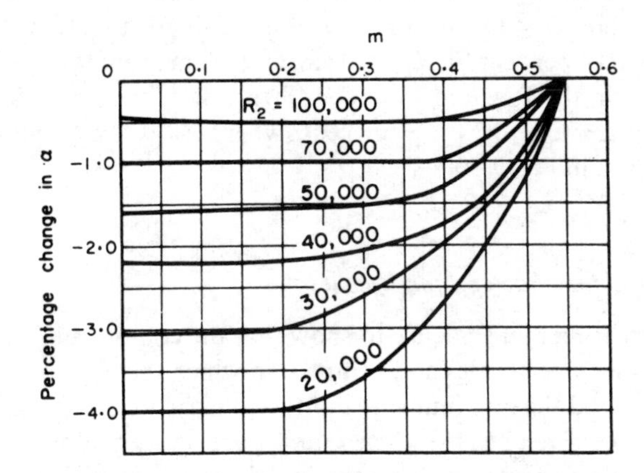

FIG. 7.10. Reynolds-number correction for standard nozzle.

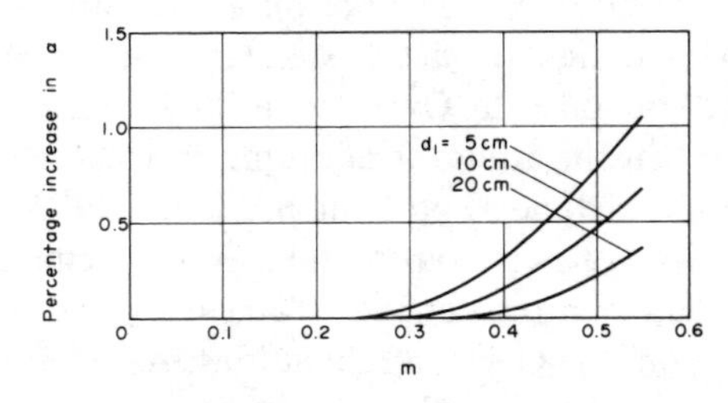

FIG. 7.11. Pipe-size correction for standard nozzle.

TABLE 7.9. MINIMUM PIPE DIAMETER FOR NEGLIGIBLE ROUGHNESS CORRECTION

Pipe material	Diameter (cm)	Pipe material	Diameter (cm)
Steel (not rusty):		Brass, copper, lead, plastic,	
cold-drawn	5	etc.	5
seamless	6·5	Cast iron:	
welded	9	not rusty	20
Steel (rusted)	20	rusty	40
Steel (galvanized)	10	bituminized	10

lengths of unobstructed pipe are the same as for the orifice (see Table 7.3 and p. 173). These lengths are in fact the same for all the constrictions standardized by the B.S.I., who emphasize that for the same tolerance the upstream lengths are twice as long for single-pressure taps as for multiple taps. The tolerances quoted in ref. 2 for the nozzle are of the same order as those for the orifice: the tolerance on basic α is 0·5 per cent up to $m = 0{\cdot}36$, rising to 1 per cent at $m = 0{\cdot}47$ and 1·5 per cent at $m = 0{\cdot}55$; the maximum pipe-size tolerance, which is zero for all pipes when $m \leq 0{\cdot}3$, occurs at $m = 0{\cdot}55$, where it is about 1·25 per cent for $d_1 = 5$ cm, 0·7 per cent for $d_1 = 10$ cm, and 0·3 per cent for $d_1 = 20$ cm; and the tolerance on the Reynolds-number correction is the same as that for the orifice, i.e. equal to $\pm 0{\cdot}33\ Z_R$ per cent, where Z_R is the Reynolds-number correction given in Fig. 7.10.

The Compressibility Factor ε for Nozzles

There is experimental evidence to show that the compressibility factor ε for constrictions such as nozzles and venturis, in which there is no contraction of the stream beyond the minimum area of the constriction itself, agrees with the theoretical value given by (18) when χ_c and χ are both equal to 1. For nozzles, the principal experiments on which this conclusion is based are those of Witte (ref. 10; 1928), which were made with superheated steam for values of m up to 0·4 and of p_2/p_1 down to the critical value (see p. 189). The 1937 edition of ref. 3 quoted some confirmatory evidence obtained with air for a nozzle that differed somewhat in design from the standard; and there are also some results (unpublished) obtained by Ower for air flowing through venturis.

There is thus some justification for adopting the theoretical value of ε for nozzles, and this has been done by both the B.S.I. and the V.D.I., in both cases for the same lower limit of p_2/p_1 (absolute) as was specified for the standard orifice, namely 0·8 and 0·75 respectively. Table 7.10 gives values of ε for air ($\gamma = 1{\cdot}4$) compiled from the German tabulations. As for the orifice, both the B.S.I. and the V.D.I. give data for other gases as well as air.

TABLE 7.10. VALUES OF THE COMPRESSIBILITY FACTOR ε FOR STANDARD NOZZLES AND VENTURIS ($\gamma = 1\cdot4$)

	Values of $(p_1 - p_2)/p_1$								
m	0·25	0·20	0·15	0·10	0·08	0·06	0·04	0·02	0
0	0·856	0·886	0·916	0·945	0·956	0·967	0·978	0·989	
0·1	0·854	0·885	0·915	0·944	0·955	0·967	0·978	0·989	
0·2	0·850	0·881	0·912	0·942	0·954	0·966	0·977	0·989	
0·3	0·842	0·875	0·907	0·938	0·951	0·964	0·975	0·988	1·000
0·4	0·830	0·864	0·899	0·932	0·946	0·960	0·973	0·987	
0·5	0·812	0·849	0·886	0·923	0·938	0·954	0·969	0·985	
0·6	0·785	0·826	0·868	0·910	0·928	0·945	0·963	0·982	

The Pressure Loss in Nozzles

It can be shown that at a given rate of flow in a pipe of given size at the same pressure difference, the pressure loss with a nozzle is the same as with an orifice.[2] For these conditions to be fulfilled, however, m for the nozzle will be less than for the orifice. Figure 7.12 shows the pressure loss in orifices, nozzles, and venturis, expressed in terms of the velocity head $\frac{1}{2}\varrho v_1^2$, where v_1 is the mean velocity in the pipe.

This diagram illustrates the large pressure drop that occurs at small values of m for both orifices and nozzles. We may compare this pressure drop with that in a length of straight pipe which, from equation (4) of Chapter V, is given by

$$p_1 - p_2 = \frac{4l}{d_1}\gamma\tfrac{1}{2}\varrho v_1^2.$$

A usual value of γ is about 0·005, for which the pressure loss in a length of pipe equal to 1 diameter is 0·02 velocity heads. The relatively high pressure loss due to nozzles and orifices is at once apparent: when $m = 0\cdot5$, for example, the loss is of the order of that of a length of pipe 50 diameters long, and it increases rapidly as m decreases. Several advantages of using as small a value of m as possible have been mentioned earlier in this chapter; here we find a disadvantage against which the other considerations must be balanced.

The Velocity Distribution in the Nozzle Jet

Reference has already been made to the uniform velocity distribution over the greater part of the jet issuing from the nozzle throat. In a nozzle specially designed to produce a uniform jet of this kind, [25] the variations of total pressure and of static pressure across the jet were well within 0·1 per cent of the velocity pressure over about 93 per cent of the full outlet diameter; and the flow through the nozzle could be calculated on the assumptions that the

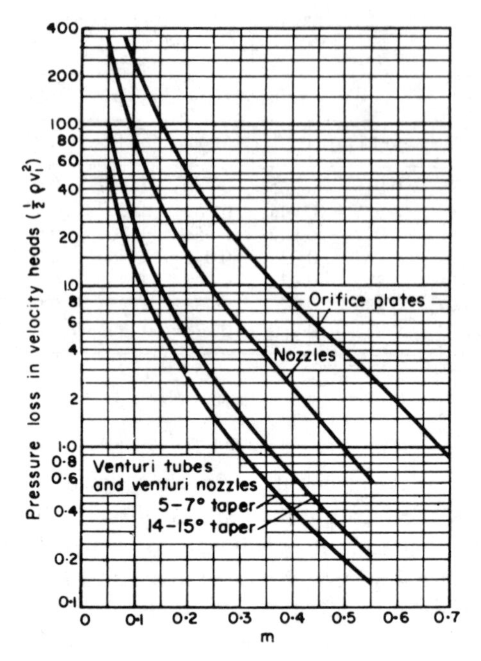

FIG. 7.12. Pressure loss in orifices, nozzles, and venturis.

effective area was 99 per cent of the true outlet area, and that the velocity was measured in the zone of uniform flow.

Special precautions were taken in this nozzle to ensure a thin boundary layer. This was achieved mainly by having a large contraction ($m = 0{\cdot}14$), and a series of gauze screens and honeycomb cells upstream to ensure uniform flow at entry. The nozzle itself did not have the abrupt entry of the standard nozzle: its profile was an easy curve from inlet to outlet, and on the downstream side there was a length of straight pipe connected directly to the nozzle throat so that the stream was confined by solid boundaries throughout.

Although the jet from the standard nozzle is perhaps not quite so uniform as from this special nozzle, it is probably still sufficiently uniform, and has a sufficiently thin boundary layer, for approximate measurements of quantity to be made from a single pitot–static observation on the axis of the jet.

The Venturi Tube

Most of the unrecoverable loss of pressure with an orifice or a nozzle is due to the sudden increase of area after the air has passed the section of minimum area: the rapid convergence of the stream on the upstream side contributes considerably less to the total loss. The loss of pressure on the downstream side occurs because the space between the boundaries of the jet and the walls of the pipe, before the stream expands again to the full pipe diameter, becomes filled with "dead air" in a violently eddying state, in which much energy is dissipated. If the space that these eddies would otherwise occupy is filled by solid material, with suitably shaped boundaries, their formation can be largely prevented. In other words, we can recover most of the pressure by guiding the stream by means of a conical length of pipe, with its smaller end of the same cross section as the jet, and gradually expanding in size along the direction of flow until the full pipe diameter is reached. An arrangement of this kind, combined with a conical entry of sharper taper preceding the throat, is known as a venturi tube;† the type standardized by the B.S.I.[(2)] is shown in Fig. 7.13. Apart from the greatly reduced loss of pressure, the device has no outstanding advantages over the plate orifice or the nozzle; and it will probably only be used when this feature is of importance. It requires considerably more space in the pipeline than either of the other two forms of constriction because, unless there is a long, gradually expanding portion after the throat, the whole purpose of the device is defeated.

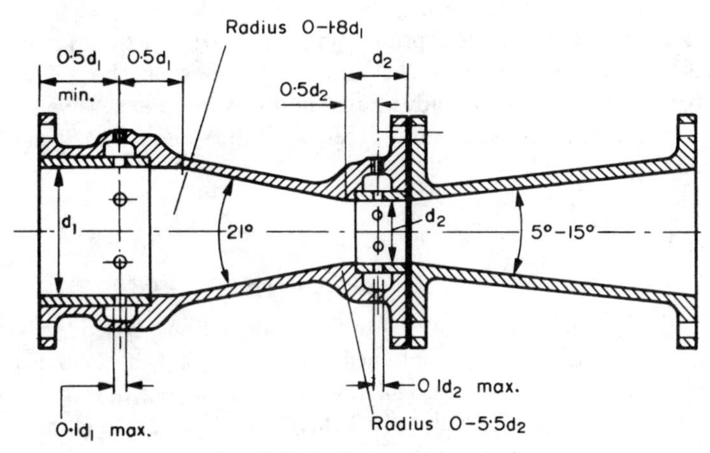

FIG. 7.13. B.S.I. standard venturi.

If the angle of the diverging outlet cone is too great, separation of the flow will occur, with a consequent loss of pressure.‡ Experience has shown that in order to ensure freedom from separation, the (total) apex angle of the cone should not exceed 6° or 7°. This, however, would result in a venturi which would be undesirably long for many practical purposes, particularly at low values of m. At $m = 0{\cdot}3$, for example, the length of the outlet cone with a 6°

† Named after G. B. Venturi (1746–1822) who first enunciated the principle on which it operates.

‡ See Diffusers, Chapter V.

apex angle would be about 4·5 pipe diameters, i.e. nearly 70 cm in a 15-cm pipe. For this reason, the venturis standardized by the B.S.I. and the V.D.I. may have outlet-cone angles up to 15°. The unrecoverable pressure loss (see below) is still considerably less than that for either the standard orifice or the standard nozzle; at the same time, however, it is considerably greater than that of the same length of straight pipe of diameter d_1.

The slope of the inlet cone can be steeper. Obviously, to reduce frictional losses, this cone should be as short as possible; but there is a limit to the slope because, if the rate of contraction of the stream is too rapid, there will be a tendency for the jet to contract after passing the smallest area of the constriction so that the throat of the venturi will not run full. Experience has shown that the total apex angle of the inlet cone should not appreciably exceed the value of 21° specified by the B.S.I.

Pressure tappings may be either single holes, or a number of holes around the section communicating with an annular chamber from which the pressure is tapped off to the manometer; the B.S.I. specify that the diameters of any pressure holes must not exceed $0{\cdot}1d_1$ on the upstream side and $0{\cdot}1d_2$ at the throat.

In an alternative type of venturi, the venturi nozzle, also specified by the B.S.I. and the V.D.I., the internal profile of the inlet up to the throat is identical with that of the standard nozzle. Beyond being somewhat shorter than the former type, it has no obvious advantage; and it is more expensive to make because of the careful shaping that is necessary if the standard coefficients are to be used.

Discharge Coefficients and Compressibility Factors for Venturis

As for the nozzle, (22) is used for flow calculations based on venturi observations, with the value of ε obtained from theory or Table 7.10. For α we use the nozzle values for the venturi having the nozzle-shaped inlet; and for the venturi with conical inlet the values in Table 7.11 are used, which have been taken from the B.S.I. data.

The values of α are subject to corrections and tolerances as for the other standard constrictions. For the venturi nozzle, the corrections, tolerances, and requisite lengths of pipe upstream and downstream are identical with those for the standard nozzle. For the type with conical inlet, the tolerances on the Reynolds-number corrections are the same as for the nozzle; and the corrections for pipe diameter (roughness) are about $-0{\cdot}8$, $-0{\cdot}5$, $-0{\cdot}3$, and 0 per cent for diameters of 10, 20, 40, and 80 cm and above respectively. For these sizes the pipe-size correction is practically independent of m: for a 5-cm pipe its values are $-1{\cdot}0$, $-1{\cdot}2$, $-1{\cdot}3$, $-1{\cdot}4$, and $-1{\cdot}3$ per cent at $m = 0{\cdot}1$, 0·2, 0·3, 0·4, and 0·5 respectively. The tolerances on these pipe-size corrections are slightly less than for the nozzle. The tolerance on basic α is 0·75 per cent for

TABLE 7.11. BASIC DISCHARGE COEFFICIENTS α AND REYNOLDS-NUMBER CORRECTIONS Z_R FOR STANDARD VENTURI WITH CONICAL INLET

$R_2 = v_2 d_2/\nu_2$	$m = 0{\cdot}1$		$m = 0{\cdot}2$		$m = 0{\cdot}3$		$m = 0{\cdot}4$		$m = 0{\cdot}5$		$m = 0{\cdot}55$	
	Basic α	Z_R (per cent)	Basic α	Z_R (per cent)	Basic α	Z_R (per cent)	Basic α	Z_R (per cent)	Basic α	Z_R (per cent)	Basic α	Z_R (per cent)
5,000		−5·0		−5·0		−4·8		−4·4		−3·4		−2·4
10,000		−3·4		−3·4		−3·3		−3·0		−2·2		−1·6
20,000		−2·3		−2·3		−2·2		−2·0		−1·5		−0·9
30,000	0·987	−1·7	0·988	−1·7	0·988	−1·6	0·987	−1·5	0·986	−1·1	0·985	−0·6
50,000		−0·9		−0·9		−0·8		−0·7		−0·3		−0·1
100,000		−0·4		−0·4		−0·4		−0·2		0		0
300,000		0		0		0		0		0		0

values of m up to 0·3; 0·85 per cent at $m = 0·4$; 1·1 per cent at $m = 0·5$; and 1·4 per cent at $m = 0·55$.

Standard venturis may not be used in smooth pipes less than 5 cm in diameter; but, as for the standard orifice and nozzle, the lower limit may have to be greater than this for rough pipes. The limits of m are 0·05 and 0·55, and the smallest permissible throat diameter is 2 cm.

Loss of Pressure in Venturis

With an outlet cone having an apex angle of 5–7°, the loss of pressure in a venturi, expressed as a percentage of the pressure difference between inlet and throat, ranges (approximately linearly with m) from about 13 per cent at $m = 0·1$ to about 6·5 per cent at $m = 0·5$, the corresponding values for the standard nozzle being about 83 and 33 per cent. For the largest outlet angle allowed by the B.S.I. the corresponding figures are about 24 per cent and 10 per cent. Expressed in terms of the pipe velocity head $\frac{1}{2}\varrho v_1^2$, these losses are shown in Fig. 7.12. Although they are still well above the losses in a straight pipe, even when its length is increased by the overall length of the venturi, they are well below those for nozzles and orifices.

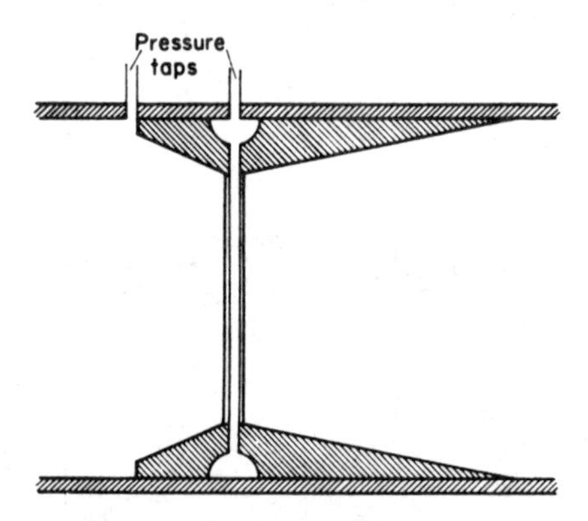

FIG. 7.14. Diagram of Dall tube.

The Dall Tube

An interesting modification of the venturi tube,[26] known as the Dall tube after its designer H. E. Dall, appeared on the market in 1946. M. L. Nathan[27] had patented a very similar device in 1937, but his design was not developed commercially. The Dall tube, shown diagrammatically in Fig. 7.14, is essentially a venturi in which the smaller ends of the inlet and outlet cones are connected together with only a very short length of parallel-walled throat between them.

In this throat there is an annular slit extending around the entire circumference and open to the air or other fluid flowing in the pipeline. There is a small shoulder† at the entrance to the inlet cone and pressure tappings are situated at the shoulder and at the annular slit as shown in Fig. 7.14. The total angles of both the inlet and outlet cones (about 40° and 15° respectively) are considerably greater than in the standard long-outlet venturi. The result is that the overall length of a Dall tube, which lies between $1{\cdot}5d_1$ and $2{\cdot}2d_1$ (d_1 = pipe diameter), is only about a quarter to one-fifth that of the long-outlet venturi producing the same differential pressure. The discharge coefficient of the Dall tube ranges from about 0·73 at small area ratios ($m < 0{\cdot}1$) to 0·62 at $m = 0{\cdot}7$[28]; and the differential pressure is between 1·8 and 2·2 times that of the long-outlet venturi of the same m. This higher differential is an important advantage, since it makes possible the achievement of a given differential, i.e. a given accuracy of measurement, by the use of a Dall tube of greater area ratio m than would be necessary with a venturi. This in itself makes for a lower irrecoverable pressure loss.

Even at the same value of m, however, the irrecoverable pressure loss of a Dall tube compares favourably with that of the long-outlet venturi. Miner[29] has shown that it is actually less for all values of m above about 0·3 and not much greater for lower values of m. It seems therefore that, despite the steep outlet cone, pressure recovery between throat and outlet is practically as good as that of a venturi; i.e. that separation of the flow is in some way prevented by the action of the annular slit, and that the pressure loss is little more than that due to wall friction.

It is stated in ref. 28 that the pressure loss in a Dall tube producing a differential pressure of four velocity heads is equal to the friction loss in 5 or 6 diameter-lengths of straight pipe. To compare this with the losses in venturis we shall assume (as before, see p. 179) that the resistance coefficient γ for the straight pipe is 0·005, which gives a resistance of 0·02 velocity heads per diameter-length of pipe. Table 7.12 shows the pressure losses in venturis producing a range of differential pressures. Since the discharge coefficient for the venturi is very nearly 1, the differential pressures are, as stated on p. 153, very close to the theoretical values calculated from (5) which are tabulated in column 2 of the table. The lengths of the venturi for different values of m have been calculated from the data for the standard long-outlet venturis given in ref. 2.

The last column of Table 7.12 shows that, for a differential presure of four velocity heads, the pressure loss due to a long-outlet venturi is equal to that in nearly three times its own length of straight pipe, i.e. to a straight pipe about 18 diameters long since (see column 3) the overall length of the venturi is 6·2 pipe diameters. As stated above, a Dall tube giving the same differential

† Nathan's earlier patent does not specifically mention a shoulder at entry, but it does emphasize the importance of the annular slit in the throat.

TABLE 7.12. PRESSURE LOSSES IN VENTURIS OF 5–7° TAPER

m	Differential pressure produced by venturi (velocity heads)	Total length of venturi (pipe diameters)	p_v Pressure loss in venturi (velocity heads)	p_s Pressure loss in same length of straight pipe (velocity heads)	Length of pipe with same pressure loss as venturi (pipe diameters)	p_y/p_v
0·1	99	11·0	13·2	0·22	660	60
0·2	24	9·8	2·7	0·20	135	13·8
0·3	10·1	8·0	0·9	0·16	45	5·6
0·4	5·2	6·8	0·4	0·14	20	2·9
0·435	4·0	6·2	0·35	0·12	17.5	2·8
0·5	3·0	5·9	0·2	0·12	10	1·7
0·55	2·3	5·4	0·13	0·11	6·5	1·2

pressure of four velocity heads has a pressure loss equal to that of 5 or 6 diameter lengths of straight pipe. Thus, at four velocity heads differential, the pressure loss due to the Dall tube is less than that due to the venturi by the friction loss in about 12 diameter-lengths of straight pipe. This seems to be due entirely to the shorter overall length of the Dall tube (about 2 diameters as against 6 for the venturi), for the figures show that each device suffers a pressure loss equal to that of about three times its own length of straight pipe.

The Dall tube's advantages of short length and high differential pressure (or low total-pressure loss) were, however, subsequently found to be accompanied by certain disadvantages: in particular, the uncalibrated accuracy proved to be less precisely defined than was originally stated, and the discharge coefficient varied considerably with Reynolds number. Detailed studies led to the improved design described in ref. 30, which includes extensive data.

The Use of Orifices and Nozzles Other than in Pipelines

It is sometimes necessary to measure the quantity of air flowing between the atmosphere and large receivers or air drums; and one of the most convenient methods of making such measurements is to install a constriction in one wall of the drum and to measure the pressure drop across it. Any form of constriction can be used for this purpose but, because of its relative simplicity, the plate orifice is usually preferred.

An example of this use of the orifice arises in the measurement of the air supply to an internal-combustion engine on test, on which comprehensive experiments have been made by Watson and Schofield,[31] Durley,[32] and

Müller.[33] The air intake of the engine is connected to a large air box or drum,† and all the air taken by the engine is sucked into this drum from the atmosphere through a small orifice in one side. Within the box, the pressure is less than the atmospheric pressure outside. All we need do, then, to measure the pressure drop across the orifice is to connect one side of a manometer, whose other side is open to the atmosphere, to a hole in the side of the box. If, as is usual in this method of measurement (see below), the orifice is small in relation to the box, the flow inside the latter is everywhere very slow and the pressure practically uniform throughout. The pressure tapping can therefore be situated at any convenient place in the box; but, if standard discharge coefficients are to be used, the tapping should be as specified for the standard orifice (or other constriction), and care should be taken to ensure that there are no burrs or other surface irregularities in its vicinity on the inner surface of the drum.

This case represents one of the two possibilities of using a constriction without either the upstream or downstream (or both) lengths of pipe that are present in the more usual applications of these devices. The other case is that of a constriction at the end of a pipe discharging into or taking air from the atmosphere or a large space. This was investigated by Stach[34] for the standard orifice and the standard nozzle. Both cases are provided for in B.S. 1042, but data are given only for the orifice with corner tappings, the nozzle, and the venturi nozzle. In these applications a space may be considered as large provided that there is no wall closer than $4d_2$ to the axis of the constriction or to the plane of the constriction in the space. In addition, when the space is on the upstream side of the constriction, the velocity at any upstream point more than $4d_2$ from the constriction must be less than 3 per cent of the velocity through the constriction; this requirement is intended to eliminate the effects of disturbances such as draughts. As regards the diameters of the constrictions allowed for this type of measurement, when there is an upstream length of pipe, the value of m must not exceed 0·15; when there is no upstream length, the definition of a large space ensures that the effective upstream diameter of the space is greater than $8d_2$, i.e. that m is less than 0·015.

Subject to these restrictions, when the flow is from a pipe into a large space the values of the discharge coefficient and of all corrections and tolerances are the same as those to be used when the device is installed in a pipeline in the normal way with m equal to 0·15. When the flow is from a large space into a pipe or into another large space, the values of α and of all corrections (except that for pipe size, which is unity) and tolerances are those corresponding to $m = 0$.

It is of some interest to note that, although the orifices used in the investigations described in refs. 31, 32, and 33 all had sharp edges, none of them conformed to the standard specifications in all respects. Yet in all cases the mean

† See also Chapter XII.

values of α obtained agreed closely with those given in B.S. 1042, as quoted above, for this type of measurement.

Some simplification of the flow calculation is usually possible when constrictions are used in this way. The basic equations to be used are (12) or (17) according to the value of p_1/p_2. For flow from or to the atmosphere, p_1 or p_2 is atmospheric, and it will usually be possible to select an orifice of such size that $(p_1 - p_2)$ is a small fraction (1 per cent or less) of the absolute value of p_1 or p_2. In that case the simpler equation (12) for incompressible flow can be used. Further, as already explained, m for flow from a space into a pipe will be less than 0·015; hence m^2 can be neglected in (12) or (17), and if the flow can be regarded as incompressible it can be calculated from the equation

$$q = \alpha a_2 \sqrt{[2\varrho(p_1 - p_2)]}.$$

For flow from a pipe into a space ($m \not> 0{\cdot}15$), the maximum error incurred by the use of the same simplified equation will be 1·1 per cent.

Special Forms of Orifice

Various forms of plate orifice have been suggested from time to time for various purposes. They include segmental and eccentric orifices in which the bottom of the pipe is unobstructed, but these are as a rule more valuable in liquid than in gas or air flow: their purpose is to prevent solid particles carried in suspension in the fluid stream from being deposited at the downstream face and blocking the pressure tappings.

An annular orifice proposed by Howell[35] was found to be less affected by upstream conditions than the standard type. Later tests for the N.E.L. on a differently supported annular orifice showed that its discharge coefficient was much less affected than that of the concentric orifice by swirl in the approaching flow, but was still sensitive to changes in the velocity distribution.

Figure 7.4 shows the considerable changes that occur in the discharge coefficient of the standard sharp-edged orifice at low Reynolds numbers. Various designs less susceptible to this effect have been proposed.[22, 23, 36, 38, 39, 40] In most of these the sharp edge of the standard orifice is replaced by a curved inlet, and there is no parallel portion or downstream bevel; in fact, they resemble short nozzles with no parallel throat. A type having the inlet curve shaped as in the standard nozzle was said to have constant α for values of m between 0·37 and 0·45 over the pipe Reynolds-number range 4000–100,000 (ref. 3, 1948). Other similar types which were stated to have constant α at low Reynolds numbers had semi-circular and quarter-circle edges. Later work[40] has shown however, that with the latter there is still a 3 per cent change in α in the Reynolds-number range 250–50,000, although most of this occurs in the unstable regime between laminar and turbulent pipe flow; between Reynolds numbers of 4000 and 50,000 the variation in α is less than $\pm 0{\cdot}5$ per cent.

Critical-flow Metering

Equation (8), the theoretical expression for the flow of a compressible fluid through a constriction in which the jet does not contract beyond the throat, shows that there is a value of the pressure ratio $\phi(=p_2/p_1)$ for which the flow is a maximum. For a given value of the upstream pressure p_1 and density ϱ_1 maximum flow occurs when

$$\frac{(1-\phi^{(\gamma-1)/\gamma})\phi^{2/\gamma}}{1-m^2\phi^{2/\gamma}} \tag{23}$$

is a maximum.

The value of ϕ for maximum flow, which we term the *critical pressure ratio* and denote by ϕ_c, is obtained by equating to zero the differential coefficient of (23) with respect to ϕ. In this way we obtain

$$2\phi_c^{(1-\gamma)/\gamma} + (\gamma-1)m^2\phi_c^{2/\gamma} = \gamma + 1. \tag{24}$$

When $m = 0$, this reduces to the well-known expression for the pressure ratio giving the maximum flow through a nozzle of the specified type discharging into a large space, namely

$$(\phi_c)_0 = \left(\frac{2}{\gamma+1}\right)^{\gamma/(\gamma-1)} \tag{25}$$

For air, $\gamma = 1{\cdot}4$ and $(\phi_c)_0$ given by (25) is 0·528. ϕ_c increases slightly with m reaching 0·550 at $m = 0{\cdot}4$. When the pressure drop through the nozzle is such that the critical pressure ratio is reached, the velocity at the throat is equal to the local speed of sound, which is also the speed at which pressure changes are propagated upstream. Hence any pressure changes that occur at the throat cannot affect entry conditions, and the rate of flow can be increased only by increasing p_1: any reduction of p_2 will not have any effect. This theoretical prediction is confirmed in practice in the case of contoured nozzles, except for small effects due to the boundary layer on the nozzle wall: since the velocity falls to zero at the surface, velocities in the innermost part of the boundary layer are subsonic, and here a route is available by which an upstream influence can be exerted.

The boundary layer has the further effect of reducing the cross-sectional area of virtually frictionless flow to slightly less than the geometric area of the nozzle. The magnitude of this effect can be calculated. Taking boundary-layer effects into account, together with the effects of the transverse pressure gradient across the throat (neglected in the simple one-dimensional analysis outlined above), Stratford[(41)] has developed an optimum shape of nozzle which gives values of the discharge coefficient very close to unity, provided that the throat Reynolds number exceeds about 10^6 and that steps are taken to ensure an adequate degree of uniformity in the entry flow. Such a nozzle can be used for calibrating other flowmeters. Reference 42 describes an experimental set-up for this purpose, in which honeycombs and gauze screens are incorporated, together with ample contraction upstream of the nozzle.

B.S. 1042[2] provides for critical-flow metering with the standard nozzle provided that $m \not> 0{\cdot}4$. This method is useful for such purposes as the measurement of the output of a compressor by discharging it to atmosphere. The basic equation for such measurement is obtained by substituting $p_1(1 - \phi)$ for $(p_1 - p_2)$ in (17), giving

$$q = \varepsilon \alpha a_2 \sqrt{\left(\frac{2\varrho_1 p_1}{1 - m^2}\right)} \sqrt{(1 - \phi)}. \tag{26}$$

Inserting the value of ε from (18), with $\chi_c = \chi = 1$ (see p. 178), eqn. (26) for critical-flow metering with the nozzle becomes

$$q = \alpha a_2 \sqrt{(p_1 \varrho_1)} \sqrt{(\gamma \phi_c^{(\gamma + 1)/\gamma})}, \tag{27}$$

where ϕ_c, the critical pressure ratio p_2/p_1, is given by (24). It will be observed that it is necessary only to measure the upstream pressure p_1, and not the throat pressure. The value of the discharge coefficient α to be used is the same as that applicable to ordinary sub-critical flow as given in Table 7.8. The upstream temperature has, of course, to be measured as well in order to determine ϱ_1. If necessary, a method of temperature measurement suitable for compressible-flow conditions (see pp. 70–2) must be employed.

The pressure loss across a nozzle when used in these "choked flow" conditions is substantially greater than in subsonic flow. Some of the loss can be recovered by incorporating a diffusing section downstream as in a venturi tube (see, for instance, ref. 41).

Plate orifices also can be used in choked-flow conditions. Their discharge coefficients, however, do vary with the downstream pressure, because they are dependent on the changing degree to which the issuing jet contracts (vena contracta: see p. 153). Data on thick-plate orifices (or "cylindrical nozzles") are reported in ref. 43. For these, too, the addition of a diffusing section (within the thickness of the plate) reduces the pressure loss.

Flow Straighteners

Reference has been made to the need for flow straighteners to improve the accuracy of flow measurement in pipes. The main purpose of the devices seems to be to remove swirl, which can have a large effect on the pressure recorded at side holes, and also on the discharge coefficients of standard constrictions.[44] Some success has met attempts to design straighteners which not only remove swirl but also, when the initial conditions are not exceptionally adverse, produce a downstream velocity distribution approaching that for fully developed pipe

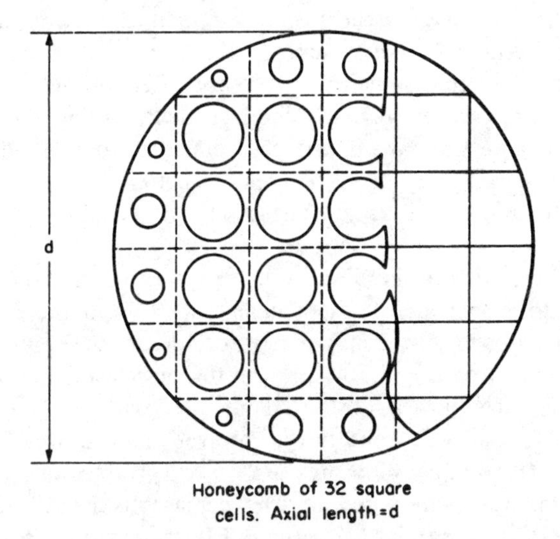

FIG. 7.15. Flow straightener.

flow, so that standard constriction discharge coefficients can be used. Zanker[45] has described a straightening device consisting of a square-celled honeycomb 1 pipe diameter d in length, to the front of which is attached a thin plate containing thirty-two small sharp-edged orifices (Fig. 7.15), each centrally in front of a square honeycomb cell. The diameters of the holes increase with their distance from the wall, the grading (particulars of which are given in ref. 45) being such that the desired velocity distribution is approached. Unfortunately, this device has a fairly high resistance. Zanker experimented with two examples, one designed to give the desired distribution at a Reynolds number of 10^5 and the other at 10^6: the pressure losses were 5·7 and 5·1 velocity heads respectively. With a settling length of 4 pipe diameters between them and a standard orifice with $m = 0{\cdot}53$, the discharge coefficient with each of these straighteners was within $\pm 0{\cdot}8$ per cent of the standard value for various types of approach-velocity distribution, including one with marked asymmetry and one with a swirl angle of 12–15° at the pipe walls.

Good results were also obtained by Sprenkle and Courtright[46] with a straightener consisting of three transverse plates spaced 1 pipe diameter apart, each containing 250 circular orifices equal in diameter and of total area (for each plate) about one-sixteenth of the pipe area. With a downstream settling length of $8d$, good velocity distributions were obtained. The pressure loss, however, was about 15 velocity heads, which could be reduced to 11, without adverse effect on the efficiency of the device, by bevelling the leading edges of the orifices. This straightener is one of two recommended in the U.S. Standard Code[6] for flow measurement with constrictions; the other consists of a number of tubes lying along the flow, of length at least $8d$, the axes of adjacent tubes being not more than $d/4$ apart.

Published data are insufficient at present to enable us to be certain that devices such as these, particularly the last mentioned in the previous paragraph, will enable standard discharge coefficients to be used in all circumstances with an accuracy of 99 per cent or better. In many cases it will be preferable to remove swirl as far as possible and to calibrate the installation *in situ*.

For the removal of swirl, Zanker(45) found the honeycomb effective, but Spencer(47) doubts whether this is so in all cases. Indeed, the B.S.I.(2) state that swirl due to a strong source may persist for 100 or more pipe diameters, and that straighteners of the honeycomb or the nest-of-tubes type may not be effective unless they cause a substantial pressure loss.

Except in very adverse conditions, however, there can be little doubt that honeycombs with relatively low pressure loss will produce an improvement, and that their effectiveness in removing swirl increases as the cell size decreases. In another one of their standards,(48) the B.S.I. recommend a square-celled type in which the length of side w of the cell walls is 7–15 per cent of the pipe diameter and the axial length of the honeycomb is $3w$. The National Engineering Laboratory use honeycombs of 9·5 mm hexagonal mesh, 76 mm long axially, irrespective of pipe diameter. The cell walls need be no thicker than is necessary for strength and avoidance of flutter. In fact, these honeycombs are made of paper, and the pressure loss is only about 0·9 velocity heads in the speed range 12–24 m/sec. Salter(49) used a honeycomb comprising horizontal rows of equilateral triangles 6·4 mm apart, the cell walls being 0·025 mm thick; the pressure drop for a 5-cm axial length was about 0·4 velocity heads at 30 m/sec. A simpler type, more easy to construct, may effect some improvement if the swirl is not too great,(50) but there are no published data on its performance. It consists of six thin, flat vanes arranged radially in the pipe and extending for 1·5 diameters along the flow direction.

Whatever the type of honeycomb used, a downstream settling length of pipe of about 10 diameters should be provided where possible before the section of pressure measurement is reached.

There are insufficient data at present available to enable more precise advice to be given regarding the design of flow straighteners to suit any specified conditions. But a simple device described by Lugt(44) (who attributed it to G. Hutarew) can be used to find out whether significant swirl exists at the measurement plane in a pipe, and how effective any particular flow straightener is in removing it. As shown in Fig. 7.16, this swirl indicator, which is temporarily inserted through the pipe wall for use, consists of a fitting with two pressure holes so arranged that any swirl at the walls will produce a differential pressure detectable by a simple manometer.

If, as distinct from reducing swirl, it is desired to correct an unfavourable

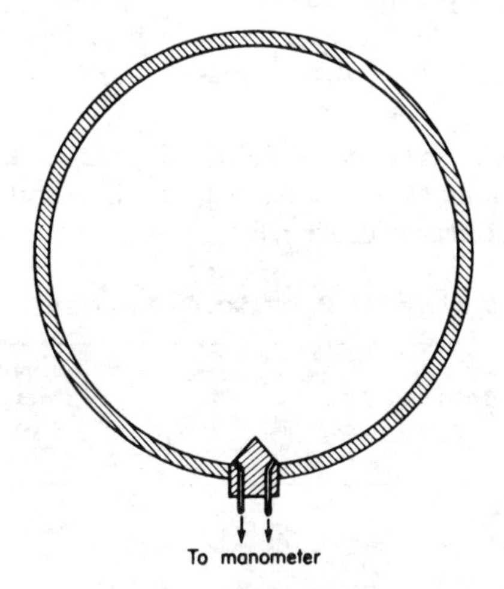

FIG 7.16. Swirl detector.

cross-sectional velocity distribution, considerable improvements can be obtained by the use of gauze screens. In fact, such screens may themselves also reduce swirl to some degree.[45] It is not possible to give any general guidance on the dimensions of these gauzes – the worse the velocity distribution the closer will be the mesh required. For reducing comparatively small disturbances in a 45-cm wind tunnel Salter[49] found 0·25-mm diameter wires spaced twelve to the centimetre effective; the pressure loss with this gauze was about 2·1 velocity heads at 9 m/sec. As with honeycombs, a settling length downstream of a gauze screen is desirable.

The Resistance of Honeycombs and Gauzes

The resistance coefficient of a honeycomb or a gauze is conveniently expressed in term of the loss of pressure, in velocity heads, suffered by the air in passing through it. In general, and particularly with gauzes, the resistance coefficient will depend on the air speed, decreasing as the latter increases.

To estimate the resistance of a gauze screen, the following formula[51] will be found to give results sufficiently accurate for practical use:

$$K = 6{\cdot}5\,\frac{1-\beta}{\beta^2}\left(\frac{vd}{\beta\nu}\right)^{-1/3},$$

where K is the pressure loss in velocity heads, β is the "open" area ratio of the gauze (see below), v is the mean speed ahead of the screen, d is the diameter of the wires of which the gauze is made, and ν is the kinematic viscosity of the air.

If there are n wires per unit length,

$$\beta = \left(\frac{1/n - d}{1/n}\right)^2 = (1 - nd)^2.$$

No general formula can be given for estimating the resistance of honeycombs. Table 7.13, supplied by Mr. C. Salter, will supplement those already quoted. Further data have been provided in ref. 52.

TABLE 7.13. RESISTANCE OF HONEYCOMBS

Type of honeycomb	Approximate resistance (velocity heads)	Velocity ahead of honeycomb (m/sec)
Hexagon cells 10 cm long made from 0·15 mm walled hexagon brass tubes 10 mm pitch	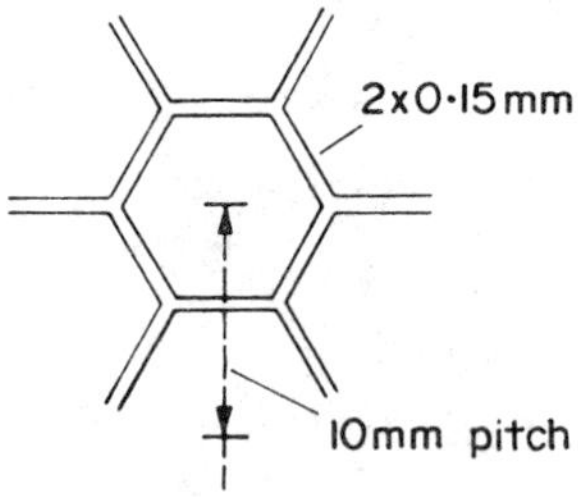 1·1	1·8
	0·85	3·0
	0·65	4·9
	0·55	6·7
Alternate flat and zig zag strips 0·25 mm thick stainless steel, cells 5 cm long	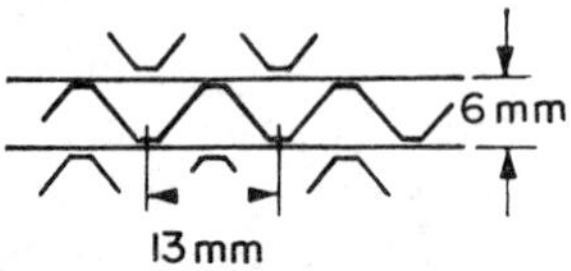 0·38	30
25 mm square cells, 0·25 mm walls, 8 cm long	0·23	9–24
13 mm square cells, 0·5 mm walls, 10 cm long	0·31	9–24
Resinated paper hexagon cell 6 mm between flats	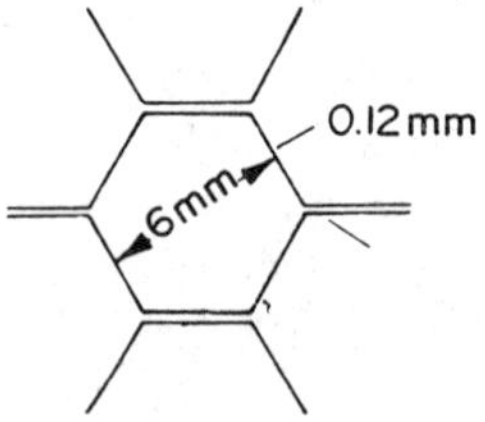 0·55 (cells 25 mm long)	3
	0·3 (cells 25 mm long)	10
	0·22 (cells 25 mm long)	15 and over
	0·6 (cells 76 mm long)	7·5
	0·5 (cells 76 mm long)	11
	0·45 (cells 76 mm long)	15
	0·44 (cells 76 mm long)	18

References

1. F. C. JOHANSEN, Flow through pipe orifices at low Reynolds numbers, *Proc. Roy. Soc.* A **126** (1930) 231.
2. B.S. 1042: Part 1: 1964, *Methods for the Measurement of Fluid Flow in Pipes*, Part 1: *Orifice Plates, Nozzles and Venturi Tubes*, British Standards Institution, London, 1964.

3. DIN 1952, V.D.I. Durchflussregeln, *Regeln für die Durchflussmessung mit genormten Düsen und Blenden*, Verein deutscher Ingenieure, Düsseldorf, 1969.
4. J. L. Hodgson, The commercial metering of air, gas, and stream, *Proc. Instn Civil Engrs* **204** (1917) 108; see also *The Orifice as a Basis of Flow Measurement*, Instn Civil Engrs, Selected Engineering Papers, 31 (1925).
5. *Fluid Meters, their Theory and Application*, A.S.M.E. Research Committee on Fluid Meters, 6th edition, 1971.
6. *Flow Measurement*, Supplement to A.S.M.E. Power Test Codes, Chapter 4, Part 5, 1959.
7. R. E. Sprenkle, The thin plate orifice for flow measurement, *Flow Measurement in Closed Conduits*, paper B-1, Vol. I, 127, H.M.S.O., Edinburgh, 1962; see also discussion on this paper.
8. F. C. Johansen, The influence of pipe diameter on orifice discharge coefficients, *Engr* **149** (1930) 679.
9. J. M. Spitglass, Orifice coefficients – data and results of tests, *Trans. A.S.M.E.* **44** (1922) 919.
10. R. Witte, Durchflussbeiwerte der I.G. Messmündungen für Wasser, Öl, Dampf und Gas, *Z.V.D.I.* **72** (1928) 1493; Durchflusszahlen von Düsen und Staurändern, *T.M.T.* **1** (1930) 34, 72, 113; Die Strömung durch Düsen und Blenden, *Forsch. IngWes* (new title of T.M.T.) **2** (1931) 245, 291; Neure Mengenstrommessungen zur Normung von Düsen und Blenden, *Forsch. IngWes* **5** (1934) 205.
11. M. Jakob and F. Kretschmer, Die Durchflusszahlen von Normaldüsen und Normalstaurändern für Rohrdurchmesser von 100 bis 600 mm, *V.D.I. Forschungsheft* 311 (1928).
12. F. Kretschmer, Einfluss der Reibung auf den Durchflussbeiwert der Drossel-Messgeräte, *Forsch. IngWes* **4** (1933) 93.
13. E. A. Spencer, H. Calame, and J. Singer, Edge sharpness and pipe roughness effects on orifice plate discharge coefficients, N.E.L. Report 427 (1969).
14. T. E. Stanton, The mechanical viscosity of fluids, *Proc. Roy. Soc.* **A 85** (1911) 366.
15. K. J. Zanker and L. Fellerman, *Some Experiments with Orifice Plate Flow Meters*, British Hydromechanics Research Association and National Engineering Laboratory, B.H.R.A.–N.E.L. Joint Report No. 2 (1961).
16. G. Ruppel, Einfluss der Expansion auf die Kontraktion hinter Staurändern, *T.M.T.* **1** (1930) 151.
17. A. Busemann (Note on Ref. 15), *T.M.T.* **1** (1930) 338.
18. E. Buckingham, Note on contraction coefficients of jets of gas, *Bureau of Standards J. of Research* **6** (1931) 765; also Notes on the orifice meter; the expansion factor for gases, *ibid.* **9** (1932) 61.
19. T. H. Redding, *A Bibliographical Survey of Flow through Orifices and Parallel-Throated Nozzles*, British Scientific Instrument Research Association Publication M 9, Chapman & Hall, London, 1952.
20. C. F. Smith (discussion on paper B-4), *Flow Measurement in Closed Conduits*, Vol. I, 248, H.M.S.O., Edinburgh, 1962.
21. E. A. Spencer and P. Harrison, *Flow Measurement by Orifice Plates of Precision Manufacture*, N.E.L. Rep. 14 (1961).
22. H.-G. Giese, Mengenmessung mit Düsen und Blenden bei kleinen Reynoldsschen Zahlen, *Forsch. Ing Wes* **4** (1933) 11.
23. M. Hansen, Düsen und Blenden bei kleinen Reynoldsschen Zahlen, *Forsch. Ing Wes* **4** (1933) 64.
24. K. Jaroschek, Vergleichende Durchflussmessungen mit Düsen und Blenden, *Z.V.D.I.* **80** (1963) 643.
25. J. C. Ascough, The development of a nozzle for absolute airflow measurement by pitot-static traverse, *R. & M.* 3384 (1964).
26. G. Kent Ltd., Low loss venturi tubes, British Patent 689474 (1950, 1953).
27. M. L. Nathan, Improvements in meters for the measurement of fluid flow, British Patent 473562 (1936, 1937).
28. H. E. Dall, Flow tubes and non-standard devices for flow measurement with some coefficient considerations, *Flow Measurement in Closed Conduits*, paper D-1, Vol. II, 385, H.M.S.O. Edinburgh, 1962.

29. I. O. MINER, The Dall flow tube, *Trans A.S.M.E.* **78** (1956) 475; also *Instrum. Engr* **27** (1957) 45.
30. D. HALMI, Metering performance investigation and substantiation of the "Universal Venturi Tube" (U.V.T.), (Parts 1 and 2), *Trans. A.S.M.E., J. Fluids Engng* **96** (1974) 124.
31. W. WATSON and H. SCHOFIELD, On the measurement of the air supply to internal combustion engines by means of a throttle plate, *Trans. Instn. Mech. Engrs*, Parts 1–2 (1912) 517.
32. R. J. DURLEY, On the measurement of air flowing into the atmosphere through circular orifices in thin plates and under small differences of pressure, *Trans. A.S.M.E.* **27** (1906) 193.
33. A. O. MÜLLER, Messung von Gasmengen mit der Drosselscheibe, *Z.V.D.I.* **52** (1909) 285.
34. E. STACH, Die Beiwerte von Normdüsen und Normbelenden im Einlauf und Auslauf, *Z.V.D.I.* **78** (1934) 187.
35. A. R. HOWELL, Note on the R.A.E. annular airflow orifice, *R. & M.* 1934 (1939).
36. H. E. DALL, Developments in differential producers for flow measurement, *Instrum. Engr* **2** (1959) 144.
37. R. WITTE, Durchflussbeiwerte der I.G. Messmündungen für Wasser, Öl, Dampf und Gas, *Z.V.D.I.* **72** (1928) 1493.
38. B. W. KOENNECKE, Neue Düsenformen für kleinere und mittlere Reynoldszahlen, *Forsch. IngWes* (1938) 109.
39. G. WÄLZHOLZ, Die Doppelblende, *Forsch. IngWes* **7** (1936) 191, 226.
40. M. BOGEMA and P. L. MONKMEYER, The quadrant edge orifice – a fluid meter for low Reynolds numbers, *Trans. A.S.M.E., J. Basic Engng* **82**D (1960) 729.
41. R. S. STRATFORD, The calculation of the discharge coefficient of profiled choked nozzles and the optimum profile for absolute air flow measurements, *J. R. Ae. S.* **68** (1964) 237.
42. D. W. SPARKES, A standard choked nozzle for absolute calibration of air flowmeters, *Ae. J. R. Ae. S.* **72** (1968) 335.
43. T. J. S. BRAIN and J. REID, Performance of small diameter cylindrical critical-flow nozzles, N.E.L. Report 546 (1973).
44. H. LUGT, Einfluss der Drallströmung auf die Durchflusszahlen genormter Drosselmessgeräte, *Brennstoff-Wärme-Kraft* **13** (1963) 121.
45. K. J. ZANKER, The development of a flow straightener for use with orifice-plate flowmeters in disturbed flow, *Flow Measurement in Closed Conduits*, paper D-2, Vol. II, 395, H.M.S.O., Edinburgh, 1962.
46. R. E. SPRENKLE and N. S. COURTRIGHT, Straightening vanes for flow measurement *Mech. Engng N.Y.* **80** (1958) 71, 92.
47. E. A. SPENCER, Developments in industrial flowmetering (Part I), *Chem. and Process Engng* **44** (1963) 297.
48. B.S. 848: Part 1: 1963, *Methods of Testing Fans for General Purposes*, British Standards Institution, London, 1963.
49. C. SALTER, *Low Speed Wind Tunnels for Special Purposes*, N.P.L. Rep. NPL/Aero/155, 1961.
50. F. M. A. Code 3: 1952, *Fan Performance Tests*, Fan Manufacturers' Association Limited London, 1952.
51. K. E. G. WIEGHART, On the resistance of screens, *Aero. Quart.* **4** (1953) 186.
52. J. WHITAKER, P. G. BEAN, and E. HAY, Measurement of losses across multi-cell flow straighteners, N.E.L. Report 461 (1970).

CHAPTER VIII

THE VANE ANEMOMETER

ALTHOUGH it is subject to certain practical limitations indicated below, the vane anemometer has been extensively employed for many years. As a means of measuring air speed, particularly in the range of low speeds where the pressures set up by the motion are so small that it is difficult to measure them accurately, its simplicity renders it an exceedingly useful instrument; and if properly used it gives all the precision ordinarily required in engineering practice.

The vane anemometer is, in effect, simply a windmill consisting of a number of light, flat vanes mounted on radial arms attached to a common, small, steel spindle, which rotates in two low-friction bearings. Eight vanes made of thin sheet aluminium alloy, or mica, are almost invariably used; and the air forces acting on the vanes cause the spindle to rotate at a rate depending mainly on the air speed.† Suitable gearing transmits the motion of the spindle to a pointer or, more usually, to a number of pointers moving over graduated dials. Normally, the dials are marked with scales of metres; and experience has taught manufacturers of these instruments how to proportion the various dimensions, vane inclinations, and gearing ratios, so that the number of metres indicated on the dials in a given time is approximately equal to the distance traversed by the air in the same time. Therefore, to determine the air speed it is necessary to observe with the aid of a stop-watch the number of "metres of air", as shown by the indicating mechanism, to pass the instrument in unit time. This gives the indicated air speed, from which the true air speed is obtained by reference to the calibration curve for the particular anemometer in use.

Instruments of this type are mainly of value for measuring air speeds in large ducts or ventilating shafts, or in smaller pipes (subject to the restriction on the limiting size mentioned below) where observations can conveniently be made at an open end, preferably an outlet end. They are probably best used mounted on light, preferably streamlined, supports and inserted into the air-stream from one side. It is important that the observer should not stand too close to the instrument whilst readings are being obtained, as his presence will disturb the flow even if he is situated downstream of the instrument.

† See p. 210 for the effect of air density.

Two types of anemometer are in common use. In one, the plane of the dials is perpendicular to the plane of rotation of the blades (Fig. 8.1); this design is not as compact as the other in which the indicating dial is concentric with the blades, but it is probably to be preferred on account of the greater facility that it offers for taking observations from the side.

In some instruments, provision is made for throwing the indicating mechanism out of mesh with the vane spindle by the movement of a small lever, and also, sometimes, for setting the reading to zero at any stage. These devices appear to be designed to obviate the use of a stop-watch, the intention being that readings should be taken by throwing the indicating mechanism into gear at a given instant and disconnecting it when a certain time has elapsed, which can be observed with any watch provided with a seconds hand. This method of using vane anemometers is not to be recommended; more accurate results will certainly be obtained by observing with a stop-watch the time taken for a given number of complete revolutions of one of the pointers. The other method should only be adopted if a stop-watch is not available.

In order to reduce starting and stopping errors to small percentages of the total readings, no observations should occupy less than 100 sec, or comprise a reading of less than 50 m on the instrument. It is advisable also to take at least two check readings of every observation, for it will often be found that the flow is subject to fluctuations, so that the average over a given period is not quite constant.

It has often been stated that vane anemometers are unreliable and that their indications are prone to unaccountable inconsistencies. This is not true if the instruments are properly used. Trouble frequently arises through the use of an anemometer to measure air speeds for which it is not designed. If the speed is too high, the vane setting may suffer a permanent distortion, which will cause a corresponding permanent alteration in the calibration curve. On the other hand, if the speed is too low, the friction of the bearings and gearing exercises an appreciable effect (see below). Instruments such as that illustrated in Fig. 8.1 generally have jewelled bearings and are made in three or four types suitable for air speeds ranging from about 15 to 250 or 300 m/min, 45 or 60 to 900 m/min, and 90 to 2000 or 3000 m/min. Each type can confidently be relied upon in its design range as specified by the maker.

Low-speed Anemometers

For low air speeds it is essential to have an instrument in which the friction is very small. In order that the starting torque due to the wind be large compared with the resisting frictional torque, it is desirable that the vanes shall be large and set at a large radius from the axis. These dimensions are, however, limited by considerations of lightness; for the greater the inertia of the moving parts, the less the ability of the instrument to follow rapid fluctuations in the speed of the air, and the greater the errors due to this cause. Moreover, the resisting frictional torque is composed partly of a torque due to the weight of the moving parts, and partly of a torque due to the end thrust of the spindle on its downstream bearing, arising from the downstream component of the air force on the vanes; the first of these factors will increase with the weight of the vanes and the second with their size. As is usual in design generally, a compromise has therefore to be effected between conflicting requirements.

A form of low-speed anemometer that gave consistent readings down to air speeds of about 0·2 m/sec was designed by Ower.[1] The vanes, which were about 3 cm deep, were of the usual pattern, inclined at an angle of 45°† to the axis, and the outer radius of the vane circle was about 5 cm. There was no gearing except a worm drive through which the rotation of the vanes was transmitted, with a reduction of 50 to 1, to a pointer moving over a dial attached to one side of the casing. The pointer was fixed directly to the worm-wheel spindle, and the only markings on the dial were four short radial lines, to indicate quarter-revolutions of the pointer. Three tapped bosses were fitted to the casing for convenience in mounting the instrument to suit various conditions of use.

Other low-speed vane anemometers, responding to about the same lower limit of air speed, are now on the market. In appearance they resemble the Ower type except that they have no gear train, the revolutions of the vane wheel being indicated by methods that produce no friction.

In one type, the passage of the vanes past a certain spot interrupts the light from a small lamp illuminating a photocell, and so causes electric pulses which are counted by a suitable circuit. In another type electric pulses are produced by a device depending on change of capacity. The advantages of both these types over the simpler and cheaper Ower type lie not so much in response to lower air speeds owing to the elimination of gearing friction – as already indicated their sensitivity is not significantly better – but in their convenience as distant-reading instruments. Some of them also have sealed ball-bearings and are reported to behave satisfactorily in dusty atmospheres.

A different type of low-speed anemometer is that designed by Rees,[2] which we have already briefly noted (see p. 4). Although perhaps not as simple to use as the vane anemometer, it has the advantage over the latter that it can be used to measure appreciably lower speeds – down to about 4·5 cm/sec. On the other hand, the instrument can only be used to measure horizontal air velocities, whereas the vane anemometer is subject to no such restriction, provided that, for accurate work, it is calibrated in the attitude in which it is to be used.

† Subsequently reduced to 40° in the standard type (see p. 215).

Calibration Curves

Every anemometer should be calibrated in an airstream whose speed can be controlled and measured by independent means, or by moving it at known speeds through still air; for it is not possible to predict the readings of an instrument accurately from its mechanical constants. As a rule, the maker will supply a calibration curve or table showing the corrections to the indicated air speeds that have to be applied all through the range for which the instrument has been designed. Some makers state that a certain number of metres per minute, constant over the whole speed range, should be added to the readings to give the true air speed. A correction of this kind may perhaps be sufficiently accurate for a few purposes. But it should not be accepted without verification if precision is required: extensive experience of calibrating vane anemometers has not included a single instrument whose error was constant over the whole range.

Excluding this form of correction, the calibration provided by the makers can generally be relied upon to limit errors to about 1 per cent; and, provided that the anemometer is reasonably well treated, in particular to avoid distortion of the vanes, the calibration need not be frequently checked. But for accurate work it is desirable that the instrument should be recalibrated at intervals.

The method of carrying out an anemometer calibration is in itself a simple matter, but considerable care must be taken to provide a suitably steady and uniform airstream of measurable speed and of sufficiently large cross-section in relation to the size of the anemometer (see p. 227). If the amount of work done with the anemometer is not sufficient to justify the expense by the user of providing his own testing equipment, he can return it to the makers or submit it to the National Maritime Institute for recalibration.

Readings in an Inclined Stream

The readings of a vane anemometer are not seriously affected by fairly large inclinations of the direction of the air stream to the axis of the vane circle. Although it is not possible, on account of the numerous differences in details of shape and construction that occur amongst existing types of vane anemometers, to draw generalized conclusions from the observed behaviour of a particular instrument, Fig. 8.2, which was obtained from tests of a calibrated anemometer, shows the order of the error that may be expected in inclined winds.

The results are plotted in the form v_ψ/v_s against angle of yaw, where v_ψ is the speed indicated by the anemometer under test at an angle of yaw $\psi°$, and v_s is the true air speed as indicated by a separate standard anemometer. It will be seen that the instrument could be yawed through about 10° or 12° before an error of 1 per cent on the indicated speed was incurred. It is probable, therefore,

FIG. 8.1. Side-reading vane anemometer.

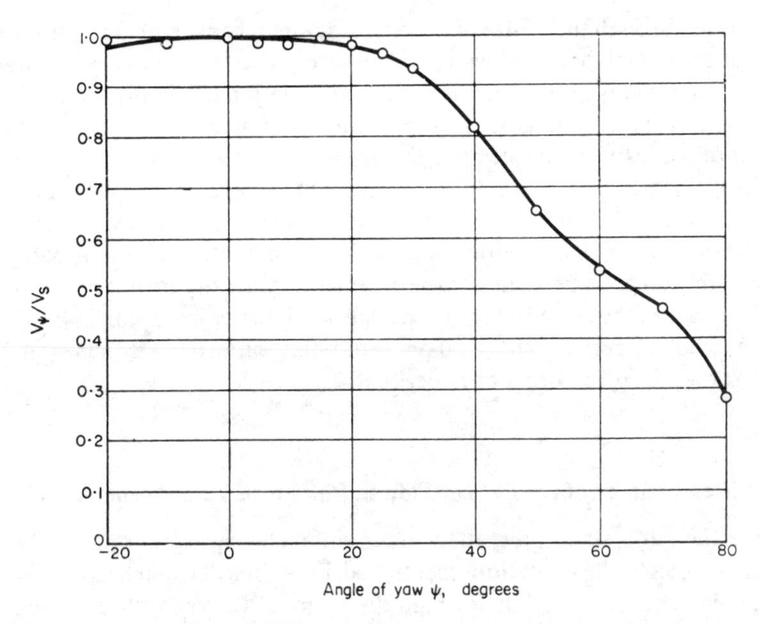

FIG. 8.2. Effect of yaw on vane anemometer.

that little apprehension need be felt about the accuracy of the readings of a vane anemometer if the mean direction of the air current is known within reasonable limits.

Characteristics Indicated by Theory

The theory of the vane anemometer will be considered in the second part of the present chapter; but, since the theoretical treatment is unavoidably somewhat involved, it will be useful at this stage to state the main deductions to which it leads. An anemometer is generally used to measure air speeds appreciably higher than the lowest speed that will just set the vanes in rotation. In these circumstances, it will be shown that the readings of a well-made anemometer, in which the frictional resistance is very small, do not usually require correction for changes in air density corresponding to normal changes in atmospheric pressure and temperature. For special low-speed work, variations of the density of the air may, however, become important, and can be taken into account by the method described on pp. 210–13.

Since there is in general a variation in velocity from point to point across a section of a pipe, it follows that, when an anemometer is used in a pipe there will usually be a variation of air speed across the anemometer disk. In these conditions the anemometer does not indicate exactly the speed of the filament of air passing through the centre of the disk, and significant errors may arise from this cause. Unless precautions are taken to equalize the velocity at different points along the pipe diameter if a marked variation originally exists, a vane anemometer should not be used in a pipe whose diameter is too small in relation to that of the instrument (see pp. 222–7).

Finally, when used in a pulsating airstream, a vane anemometer overestimates the average speed; but, provided that the extremes of speed are not more than about 15 per cent on either side of the mean, the error will probably not appreciably exceed 1 per cent. On the other hand, since the error is proportional to the square of the amplitude of the pulsations, it increases rapidly if the amplitude exceeds 15 per cent: for a 50 per cent pulsation with a frequency of 2 Hz the error will be of the order of 12 per cent.

Measurement of the Rate of Flow in Pipes with Vane Anemometers

Like the pitot–static tube, the vane anemometer measures a local velocity which, subject to the limitations mentioned above, may be taken as the velocity of the filament of air passing through the centre of the vane wheel. Therefore, to determine the quantity of air flowing along a pipe by the use of this instrument, readings must be taken with the axis of the vane wheel at the points at which the velocity head would be observed if a pitot–static tube were being employed for the purpose (see Chapter VI). If v_1, v_2, v_3, etc., are the air speeds in metres per second thus observed, the mean speed is given by equation (2) of Chapter VI, or, if Aichelen's method is used, by the means of a number of readings taken at a radius of 0·762a, where a is the full pipe radius.

From the mean velocity v_m thus obtained, the methods of calculation set out in detail on pp. 120–3 give the following expressions for the mass of air flowing in kilograms per minute in a pipe of area S square metres:

$$Q = 27{\cdot}85 S v_m \frac{b}{273 + t},$$

when temperature t is measured in degrees Centigrade and the barometric pressure b in millimetres of mercury.

In using this equation b must be increased or decreased by the static pressure if this exceeds 10 cm of water (see p. 149) to limit the error the error on Q to 1 per cent.

A word of caution is necessary regarding the recommendation sometimes made that, when being used to measure the rate of flow in a large duct, the anemometer should be moved slowly across the section so that the complete area is covered evenly during the period of observation. This will not, in general, give the true mean velocity, even apart from the fact that if the rate of lateral movement of the anemometer is more than about one-third of the air speed, the relative wind will be inclined at angles of more than 18° to the axis of the instrument, which may cause appreciable errors in the indicated air speed (see Fig. 8.2).

The Theory of the Vane Anemometer

In order to apply the results of the theoretical analysis that follows to predict the behaviour of the vane anemometer in various working conditions, a knowledge is required of the aerodynamic forces acting on small flat plates. Before the anemometer theory[3–5] was developed the only published information of this nature was that of Eiffel.[6] This was obtained in the early 1900s, and the form in which the force coefficients were given is different from that used in modern aerodynamics. To save labour, however, Eiffel's form of coefficient was used when the theory was first developed,[3] and will be adopted here. It may be noted that later on Ower and Duncan[4] measured the forces on actual anemometer blades; and these results, which were in fair agreement with Eiffel's, have been used in the analysis but expressed in Eiffel's coefficients.

Aerodynamic Characteristics of Flat Plates

The vane anemometer consists of a number of small flat plates inclined at an angle to the direction of the wind; we therefore first consider the system of forces that acts on a flat plate when exposed to an air current. In Fig. 8.3, AB represents an end view of the plate inclined at an angle φ to the wind direction OQ, this angle being usually termed the angle of incidence of the plate.

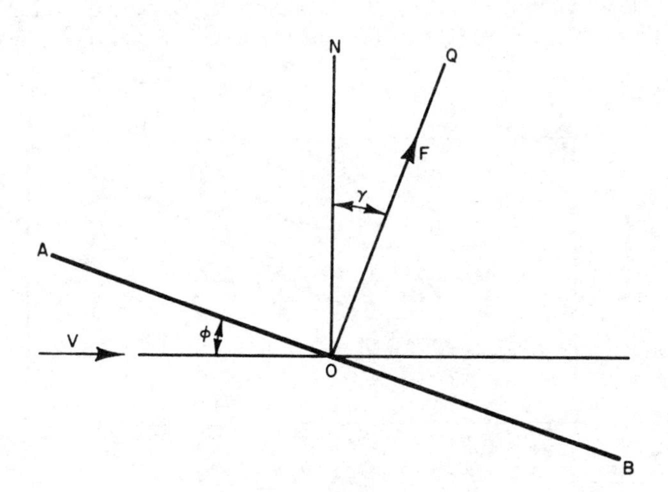

FIG. 8.3. Aerodynamic reactions on a flat plate.

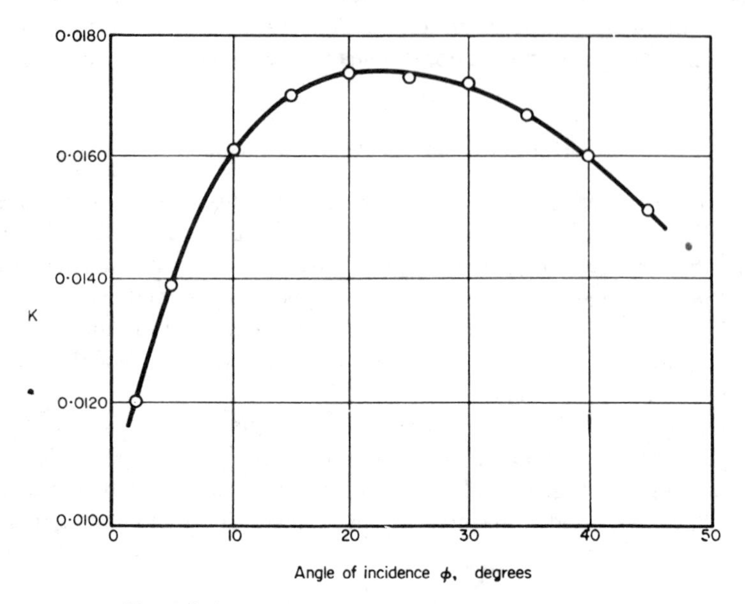

FIG. 8.4. Aerodynamic-force coefficient on a flat plate.

Let V be the wind speed, ϱ the density of the air, and S the area of the plate. The aerodynamic reactions on the plate may be reduced to a single force F acting along OQ, the point O, whose position along AB depends upon the incidence φ, being known as the centre of pressure. It is found by experiment that OQ makes an angle γ with ON, the normal to the wind direction through O, such that γ is nearly equal to φ (i.e. the force is almost at right angles to the

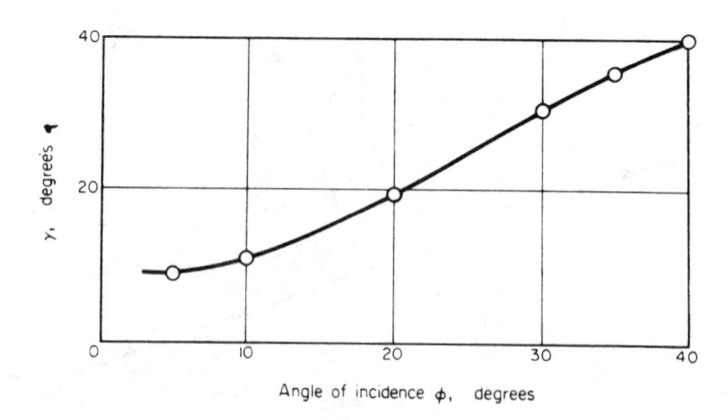

FIG. 8.5. Angle of aerodynamic reaction on a flat plate.

plate) except at small values of φ. Eiffel expressed the magnitude of this force F by the equation

$$F = K\varrho V^2 S\varphi, \qquad (1)$$

where K is a numerical coefficient whose value depends upon φ.

Values of K at different angles of incidence have been obtained experimentally[4] for typical anemometer blades, and are given in Table 8.1 and in Fig. 8.4. Values of γ deduced from Eiffel's experimental results[6] for small square flat plates are shown in the same table and are plotted in Fig. 8.5.

TABLE 8.1. AERODYNAMIC DATA FOR ANEMOMETER BLADES

Angle of incidence (degrees)	K	γ (degrees)
5	0·0123	9·0
10	0·0139	11·0
20	0·0174	19·5
30	0·0172	30·7
35	0·0167	35·8
40	0·0160	40·3
45	0·0151	—

Application of Flat-plate Characteristics to the Anemometer

We now apply the aerodynamic characteristics of flat plates to the vane anemometer. The following notation will be used:

F = resultant wind force on one blade
T = frictional resisting torque
Q = wind torque on blades
V = wind velocity
v = linear tangential velocity of centre of pressure of blade
$w = \sqrt{(V^2 + v^2)}$ = velocity of wind relative to blade
K = wind-force coefficient (see above)
ϱ = air density
θ = inclination of blade (when at rest) to wind direction
φ = inclination of blade (when in motion) to relative wind
= effective incidence of blade
γ = inclination of resultant wind force on blade to perpendicular to direction of relative wind
n = rotational speed of blades
N = rotational speed of pointer
C = instrument constant, such that $N = Cv$
S = area of one blade

$D =$ diameter over blade tips
$m =$ number of blades
$r =$ distance of centre of pressure of blades from axis of rotation
$I =$ total moment of inertia of all blades about axis of rotation

Case I: Mechanical friction neglected. Let AB (Fig. 8.6) represent one blade inclined at an angle θ to the wind direction RO, and let O be the centre of pressure of the blade. It should be noted that the blade is rotating in a circular path perpendicular to the plane of the paper, and that, at the instant under

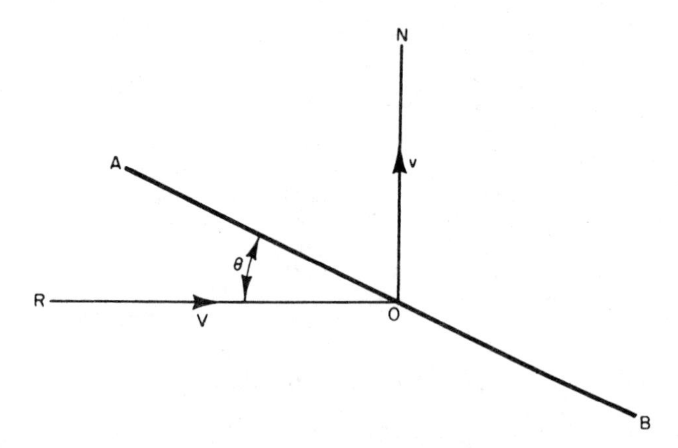

FIG. 8.7. Velocity drag ram for anemometer blade, mechanical friction neglected.

consideration, the blade is at the top of its path and moving towards the top of the page with a velocity v along ON, perpendicular to the wind direction. The blade speed v will be such that the direction of the resultant of v and V lies along AB, and there will be no resultant force perpendicular to the plane of the blade,

i.e. $$v = V \tan \theta \quad \text{and} \quad N = Cv = CV \tan \theta. \tag{2}$$

Case 2: Mechanical friction included. When mechanical friction is taken into account, the velocity diagram becomes more complicated; and it is necessary to consider the forces acting on the blade. AB (Fig. 8.7) will still move along ON, perpendicular to the wind direction, but at a lower speed v, such that the resultant wind speed w will now, when the motion has become steady, be inclined at an angle φ to the plane of the blade. The blade thus becomes a flat plate inclined to the wind at an effective incidence φ; and there will be a resultant force F on the blade acting along OQ, inclined at an angle γ to ON', the normal to the relative wind direction. The component of F along ON, multiplied by the distance of O from the axis of rotation and by the number of blades, provides the torque that overcomes the resisting torque due to mechanical friction.

If, therefore, we neglect any change in velocity of the air during its passage through the vane circle,† and any interference effect that one blade may exercise on its neighbours, we may use (1) to give us, for the blade under consideration, the following equation for F:

$$F = K\varrho w^2 S\varphi. \tag{3}$$

Resolving along ON, putting $w^2 = V^2 + v^2$, and equating the wind torque to the frictional resisting torque, we therefore have for steady motion,

$$T = K\varrho\varphi S(V^2 + v^2)mr\cos(\theta - \varphi + \gamma). \tag{4}$$

Also, from the velocity diagram,

$$v = V\tan(\theta - \varphi). \tag{5}$$

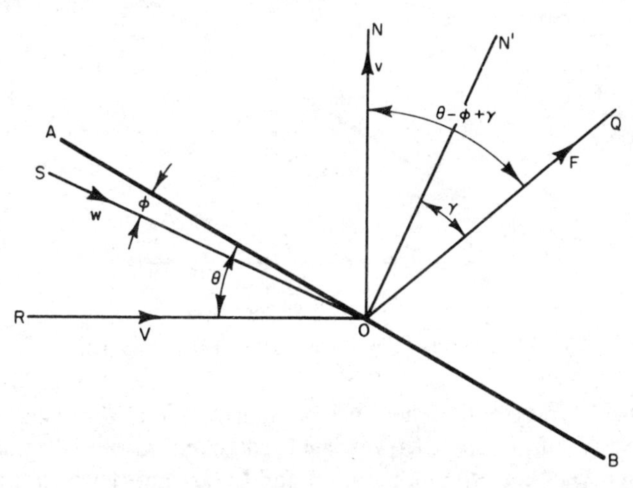

FIG. 8.7. Aerodynamic reactions on anemometer blade, mechanical friction included.

We shall find it convenient at this stage to consider the practical aspects of the equations so far derived. It will be apparent that the speed of rotation of a given set of vanes will be determined not only by the wind speed but also by the mechanical friction introduced by the spindle bearings and by the gearing. In the ideal case where friction is entirely absent, the calibration curve of the instrument, i.e. the curve of indicated wind speed (represented by v) against true

† See p. 215.

wind speed, would be a straight line passing through the origin – see (2) – whose slope would be determined (once the gearing ratio is fixed) by the inclination of the vanes to the wind direction. In practice, however, friction exercises an effect which is felt mainly at low speeds when the aerodynamic forces on the vanes are small. At high speeds, when the relative effect of friction is small, φ is found to become small compared with θ, and we see from (5) that the calibration curve of the instrument will be practically linear.† At low speeds φ is comparable in magnitude with θ, and v will be zero at a certain low value of V when φ becomes equal to θ.

The behaviour of all well-made anemometers substantiates the above reasoning. The usual type of calibration curve for such instruments is illustrated in Fig. 8.8. At high speeds the curve exhibits no significant departure from linearity; but at low speeds it bends over somewhat in the manner shown,

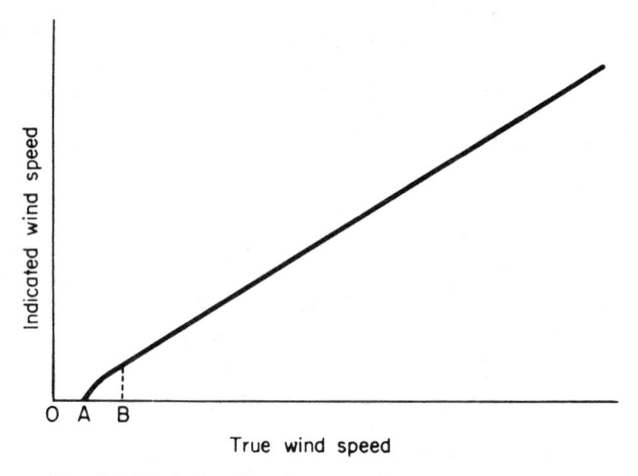

FIG. 8.8. Typical calibration curve for vane anemometer.

the abscissa OA representing the true wind speed at which the vanes just begin to turn, i.e. the limiting speed at which φ becomes equal to θ. The magnitudes of the wind speeds corresponding to OA and to OB (above which the curve is linear) vary, as is naturally to be expected, with different instruments; in the Ower low-speed anemometer to which reference has been made above, OA was found to be about 0·2 and OB about 0·3 m/sec.

The friction torque T in (4) cannot be computed from first principles; it will obviously depend upon the construction of the instrument, and so will have to be determined separately for each anemometer by experiment if a knowledge of its magnitude is required. The necessary tests, however, are simple: they consist only of a calibration to establish the relation between v and V for a

† See also footnote on p. 219.

number of values of the wind speed V. The torque T for a given speed V can then be calculated by the use of (4) and (5), v being taken from the calibration curve, and the values of K and γ being obtained from Figs. 8.4 and 8.5 at the values of φ calculated from (5).

Now the torque T is made up of two components, one due to the weight W of the moving parts, and one due to the end-thrust of the spindle on the rear bearing, arising from the component P of the resultant wind force acting along the wind direction. Hence we may write

$$T = \mu W r_1 + \mu P r_2, \tag{6}$$

where μ is the coefficient of friction, and r_1 and r_2 are the effective radii, whose actual values do not concern us, at which the forces μW and μP act.

If, as is usual, the bearings are cylindrical, it is justifiable to regard r_1 and r_2 as constant. If, further, we assume that μ is constant, we may, since W is constant, write (6) in the form

$$T = k + hP, \tag{7}$$

where k and h are constants.

T is given by (4) and can be calculated from the calibration curve, as already shown. Similarly, P will be given by

$$P = K\varphi\varrho S(V^2 + v^2)m \sin(\theta - \varphi + \gamma), \tag{8}$$

and is calculable in an analogous manner. From (7) we may expect a curve of T against P to be a straight line, from which the values of k and h can be determined. Values of T and P for the low-speed anemometer previously mentioned are plotted in Fig. 8.19. It will be seen that, except for the small values near the origin (corresponding to values of V below 0·3 m/sec), the points lie well on a straight line, which gives values of k and h of $2{\cdot}2 \times 10^{-6}$ N m and $2{\cdot}4 \times 10^{-2}$m respectively.

TABLE 8.2

Wind speed V (m/sec)	$T \times 10^6$ (N m)	$P \times 10^4$ (N)
0·18	3·34	0·96
0·21	4·03	1·19
0·24	4·64	1·34
0·27	5·22	1·49
0·30	5·67	1·59
0·46	7·31	2·16
0·61	9·07	2·85
1·22	16·03	5·60
1·83	22·98	8·44
2·44	30·10	11·39

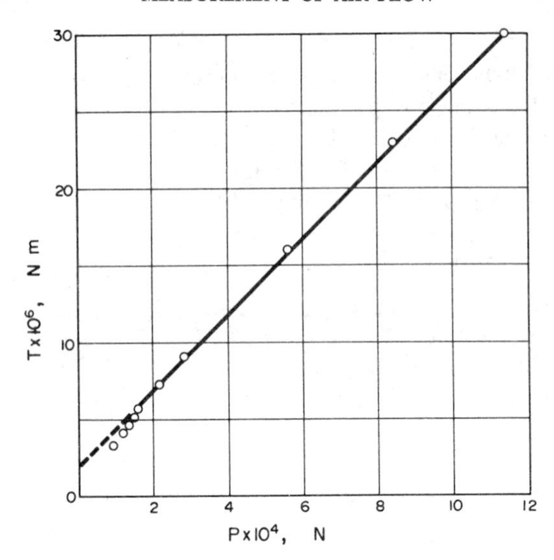

FIG. 8.9. Thrust and torque for a vane anemometer.

The values of T and P plotted in Fig. 8.9 are given in Table 8.2. The mechanical constants of the instrument to which they relate are as follows:

$S = 6{\cdot}3\ \text{cm}^2$; $r = 3{\cdot}5$ cm; $m = 8$; $D = 9{\cdot}5$ cm,
$\theta = 45°$; gear reduction (vanes to pointer) $= \frac{1}{50}$;
$I = 61\ \text{g cm}^2$.

Effect of Variations of Air Density

We now consider the effects of variations of atmospheric density on the readings of a vane anemometer. If an anemometer is employed for the measurement of low speeds, for which purpose it has advantages over certain other types of air-speed measuring instruments, errors will arise if, when measurements are being taken, the air density differs from that existing at the time when the instrument was calibrated, for which conditions alone the calibration curve is valid. The calibration curve or certificate accompanying an anemometer intended for low-speed work should bear a note recording the atmospheric density, or at least the atmospheric temperature and pressure, at the time of calibration.

The problem we are now considering is the determination of the effect of a variation in ϱ on the value of the blade speed v. We shall assume for the purpose of the analysis that the linear speed of the vanes v is known for each value of the wind speed V; although the calibration will not as a rule be given in terms of v, it is a simple matter to deduce it from the gearing ratio and the dimensions of the instrument. Thus, if N is the rotational speed of the pointer at a wind speed V,

$$v = N/C = 2\pi rGn, \tag{9}$$

where G is the gear ratio of reduction between the vanes and the pointer.

The radius r (see list of symbols, p. 205) may be taken with sufficient accuracy as the radius of the centre of area of a blade from the axis of rotation.

Let suffixes 0 refer to the conditions under which the instrument was calibrated, and suffixes 1 to the conditions under which measurements are being made, i.e. when the air density has changed from ϱ_0 to ϱ_1. For a given wind speed V_0, the linear blade speed v_0 can be obtained from the calibration curve when the density is ϱ_0, and the resisting torque is given by

$$T_0 = K_0\varphi_0\varrho_0 S(V_0^2 + v_0^2)mr\cos(\theta - \varphi_0 + \gamma_0),$$

and the end thrust by

$$P_0 = K_0\varphi_0\varrho_0 S(V_0^2 + v_0^2)m\sin(\theta_0 - \varphi_0 + \gamma_0).$$

Also, from (7),

$$T_0 = k + hP_0.$$

Now suppose the air density alters to ϱ_1. At a certain value V_1 of the wind speed and a corresponding value v_1 of the blade speed, the torque T_1 will be the same as before.

That is,

$$T_0 = k + hP_0 = T_1 = k + hP_1,$$

from which it follows that

$$P_0 = P_1.$$

Hence

$$K_0\varphi_0\varrho_0 S(V_0^2 + v_0^2)mr\cos(\theta - \varphi_0 + \gamma_0) = K_1\varphi_1\varrho_1 S(V_1^2 + v_1^2)mr\cos(\theta - \varphi_1 + \gamma_1) \tag{10}$$

and

$$K_0\varphi_0\varrho_0 S(V_0^2 + v_0^2)m\sin(\theta - \varphi_0 + \gamma_0) = K_1\varphi_1\varrho_1 S(V_1^2 + v_1^2)m\sin(\theta - \varphi_1 + \gamma_1). \tag{11}$$

Dividing (11) by (10), we obtain

$$\tan(\theta - \varphi_0 + \gamma_0) = \tan(\theta - \varphi_1 + \gamma_1),$$

so that

$$\varphi_0 - \gamma_0 = \varphi_1 - \gamma_1. \tag{12}$$

Now γ is a definite function of φ, so that (12) can only be true if $\varphi_0 = \varphi_1$; and since

$$\frac{v_0}{V_0} = \tan(\theta - \varphi_0) \quad \text{and} \quad \frac{v_1}{V_1} = \tan(\theta - \varphi_1),$$

the equality of φ_0 and φ_1 leads to the relation

$$\frac{v_0}{V_0} = \frac{v_1}{V_1}. \tag{13}$$

Also, since K is a function of φ, $K_0 = K_1$. Hence (10) reduces to

$$\varrho_0(V_0^2 + v_0^2) = \varrho_1(V_1^2 + v_1^2),$$

and, by virtue of (13), we find that

$$\frac{v_1}{v_0} = \frac{V_1}{V_0} = \sqrt{\left(\frac{\varrho_0}{\varrho_1}\right)}. \qquad (14)†$$

Equation (14) indicates a ready means for deriving the calibration curve of the anemometer at density ϱ_1 from that given at density ϱ_0, since the interpretation of (14) is that a point (v_0, V_0) on the ϱ_0 curve becomes the point $[v_0\sqrt{(\varrho_0/\varrho_1)}, V_0\sqrt{(\varrho_0/\varrho_1)}]$ on the ϱ_1 curve. Hence a point P' on the new curve can be obtained from any point P on the original curve by joining P to the origin O and producing the line OP to P', so that $OP'/OP = \sqrt{(\varrho_0/\varrho_1)}$ (see Fig. 8.10).

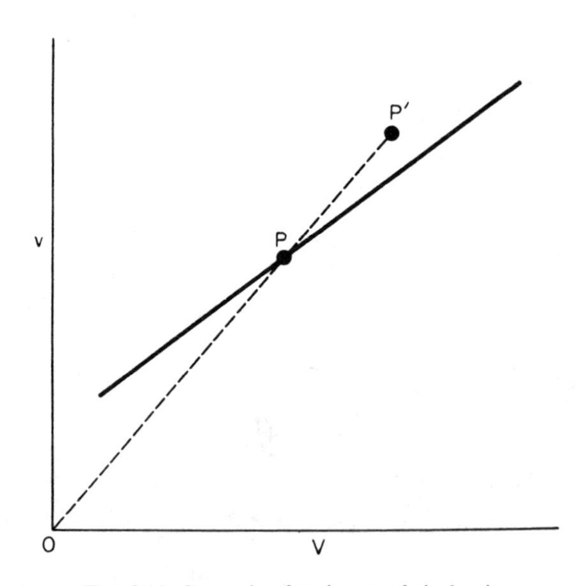

FIG. 8.10. Correcting for change of air density.

† This equation can also be deduced directly from dimensional considerations together with the assumption (implicit above) that K does not change with Reynolds number within the range considered.

Except at the lowest speeds, the original calibration curve will generally be a straight line whose equation may be expressed in the form

$$v = pV + q, \tag{15}$$

where p and q are constants.

From the foregoing analysis it follows that the corresponding portion of the calibration curve for density ϱ_1 will be a parallel straight line whose equation will be

$$v = pV + q\sqrt{\left(\frac{\varrho_0}{\varrho_1}\right)}. \tag{16}$$

Hence the linear portion of the new curve can be derived from that of the original curve by obtaining a single point in the manner indicated above and drawing a parallel line through this point.

Generally, the calibration curve will not be expressed in terms of v, the linear blade speed, but of N, the rotational speed of the pointer. This will not, however, invalidate the above argument; its only effect will be to alter the values of the constants p and q in (15) and (16).

In practice, it is not necessary always to draw a new calibration curve each time the value of the density changes. The air speed corresponding to an observed rotational speed can be obtained by calculation as follows. The calibration curve, obtained at density ϱ_0, will be given in terms of the true wind speed V and an observed quantity, which we may denote by V_i, which will be the rotational speed of the pointer, or the indicated air speed if the dial is graduated in metres. In the latter case, V_i will simply bear a constant ratio to the rotational speed, so that the two cases are in effect identical. Suppose now that the quantity V_i is observed when the density is ϱ_1. The corresponding value of V_i when the density is ϱ_0 will be $V_i\sqrt{(\varrho_1/\varrho_0)}$, and this will correspond to a value V of the wind speed, as determined from the calibration curve for density ϱ_0. The value of the wind speed required can then be shown to be equal to $V\sqrt{(\varrho_0/\varrho_1)}$.† To summarize, therefore, the procedure is to multiply the observed value of V_i by $\sqrt{(\varrho_1/\varrho_0)}$; to read off the value of V corresponding to this from the calibration curve; and, finally, to multiply this value of V by $\sqrt{(\varrho_0/\varrho_1)}$.

The above methods of correction for density changes have been derived on the assumption that the friction torque can be expressed in the linear form of (7). This is probably not quite true at the lowest speeds, but the method of correction will be found, even in this region, to give sufficiently close approximations to actual conditions.

It is of interest that the same theoretical relationship between rate of flow and density was obtained independently by Gehre and Smits,[7] and checked

† Cf. eqn. (14).

experimentally by tests of a large turbine-type gas flowmeter – in effect a vane anemometer that fills the whole pipe (see also p. 302).

Practical importance of the density correction. On the assumption that the calibration curve is a straight line, which appears to be generally true for vane anemometers if we disregard that portion of the curve in the vicinity of the lowest wind speed at which the vanes begin to turn, we can examine analytically the practical importance of the density correction. As already remarked, the calibration curves for different air densities are parallel straight lines; and the displacement along the V axis of two such curves, one relating to density ϱ_0 and the other to density ϱ_1, is equal to $q/p[\sqrt{(\varrho_0/\varrho_1)} - 1]$, where p and q are the constants of the calibration curve for density ϱ_0 (see (15)). It follows from this that if ϱ_0 is the air density for which the anemometer was calibrated and we use that calibration curve (instead of the ϱ_1 curve) to determine the value of the wind speed V corresponding to an indicated wind speed V_i when the density is ϱ_1, the correction to V will be the displacement of the ϱ_1 curve along the V axis,

i.e.
$$\text{correction} = \frac{q}{p}\left[\sqrt{\left(\frac{\varrho_0}{\varrho_1}\right)} - 1\right]. \tag{17}$$

Now an examination of the characteristics of numerous vane anemometers indicates that the value of p is nearly always of the order of 1 while q only rarely exceeds 0·6 numerically, although its sign may be positive or negative. These values of p and q relate to calibration curves in which V and V_i are expressed in metres per second. Thus we may say that the error on true speed V due to neglect of change in density from ϱ_0 to ϱ_1 is of the order of $0{\cdot}6[\sqrt{(\varrho_0/\varrho_1)} - 1]$. The percentage error on true speed is therefore $(60/V)[\sqrt{(\varrho_0/\varrho_1)} - 1]$; and if this error is not to exceed ± 1 per cent we have

$$\left|\frac{60}{V}\left[\sqrt{\left(\frac{\varrho_0}{\varrho_1}\right)} - 1\right]\right| < 1,$$

i.e.
$$\left|\sqrt{\left(\frac{\varrho_0}{\varrho_1}\right)} - 1\right| < \frac{V}{60}.$$

From this relation, Table 8.3 has been prepared, which shows, for a series of values of the true air speed V, the limits of the variation of the density ratio ϱ_0/ϱ_1 that can be allowed if the error on V is to be limited to 1 per cent.

Table 8.3 shows that the higher the wind speed the greater is the variation of density that can be neglected. Thus, when the wind speed is 1·5 m/sec, a 5 per cent density change can be neglected without causing an error of more than 1 per cent on V for an average vane anemometer. But at 10 m/sec a density of over 30 per cent can be neglected for a 1 per cent limiting error. Now density

TABLE 8.3. PERMISSIBLE VARIATION IN DENSITY RATIO FOR ERROR ON V TO BE LIMITED TO 1 PER CENT

V (m/sec)	Limits of ϱ_0/ϱ_1
1·5	0·95–1·05
3·0	0·90–1·10
6·0	0·81–1·21
10·0	0·69–1·36

changes of the order of 5 per cent will be extremely rare under ordinary conditions of use, so that we may say that usually no attention need be paid to density corrections at speeds above 1·5 m/sec.

But there are circumstances, as for example in the mines in South Africa, where values of the air density may differ by as much as 20 per cent from the average value in England. In that case, the density correction would have to be taken into account at all speeds below about 6 m/sec if the anemometer in use had been calibrated in the United Kingdom. Other instances of large density changes might occur if the anemometer were to be used for measuring the flow of hot air.

It should be remarked that the figures given in Table 8.3 are based on the values of p and q stated above, namely 1 and $\pm 0{\cdot}6$ respectively. It will usually be found that p is within 10 per cent of unity, but the value given for q seems to be rather higher than the average. The values given in Table 8.3 are therefore rather conservative; with many vane anemometers somewhat larger density variations may be neglected than the table indicates. In any case, the corrections, when necessary, can be calculated by the use of (17).

Best Blade Angle

The simple theory developed in the preceding pages can be applied to the problem of determining the angle at which the blades should be set for rotation to begin at the lowest air speed. In this way, as was shown in earlier editions of this book, the best blade angle was found to be about 31°. The changes in speed that occur as the air approaches the vane circle were neglected in this theory. Taking these inflow effects into account, van der Hegge Zijnen[(8)] showed that the best blade angle is 40°, a value that agrees exactly with results obtained by Ower† from tests of a special anemometer with variable blade settings. It is of some interest that the Ower low-speed anemometer was designed before either theory had been developed, and its blade angle of 45° was set by a fortunate

† Unpublished.

choice. A similar instrument subsequently constructed with a 40° blade angle began to rotate at an air speed slightly lower than that at which motion of the original anemometer began.

It should be remarked that neglect of the inflow effects causes errors in the calculated values of φ, the effective blade incidence. These errors are greatest when the blades are on the point of beginning to turn, i.e. the conditions that have to be investigated when applying the theory to the determination of the best blade angle. The conclusions drawn from the application of the simplified theory (neglecting inflow) to the other conditions examined in this chapter will be subject to much smaller errors; in any case, these conclusions are more qualitative than quantitative. Hence the use of the simplified theory is considered to be justified, particularly as the extended theory is complicated and laborious to apply.

Effect of a Fluctuating Wind

We have hitherto assumed that the anemometer is placed in a steady wind whose speed is constant and uniform across the whole vane circle. In practice, however, the air speed is usually more or less unsteady in magnitude and also variable from point to point across the vanes; and it is necessary to consider the errors that may be incurred from each of these causes. We shall deal firstly with the effect of a fluctuating wind speed, which we shall assume at any instant to be the same at all points over the vane circle, but to vary with time over the whole circle. We shall exclude winds in which the fluctuations are entirely random, e.g. natural winds, and confine our attention to those cases in which an anemometer is used to measure the speed of an artificially produced air current. Such a current will usually be obtained by means of a fan running at a speed which is intended to be constant, but which may fluctuate in a more or less periodic manner about a mean value; and for the purpose of analysis we shall assume that the wind speed V at any time t is given by the simple harmonic relation

$$V = V_0(1 + \lambda \sin pt), \tag{18}$$

where V_0, the mean speed about which fluctuations occur, is the value of V when $t = 0$, λ is the amplitude of the fluctuations, and $2\pi/p$ is the time for one complete cycle.

It can be shown by dimensional analysis that the wind torque Q is given by an equation of the form

$$Q/\varrho V^2 D^3 = f\left(\frac{nD}{V}\right), \tag{19}$$

where $f(nD/V)$ represents some function of nD/V. When the anemometer is in steady motion $Q = T$;† and therefore, by plotting $T/\varrho v^2 D^3$ against nD/V, the form of the function $f(nD/V)$ can be determined. To take a concrete case, we may consider once more the low-speed anemometer,† for which the values of T have been calculated (see Table 8.4). Values of $Q/\varrho v^2 D^3$ are plotted against nD/V in Fig. 8.11, and it will be seen that all the points for values of $V = 0{\cdot}3$ and upwards lie on a straight line. It should be noted that this range of V corresponds to the linear portion of the calibration curve of the instrument, so that over this region we may write

$$\frac{Q}{\varrho V^2 D^3} = a + b\frac{nD}{V}, \tag{20}$$

where a and b are constants; and this result may be expected to apply, over the linear range, to all vane anemometers of normal design, the values of a and b varying from one to another. For the particular instrument under consideration the values of a and b taken from Fig. 8.11 are 0·118 and −0·283 respectively. In a fluctuating wind, the driving torque Q due to the air no longer necessarily balances the resisting frictional torque T. The former depends on the instantaneous value of nD/V, in accordance with (20), while we may expect T to depend only on the relative rubbing velocity at the bearings. This expectation is borne out by the values given for T in Table 8.2; if these are plotted against V, to which n, and consequently also the rubbing velocity, bears a linear relation over most of the speed range, it is found that from about 0·3 m/sec upwards the points lie on a straight line

$$T = K_1 V + K_2, \tag{21}$$

in which $K_1 = 1{\cdot}15 \times 10^{-5}$ N sec

and $K_2 = 2{\cdot}07 \times 10^{-6}$ N m

We now imagine the anemometer subjected to a wind fluctuating in speed according to (18) about a mean speed V_0. At any time t when the air speed is V, let the rotational speed of the anemometer be n. Then, equating the rate of change of angular momentum of the vanes to the total torque (air minus friction), we get for the equation of motion of the vanes

$$2\pi I\frac{\mathrm{d}n}{\mathrm{d}t} - Q + T = 0, \tag{22}$$

where I is the moment of inertia of the rotor. Substituting for V, Q, and T their equivalents from (18), (20), and (21), we get from (22), after simplification,

$$\frac{\mathrm{d}n}{\mathrm{d}t} + \alpha(1 + \lambda \sin pt)n = \beta(1 + \lambda \sin pt) + \gamma(1 + \lambda \sin pt)^2 + \delta, \tag{23}$$

† Continuing reference to this anemometer is unfortunately unavoidable, since it is the only instrument whose complete characteristics are available to the authors.

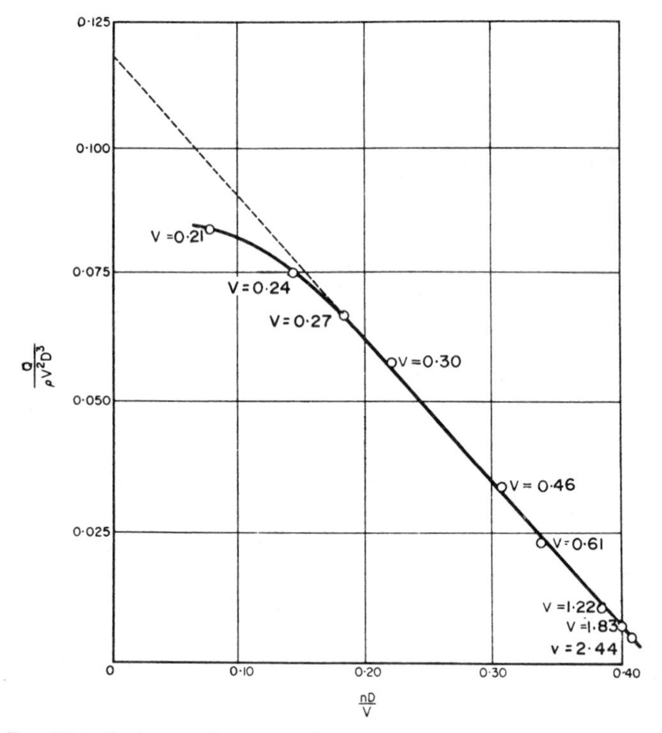

FIG. 8.11. Variation of torque with rotational speed for vane anemometer. (Values of V in metres per second)

where

$$\alpha = -\frac{b\varrho D^4}{2\pi I} V_0; \quad \beta = -\frac{K_1}{2\pi I} V_0; \quad \gamma = \frac{a\varrho D^3}{2\pi I} V_0^2; \quad \delta = -\frac{K_2}{2\pi I}. \quad (24)$$

Now (23) is a differential equation of the standard form $dy/dx + Py = Q$, where P and Q are functions of x or constants. Applying the well-known solution, we thus obtain from (23)

$$n \exp[\int\alpha(1 + \lambda \sin pt)\, dt] = \int[[\beta(1 + \lambda \sin pt) + \gamma(1 + \lambda \sin pt)^2 + \delta]\exp[\int\alpha(1 + \lambda \sin pt)\, dt]\, dt + C,$$

i.e.

$$n \exp\{\alpha[t - (\lambda/p)\cos pt]\} = \int[(\beta + \gamma + \delta) + (\beta + 2\gamma)\lambda \sin pt + \gamma\lambda^2 \sin^2 pt]\exp\{\alpha[t - (\lambda/p)\cos pt]\}\, dt + C. \quad (25)$$

This equation can be integrated if it is assumed that the term $\lambda/p \cos pt$ can be neglected. When the theory was originally published[(3)] it was shown that this term is as a rule very small in practice and it was accordingly neglected. A later investigation[(5)] confirmed the validity of this assumption even when $\lambda = 0{\cdot}5$, i.e. when the amplitude of the velocity pulsations is as much as 50 per cent on either side of the mean value V_0, and the frequency is 2 Hz; and it follows from the form of the term that, for a given λ, the error due to its neglect decreases the faster the pulsations. There seems therefore to be little doubt that the term can be neglected in the great majority of practical cases.†

We can thus reduce (25) to

$$n\,e^{\alpha t} = \int[(\beta + \gamma + \delta) + (\beta + 2\gamma)\lambda \sin pt + \gamma\lambda^2 \sin^2 pt]e^{\alpha t}\,dt + C, \tag{26}$$

which can be integrated, giving

$$n = \frac{\beta + \gamma + \delta}{\alpha} + \frac{\lambda(\beta + 2\gamma)}{\alpha^2 + p^2}(\alpha \sin pt - p \cos pt)$$
$$+ \frac{\gamma\lambda^2}{2\alpha(\alpha^2 + 4p^2)}(\alpha^2 + 4p^2 - \alpha^2 \cos 2pt - 2\alpha p \sin pt) + C\,e^{\alpha t}. \tag{27}$$

Substituting for α from (24) and using the values of the mechanical constants given on p. 210, we find that the term $C\,e^{-\alpha t}$ is reduced to one-tenth of its initial amplitude in under 5 sec, so that we can ignore it after the motion has settled down soon after the start. What we are concerned with is the average value of n over a complete cycle of the pulsation once the motion of the anemometer has settled down. Denoting this average value of n by $\bar{n}$, we deduce from (27), which gives the *instantaneous* value of n at any specified instant, that

$$\bar{n} = \frac{\beta + \gamma + \delta}{\alpha} + \frac{\gamma\lambda^2}{2\alpha}. \tag{28}$$

Substitute for α, β, and γ from (24); then

$$\bar{n} = \frac{K_1}{\varrho b D^4} - \frac{a}{bD}V_0 + \frac{K_2}{b\varrho D^4 V_0} - \frac{a}{bD}V_0\frac{\lambda^2}{2}. \tag{29}$$

In a steady stream $\lambda = 0$, so that the last term on the right-hand side of (29) vanishes, and the residual part of this equation is the equation of the ordinary calibration curve of the anemometer for steady flow.‡ In a pulsating stream, λ is not zero, and the term $-(a\lambda^2/2bD)V_0$ represents the amount by which the anemometer overestimates (overestimates, because b is negative) the mean air speed V_0.

We can now establish the error incurred if we use the steady-flow calibration

† This is not necessarily true for the measurement of natural winds, in which there may be no dominant single natural frequency as there is in the case under discussion.

‡ It is of interest to note that if the value of the various constants for the Ower low-speed anemometer already given are inserted in this equation it becomes $n = 4.39\ V - 0{\cdot}404 - (0{\cdot}0738/V)$. This is linear if the last term is negligibly small, as it tends to become at high speeds. In fact, if this equation is plotted, the curve is practically indistinguishable from a straight line at speeds from about 0·6 m/sec upwards, when it gives values of V that differ by less than 3 per cent from those given by the experimentally determined calibration curve of this instrument which, except at low speeds, could be expressed by the equation $n = 4{\cdot}51\,V - 0{\cdot}59$.

curve to deduce the air speed without making any allowance for the effects of pulsation. As before, let $\bar{n}$ be the observed rate of rotation of the vanes (or of the pointer; it is immaterial which because they are related merely by the gearing factor). Then, if V_0 is the mean about which the air speed is pulsating, and if n is the rotational speed of the vanes corresponding to this air speed *under steady conditions* we know from (29) that

$$\bar{n} = n - \frac{a}{bD}\frac{\lambda^2}{2}V_0. \tag{30}$$

By hypothesis, we use the calibration curve for steady conditions, whose equation is (29) with the last term zero, to determine from the observed $\bar{n}$ a value of the indicated speed V_i in the unsteady conditions. Thus

$$\bar{n} = \frac{K_1}{\varrho b D^4} - \frac{a}{bD}V_i + \frac{K_2}{\varrho b D^4 V_i}. \tag{31}$$

Also we know that n in (30) is given by

$$n = \frac{K_1}{\varrho b D^4} - \frac{a}{bD}V_0 + \frac{K_2}{\varrho b D^4 V_0} \tag{32}$$

Hence, substituting (31) and (32) in (30), we obtain

$$\frac{a}{bD}V_i = \frac{a}{bD}V_0 + \frac{a}{bD}\frac{\lambda^2}{2}V_0 + \frac{K_2}{\varrho b D^4}\left(\frac{1}{V_i} - \frac{1}{V_0}\right) \tag{33}$$

or

$$V_i = V_0\left(1 + \frac{\lambda^2}{2}\right) + \frac{K_2}{a\varrho D^3}\left(\frac{1}{V_i} - \frac{1}{V_0}\right). \tag{34}$$

Since the anemometer overestimates in a pulsating wind, $V_i > V_0$. Hence the last term in (34) is negative,† and the first term, i.e. $V_0[1 + (\lambda^2/2)]$, represents the maximum possible value of V_i.

Inserting this value of V_i in the last term of (34), and using the values already given for K_2, a, and D for the Ower anemometer, we find that for this instrument, when ϱ has its standard value of 1·225 kg/m^3, the last term becomes

$$\frac{0{\cdot}016}{V_0}\left[\left(1 + \frac{\lambda^2}{2}\right)^{-1} - 1\right].$$

If we take a very adverse case in which the amplitude of the pulsations is 100 per cent ($\lambda = 1$) and the speed V_0 is only 0·3 m/sec, the last term becomes 0·017 m/sec, which is less than 4 per cent of the first term. This percentage will evidently decrease as the air speed increases; and in view of its small value for the adverse example chosen, it is reasonable to assume that it can be neglected in practice. It is unlikely that any vane anemometer of the normal type will have

† For the anemometer under consideration, the constants occurring in the last term are all positive.

characteristics markedly dissimilar from the Ower type on whose behaviour this assumption is based.

Thus we deduce that the speed indicated by an anemometer in a pulsating wind is given by

$$V_i = V_0\left(1 + \frac{\lambda^2}{2}\right). \tag{35}$$

The overestimation will never exceed this, but may be somewhat less at very low speeds and excessively high amplitudes of pulsation.

The theory developed in this section is essentially quasi-stationary in that it assumes that the torque-speed relation is the same in fluctuating as in steady flow. In addition, the final result expressed in (34) is independent of frequency and so cannot be true for low-frequency pulsations in which the periodic time is large in relation to the time constant of the anemometer. Obviously, if the rotor followed changes of air speed instantly, and if the quasi-stationary assumption were true, there would be no overestimation: the true average air speed would be given by the average rotational speed in accordance with the steady-flow calibration.

This anomaly appears to be due to the neglect of the term $(\lambda/p)\cos pt$ in arriving at (34) which seems therefore, as does (35), to be strictly true only for high-frequency pulsations.

However, the work recorded in ref. 5 gave experimental results for a low-speed vane anemometer which were in good agreement with the analytical integration neglecting the term $(\lambda/p)\cos pt$ and with graphical integration of the complete equation (25) for frequencies as low as 2 Hz. In addition, independent confirmation was obtained, many years after, by van Mill,[9] who measured the quantity of air passing through a turbine flowmeter in a given time by allowing it to pass into a large plastic bag from which it was subsequently put through a wet gas meter. While passing through the turbine meter, the flow was caused to pulsate approximately sinusoidally with an amplitude of 40·3 per cent of the mean rate ($\lambda = 0{\cdot}403$), and a frequency of 5 Hz. According to (35), the pulsation error is 8·1 per cent. The mean measured error obtained from eleven separate measurements was 7·5 per cent. The error was also computed by numerical integration of the equation of motion, which gave a value of 7·2 per cent. The agreement with the theory is therefore good.

Thus we may conclude that, from the practical standpoint, (35) does give a reliable estimate of the error of a vane anemometer when the air speed pulsates sinusoidally.

Other types of pulsation were discussed by Head,[10] who derived an "approximate pulsation factor" for various types of flowmeter (see also Chapter XII) which, for a turbine meter, and therefore also for a vane anemometer, agrees with (35). For the conditions for which (35) is valid, Head's formula may be written, provided that the flow fluctuations are rapid compared with the response of the instrument,

$$V_i = V_0(1 + f\lambda^2),$$

where f is a "waveform factor" for which Head gives the following values:

$f = \frac{1}{2}$ for a sinusoidal wave form (which agrees with (35),
$f = \frac{1}{3}$ for a triangular wave form,
$f = 4x(1 - x)$ for a rectangular wave form when the maximum amplitude persists for a fraction x of the period and the minimum amplitude for the remaining $(1 - x)$ of the period.

The response of an anemometer in a fluctuating wind was also studied theoretically by Ramachandran,[11] who found that when the fluctuations follow a rectangular wave form with $\lambda = 0{\cdot}45$ and x in Head's formula equal to 0·5, the mean speed is overestimated by nearly 20 per cent. This is considerably less than that given by Head's formula, which, however, was not received without criticism (see p. 319).

Effect of a Non-uniform Distribution of Air Speed Across the Vane Circle

Because the velocities at different points in a section of a pipe are not as a rule the same, it is necessary (as explained in Chapter VI) to determine velocities at a number of points across a section if we wish to measure the total quantity flowing. The question therefore arises whether, when measurements are made with an anemometer with its axis at the proper radii, the blades will rotate at the speed appropriate to the velocity of the filament of air passing the centre of the anemometer disk, or at a rate corresponding to some other air speed.

To define the problem, we shall assume that at the section of measurement there is normal smooth-pipe velocity distribution which, as shown on p. 87, can be represented sufficiently closely by the relation

$$V_s = V_0(1 - As^2), \tag{36}$$

where V_s is the velocity at any radius s less than the full pipe radius R, V_0 is the velocity at the axis, and A is a constant having the value $0{\cdot}35/R^2$.

Consider an eight-bladed anemometer (see Fig. 8.12), with the blades uniformly spaced, placed in a pipe with the centre of the vane circle at the point O distant R_0 from the centre of the pipe on the radius PQ. Let 01, 02, 03, etc., represent the median lines of the blades, whose centres of pressure are at a radius r from the centre of the vane circle. Then Y in Fig. 8.12 is the centre of pressure of blade 01, and $OY = r$.

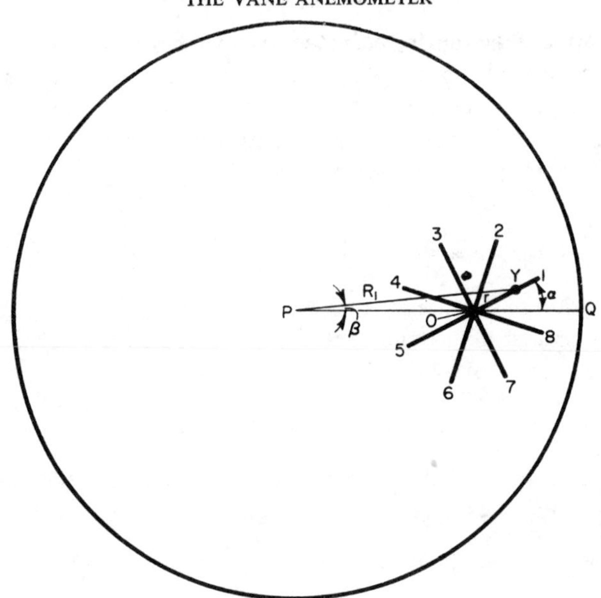

FIG. 8.12. Anemometer in non-uniform velocity field in a pipe.

We have first to find the radii R_1, R_2, R_3, etc. of the centres of pressure of the blades referred to the centre of the pipe.

For R_1 we have, by projection along and perpendicular to PQ,

$$R_0 + r\cos\alpha = R_1\cos\beta \quad \text{and} \quad r\sin\alpha = R_1\sin\beta.$$

Hence, by squaring and adding,

$$\left.\begin{aligned} R_1^2 &= R_0^2 + r^2 + 2R_0 r\cos\alpha. \\ \text{Similarly,} \quad R_2^2 &= R_0^2 + r^2 + 2R_0 r\cos\left(\alpha + \frac{\pi}{4}\right), \\ &\text{etc.,} \\ R_8^2 &= R_0^2 + r^2 + 2R_0 r\cos\left(\alpha + \frac{7\pi}{4}\right). \end{aligned}\right\} \tag{37}$$

In accordance with (36), we therefore have for the velocities of the air impinging on the centres of pressure of the blades

$$\left.\begin{aligned} V_1 &= V_0[1 - A(R_0^2 + r^2 + 2R_0 r\cos\alpha)], \\ V_2 &= V_0\left\{1 - A\left[R_0^2 + r^2 + 2R_0 r\cos\left(\alpha + \frac{\pi}{4}\right)\right]\right\}, \\ &\text{etc.,} \\ V_8 &= V_0\left\{1 - A\left[R_0^2 + r^2 + 2R_0 r\cos\left(\alpha + \frac{7\pi}{4}\right)\right]\right\}. \end{aligned}\right\} \tag{38}$$

Let the centres of pressure of the blades move under the effect of this variable velocity at a tangential linear speed v, which would correspond, according to the calibration curve, to an air speed V. Then V and v will be such that the torque on the whole instrument rotating at this speed will be equal to the sum of the individual torques contributed by each blade.

Now we have already shown (4) that

$$\text{torque} = K\varrho\varphi S(V^2 + v^2)mr\cos(\theta - \varphi + \gamma).$$

If the variation of air speed across the vane circle is not too large, we may, to a first approximation, assume that φ, and hence also K and γ, are constant for all the blades, so that we may write

$$\text{torque} = K'm(V^2 + v^2).$$

The condition that the total torque is equal to the sum of the torques of all the blades is therefore expressed by the equation

$$8K'(V^2 + v^2) = K'[(V_1^2 + v^2) + (V_2^2 + v^2) + \ldots + (V_8^2 + v^2)],$$

i.e.

$$8V^2 = V_1^2 + V_2^2 + V_3^2 + \ldots + V_8^2 = \sum V_1^2.$$

From eqns. (38) it will be found, after reduction, that

$$\sum V_1^2 = 8V_0^2[(1 - AR_0^2)^2 + Ar^2(4AR_0^2 + Ar^2 - 2)],$$

which is independent of α, and hence constant for all positions of the blades, so that

$$V = V_0\sqrt{[(1 - AR_0^2)^2 + Ar^2(4AR_0^2 + Ar^2 - 2)]}. \quad (39)$$

In this equation V is the air speed indicated by the anemometer. If the instrument indicated the speed of the air V_c passing through the centre of its disk, the value of V would be $V_0(1 - AR_0^2)$.

Hence
$$\frac{V}{V_c} = \sqrt{\left[1 + \frac{Ar^2(4AR_0^2 + Ar^2 - 2)}{(1 - AR_0^2)^2}\right]},$$

and, putting $A = 0{\cdot}35/R^2$, we obtain

$$\frac{V}{V_c} = \sqrt{\left[1 + \frac{0{\cdot}35r^2(1{\cdot}40R_0^2 + 0{\cdot}35r^2 - 2R^2)}{(R^2 - 0{\cdot}35R_0^2)^2}\right]}. \quad (40)$$

Obviously, neither R_0 nor r can be greater than R, the radius of the pipe, so that $1{\cdot}40R_0^2 + 0{\cdot}35r^2$ cannot exceed $1{\cdot}65R^2$. Hence the term $1{\cdot}40R_0^2 + 0{\cdot}35r^2 - 2R^2$ is always negative; and therefore it follows from (40) that V is always less than V_c, i.e. the speed indicated by the anemometer is always less than the speed of the air passing the centre of its disk. Now suppose we wish to limit the error to 1 per cent; i.e. we make $V = 0{\cdot}99V_c$. Then (40) becomes

$$0{\cdot}98 = 1 + \frac{0{\cdot}35r^2(1{\cdot}40R_0^2 + 0{\cdot}35r^2 - 2R^2)}{(R^2 - 0{\cdot}35R_0^2)^2}. \tag{41}$$

Writing μ for the ratio R_0/R, i.e. the ratio of the distance from the centre of the pipe at which the anemometer is placed to the pipe radius, we obtain from (41)

$$\frac{r^2}{R^2} = \frac{(1 - 0{\cdot}7\mu^2) - \sqrt{[(1 - 0{\cdot}7\mu^2)^2 - 0{\cdot}02(1 - 0{\cdot}35\mu^2)^2]}}{0{\cdot}35}. \tag{42}$$

The second term under the radical may be considered as small compared with the first up to a value of μ of about 0·7. Hence over the range of μ from 0 to 0·7, (42) reduces to the following approximate form:

$$\frac{r^2}{R^2} = \left\{(1 - 0{\cdot}7\mu^2) - (1 - 0{\cdot}7\mu^2)\left[1 - 0{\cdot}01\,\frac{(1 - 0{\cdot}35\mu^2)^2}{(1 - 0{\cdot}7\mu^2)^2}\right]\right\}\frac{1}{0{\cdot}35}$$

$$= 0{\cdot}0286\,\frac{(1 - 0{\cdot}35\mu^2)^2}{(1 - 0{\cdot}7\mu^2)^2}.$$

Table 8.4 gives values of r/R calculated from this equation for different values of μ.

TABLE 8.4

μ	r^2/R^2	r/R
0	0·0286	0·169
0·1	0·0286	0·169
0·2	0·0286	0·169
0·3	0·0286	0·169
0·4	0·0287	0·169
0·5	0·0289	0·170
0·6	0·0291	0·171

This table shows that, if the error in true air speed as measured is to be limited to 1 per cent of the speed at the axis of the anemometer, the radius of the centre of pressure of the anemometer blades must not exceed 0·17 times the radius of the pipe. In anemometers of the usual types the full radius of the vane circle is about 1·3 times the radius of the centre of pressure. The above argument therefore leads to the conclusion that an anemometer should not be used in a pipe less than $1/1{\cdot}3 \times 0{\cdot}17$, i.e. about 4·5 times its own diameter. There is, however, experimental evidence that if the pipe is appreciably less than 6 times the anemometer diameter significant errors may occur, because the anemometer tends to retard the overall rate of flow. Moreover, instruments such as that shown in Fig. 8.1, in which the indicating mechanism constitutes a large central or downstream obstruction, must introduce additional resistance. There is no published information to show what error this causes, but there can be no doubt that it will be greater the larger the anemometer in relation to the pipe. It would therefore seem prudent not to use an anemometer of this type in a pipe less than 6 times, preferably 8 times, the overall diameter of the vanes if errors greater than 1 per cent are to be avoided. If the velocity distribution in the pipe is more peaked than that for fully developed pipe flow, on which the above analysis was based, the restriction on anemometer size becomes more severe.

The foregoing arguments relate to the measurement of local velocity only. For the determination of the volumetric or mass rate of flow in the pipe, velocity observations must be made at various points in the cross-section whose positions are calculated in accordance with the rules formulated in Chapter VI. It will then be found that, in a large number of cases, the limitation on anemometer diameter in relation to that of the pipe will be determined by the prescribed position of the measuring point nearest to the pipe wall. The distance l of this point from the wall can obviously not be less than $d/2$, where d is the overall diameter of the anemometer vane circle. Table 8.5 shows how this condition limits the anemometer size for various numbers of measuring points located according to the requirements of the log-linear method explained in Chapter VI. The second column gives l in terms of the pipe diameter D.

TABLE 8.5

No. of measuring points per diameter	Distance l of nearest point to wall	Minimum D/d ratio
4	0·043D	11·6
6	0·032D	15·6
8	0·021D	23·8
10	0·019D	26·3

If the pipe diameter is 6 times the anemometer diameter – the minimum ratio for 99 per cent accuracy on local velocity measurements – half the anemometer diameter is 0·083D; and it is clear from Table 8.5 that an anemometer of this size is too large to enable the reading nearest the wall to be taken even for the four-point traverse. For accurate determinations of the volumetric flowrate by anemometer a considerably higher value of the ratio D/d is necessary; the last two columns of Table 8.5 show the minimum value of this ratio which just enables the innermost reading to be taken.

We conclude therefore that for accurate determinations of volumetric rate of flow in a pipe by vane anemometer, the anemometer diameter should be not more than one-twentieth to one-fifteenth of the pipe diameter. The corresponding ratio for rectangular ducts can be deduced from the requirements stated on p. 119.

Finally, the question of wall-proximity effect should be mentioned. There is no published information on this for vane anemometers, but some work has been done in connexion with the fundamentally similar instrument known as the propeller current meter used in hydraulic measurements. A paper by Benini,[12] in which other references are given, shows that these effects are usually small; it is therefore reasonable to assume that they can be disregarded in measurements by vane anemometer.

References

1. E. Ower, A low-speed vane anemometer, *J. Scient. Instrum.* **3** (1926) 109.
2. J. P. Rees, A torsion anemometer, *J. Scient. Instrum.* **4** (1927) 311; see also The measurement of low air velocities in mines, *Trans. Instn Mining Engrs* **74** (1927–8) 359.
3. E. Ower, The theory of the vane anemometer, *Phil. Mag.* (7) **2** (1926) 881.
4. E. Ower and W. J. Duncan, Note on anemometer theory, *J. Scient. Instrum.* **4** (1927) 470.
5. E. Ower, On the response of a vane anemometer to an airstream of pulsating speed, *Phil. Mag.* (7) **23** (1937) 992.
6. G. Eiffel, *La Résistance de l'Air et l'Aviation*, Dunod et Pinat, Paris, 1910.
7. H. Gehre and J. M. A. Smits, A turbine-type gas flowmeter, *Flow Measurement in Closed Conduits*, paper G-3, Vol. II, 701, H.M.S.O., Edinburgh, 1962.
8. B. G. van der Hegge Zijnen, Contribution to the theory of the vane anemometer, *Proc. Koninklijke Akademie van Wettenschappen te Amsterdam* **35** (1932) 1004.
9. C. H. van Mill, The dynamic behaviour of a turbine-flowmeter, *Revue A* **6** (3) (1964), 169.
10. V. P. Head, A practical pulsation threshold for flowmeters, *Trans. A.S.M.E.* **78** (1956) 1471.
11. S. Ramachandran, A theoretical study of cup and vane anemometers, *Quart. J. R. Met. Soc.* **95** (1969) 163; and *ibid.* **96** (1970) 115.
12. G. Benini, Researches on mutual interference and wall proximity effects on current meter readings, *Flow Measurement in Closed Conduits*, paper A-2, Vol. I, 13, H.M.S.O., Edinburgh, 1962.

CHAPTER IX

METHODS OF FLOW MEASUREMENT BASED UPON THE RATES OF COOLING OF HOT BODIES

THE relation between the rate of loss of heat from a heated body and the speed of flow of a fluid in which it is immersed has been studied extensively. Attention has mainly been concentrated on the use of electrically heated wires as anemometers; but there have also been investigations of the behaviour of a specially constructed alcohol thermometer when used for the same purpose.

The laws governing the convective cooling of a hot cylindrical wire by a fluid stream are now well established; and the device is widely used for flow measurements when pressure tubes are inappropriate for one reason or another. Heated-body methods possess several distinct advantages, particularly their high sensitivity at low rates of flow and – in the case of a hot wire – small size and rapid response. The wire diameter, however, needs to be very small – of the order of 0·02 mm or less. Consequently, the probe is fragile and its characteristics are seriously affected by dust deposition, oil, or other surface contamination. Its principal application is the measurement of rapid fluctuations, particularly the study of turbulent flow; in this field it is the only instrument with sufficiently rapid response, and the associated electronic equipment lends itself readily to the signal processing needed to record directly such properties of a turbulence field as r.m.s. values, correlation functions, and spectral distributions. These measurements, however, lie beyond the scope of this book; details may be found in ref. 1. Principles and practical considerations in thermal methods of flow measurement are reviewed in ref. 2.

The shielded hot wire described on pp. 232–7 overcomes the difficulties of fragility and surface contamination (although at the expense of speed of response) and finds extensive practical application in the measurement of low air-speeds. Instruments depending on the cooling of a thin film deposited on a suitable probe have greater mechanical strength, but this is of little advantage in air-flow measurements, for which wires are adequate. The much higher thermal inertia of film instruments, however, is not a serious handicap in steady-flow measurements. Shielded hot-wire instruments and thin-film probes are now available commercially, as are also unshielded hot-wire instruments, together with their associated electronic equipment.

The Cooling of Heated Wires

The extensive researches of L. V. King[3] remain the classic investigation in this field, and the relation he derived between heat-transfer rate and flow velocity has proved to be adequate for most practical purposes.† This states that the rate of heat loss H per unit length of a cylinder of diameter d, maintained at a temperature θ above that of the fluid in which it is immersed, is represented very nearly by the equation

$$H = k\theta + (2\pi k C_v \varrho d v)^{1/2}\theta, \tag{1}$$

where k is the thermal conductivity of the fluid, C_v its specific heat capacity at constant volume, ϱ its density, and v its velocity. The fluid is assumed to be flowing in a direction perpendicular to the axis of the cylinder.

King showed that, in air, this equation might be expected to hold down to a value of vd of 0·0187, v being expressed in centimetres per second and d in centimetres; this is equivalent to a speed of about 9 cm/sec for a wire of diameter 0·02 mm. For lower values of vd, King gave a different approximate formula.

In (1), which has come to be known as King's law in hot-wire anemometry, v is the total speed of the fluid past the wire. At very low velocities, however, the natural convection currents set up by the hot wire itself will exercise a cooling effect additional to that due to the translational speed v of the wire, and (1) will no longer hold. Collis and Williams[4] found that an approximation to the Reynolds number R, below which natural convection becomes important, can be obtained from the formula

$$R < \left(\frac{g d^3 \theta}{T \nu^2}\right)^{1/3},\ ‡$$

where T is the absolute temperature of the air, and the other symbols are as already defined.

For a 0·02-mm diameter wire heated to twice the (absolute) air temperature, this condition gives a Reynolds number of about 0·07.

Equation (1) may be written in the form

$$H = A + Bv^{0.5} \tag{2}$$

where A and B are functions of temperature and the properties of the fluid and

† A more accurate relation was given subsequently by Collis and Williams.[4]

‡ Those familiar with the theory of heat transfer in fluid media will recognize the parameter in brackets as the Grashof number.

the wire mentioned above. Later work has shown[1] that the relation between H and v is more accurately represented by the equation

$$H = A + Bv^{0 \cdot 45}. \tag{2a}$$

For a wire of given dimensions, maintained at a constant excess temperature and always used in the same fluid (air, for example), A and B are constants, and may be calculated approximately for the prescribed conditions. In order to verify his theoretical deductions, King undertook a comprehensive experimental investigation of the heat losses from electrically heated platinum wires ranging in diameter from about 0·03 mm to 0·15 mm, when moved on a rotating arm through still air at speeds between about 0·15 and 9 m/sec. He found that, within these limits, the results could be accurately expressed by an equation of the form of (2); and further work carried out by him and other investigators has shown that this equation holds also for speeds up to at least 30 m/sec. The experimental values of A and B determined by King were in moderately good agreement with those calculated from theory.

Hot-wire Anemometers

A typical form of simple (unshielded) hot-wire anemometer is shown in Fig. 9.1, in which A is the wire forming the heated element. A length-diameter ratio of at least 200 is usually recommended in order to minimize the effects of heat conduction from the ends of the wire, which are soldered or welded to rigid prongs BB held in an insulating base C. The leads DD connecting the wire to the measuring circuit are soldered to the prongs, for which fine (steel) sewing needles are often used, the wire being attached to the pointed ends. A suitable

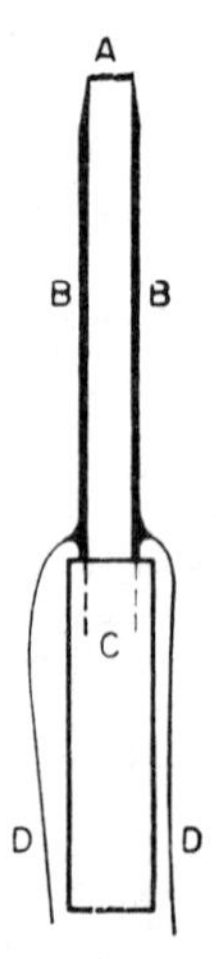

FIG. 9.1. Hot-wire anemometer.

material for the wire itself is silver-coated platinum (known as Wollaston wire), from which the silver is removed by etching locally with nitric acid after the wire has been soft-soldered to the prongs. When an extremely small diameter is required, tungsten is used, as this is 5 or 10 times as strong as platinum; but tungsten needs to be welded unless it is first copper-plated to 10 or 20 times its original diameter. Anemometer elements should be annealed before use in order to minimize drift after calibration: in any case, the wire calibration should be checked frequently. Operating temperatures are normally about 900°C for platinum, but should not exceed about 250°C for tungsten.

If the wire is heated by a current I and the resistance of the wire is R per unit length, then the heat generated per second per unit length is I^2R. Hence for an electrically heated wire (2a) becomes

$$I^2R = A + Bv^{0\cdot45}. \quad (3)$$

The heating current may either be maintained constant (Fig. 9.2) or adjusted so that the wire is maintained at constant temperature (and therefore constant

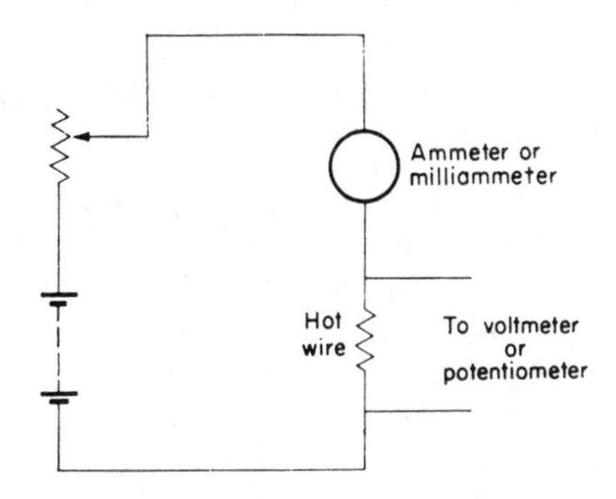

FIG. 9.2. Principle of constant-current operation.

resistance (Fig. 9.3). Calibration consists, in constant-current operation, of establishing the relationship between wind speed and the resistance of the wire or the voltage drop across it, and in constant-temperature operation, between wind speed and current. Early experiments by Simmons and Bailey[5] showed that, except at low wire-temperatures (less than about 150°C), the constant-current method led to higher sensitivity. In practice, however, the constant-resistance method has often been preferred, partly because it is somewhat easier to maintain accurately a constant resistance than a constant current, and partly because (3) may then be written

$$I^2 = I_0^2 rKv^{0\cdot45}, \quad (4)$$

where I_0 is the current needed to bring the wire to the specified resistance in still air, so that the constant K can be determined by calibration at a single air

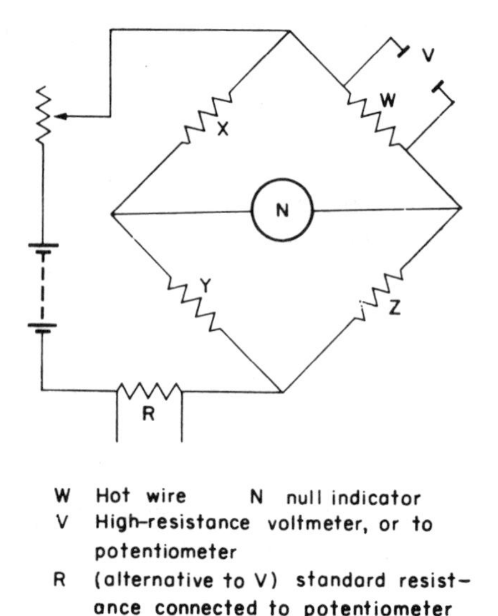

FIG. 9.3. Principle of constant temperature (resistance) operation.

speed. Details of both methods of operation, and a discussion of the effects of variation in atmospheric conditions, may be found in Chapter X of the third edition of this book, together with remarks on the choice of working temperature and notes on probe construction† (for which see also ref. 1). Apart from the measurement of rapid fluctuations and turbulence, however, the use of simple (unshielded) hot wires has now been superseded, in practical engineering, by the shielded hot-wire instrument described below, or by instruments of the type described on pp. 237–9.

A substantial increase in sensitivity can be obtained by replacing the hot wire by a thermistor bead, since a thermistor has a high rate of change of resistance with temperature. (The change is of opposite sign to that of a wire, but this does not affect the principle of the anemometer.) A thermistor anemometer responds less rapidly than a wire or a film to velocity fluctuations, but its advantages may be usefully exploited in steady-flow conditions. A commercial instrument using thermistors is described on p. 239.

The Shielded Hot-wire Anemometer

This instrument is extensively used for ventilation and other air-speed measurements at normal air temperatures. It was developed by Simmons[6]

† These are summarized on pp. 239–42.

and has a much more permanent calibration and greater mechanical strength than the unshielded hot wires so far described. The wandering calibration of an unshielded wire is largely due to the deposition of dust and possibly also to the change of specific electrical resistance with the strain caused by the air forces acting on the wire. The latter effect can be appreciable with the fine, mechanically weak wires used in ordinary hot-wire anemometry; and obviously the cleaning of these wires to remove the dust is difficult. Simmons therefore decided to try the effect of shielding the wire by enclosing it in a fine tube. Besides protecting the wire from the air forces, the shield served to increase the overall diameter of the probe – without changing the wire diameter – so that the dust deposited produced a smaller effect.

The wire of the new instrument was of nichrome, 0·1 mm in diameter and about 20 mm long; it was inserted into one of the bores of a short length of silica twin-bore tubing such as is commonly used to insulate thermocouple leads. Thus shielded from the full cooling effect of the air-stream, this hot wire would constitute too insensitive an anemometer if it were used in the ordinary way. Simmons overcame the difficulty by inserting into the other bore of the silica tubing the hot junction of a nichrome-constantan thermocouple, the two free ends forming the cold junction being fused to two copper supports outside the silica tube, one at each end, where they were maintained at the temperature of the airstream. The thermocouple thus produced an e.m.f. appropriate to the excess of the temperature of its hot junction (which was heated by the hot wire) over the temperature of the airstream; and, by using the change in this e.m.f. produced by the motion of the air as a measure of the air speed, Simmons obtained an instrument of adequate sensitivity. The hot wire was heated to about 120°C in still air by a current of about 0·6 A, which was maintained constant throughout.

Figure 9.4 shows an instrument of this type made by H. Tinsley & Co. Ltd., which has proved to be highly satisfactory. With periodic cleaning of the silica tube to remove dust, the calibration remains constant for indefinite periods, as compared with the days, or often hours, of the unshielded types using fine wires. Cleaning is easily carried out by the application with a camel-hair brush of a solvent such as carbon tetrachloride or acetone. The instrument is much more robust than the ordinary type with a wire of the same diameter; and the effect of variations of air temperature of as much as 30° or 40°C is very small. Finally, the sensitivity is ample: a three-range electrical circuit is used, with which useful ranges of 0–0·1, 0·6, and 1·5 m/sec can be obtained, the maximum speed in each range being measurable with an error of less than $\pm$ 0·5 per cent. Simmons recommended that, in horizontal flow, the instrument be used with the wire vertical, as this gives the highest sensitivity at low speeds, below about 0·08 m/sec. At higher speeds the performance is the same whether the wire is vertical or horizontal, provided that it is at right angles to the stream; if it lies along the stream, the sensitivity is much less. Typical calibration curves are given in

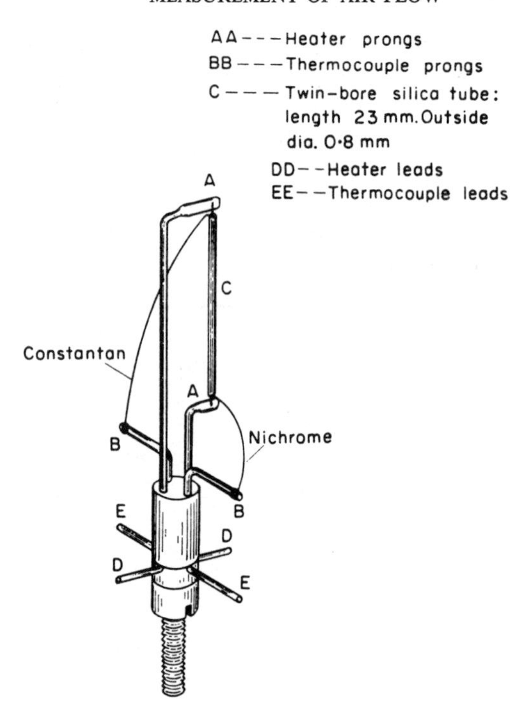

FIG. 9.4. Shielded hot-wire anemometer.

Fig. 9.5, which shows the variation of thermocouple e.m.f. e with wind speed v for two values of the wire-heating current. King's law indicates that, over ranges in which the character of the flow past the silica tube remains unchanged, $\sqrt{v}$ may be expected to vary approximately linearly with $1/\theta$ and therefore with $1/e$. The extent to which this is so is shown in Fig. 9.6. The marked departure from linearity at very low speeds is due to the effects of free convection.

Effect of variations of ambient conditions. The effects of variations of ambient temperature and pressure on the readings of a shielded hot-wire anemometer, at speeds for which the cooling due to natural convection is negligible, have been examined by Cowdrey[(7)] on the basis of an empirical relation between heat loss rate and wind speed given by Hilpert:[(8)]

$$\frac{H}{k\theta} = a\left(\frac{\varrho v d}{\mu}\right)^m, \tag{5}$$

where, as before, H is the rate of heat loss per unit length and θ is the temperature excess over that of the fluid; d is the diameter of the silica tube. The values

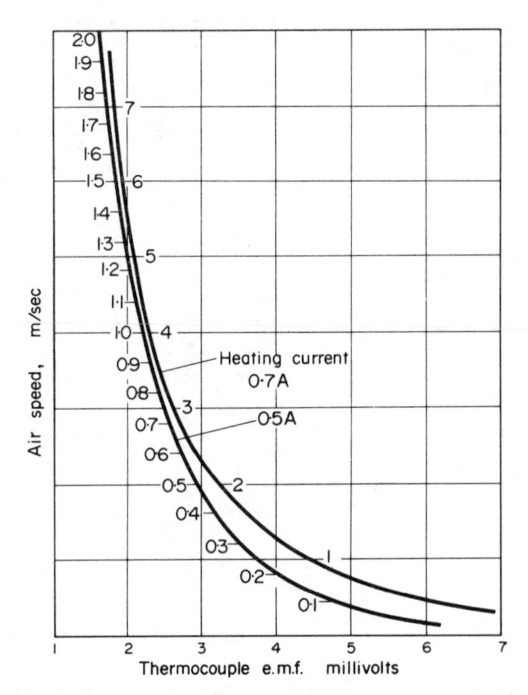

FIG. 9.5. Calibration curves for shielded hot-wire anemometer.

of a and m, for the values of $\varrho vd/\mu$ appropriate to a shielded hot-wire anemometer, may for practical purposes be taken as constant, with m equal to 0·466. Since for practical purposes k may be taken to be proportional to μ, we may write (for a given diameter d)

$$v = b\mu^{(m-1)/m}/\varrho\theta^{1/m},$$

(where b is another constant) provided that the wire current, and therefore H, is maintained constant. (We assume that the wire resistance does not vary appreciably with temperature, which is permissible when the wire is of nichrome.)

For different values of ambient temperature (t_1, t_2) and pressure, therefore, such that the corresponding air densities are ϱ_1 and ϱ_2, the same value of thermocouple e.m.f. e corresponds to different values of temperature differences (θ_1, θ_2) and of wind speed (v_1, v_2) such that

$$v_2 = G\frac{\varrho_1}{\varrho_2}v_1, \tag{6}$$

where

$$G = \left(\frac{\mu_2}{\mu_1}\right)^{(m-1)/m}\left(\frac{\theta_1}{\theta_2}\right)^{1/m}.$$

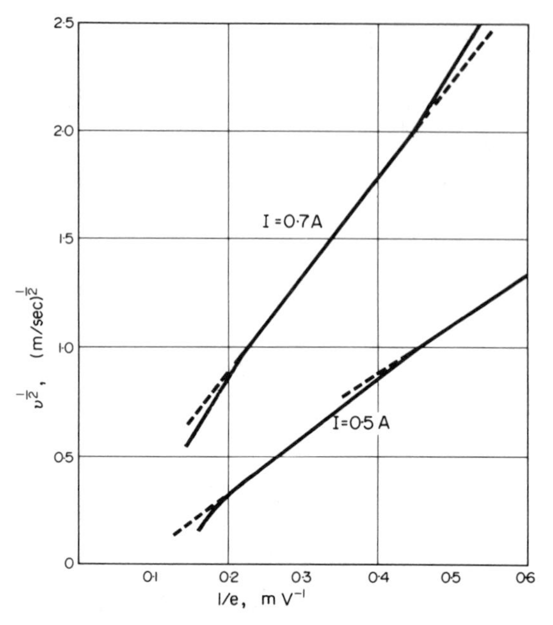

FIG. 9.6. Relation between e.m.f. and airspeed for shielded hot-wire anemometer.

Curves of G for use when the thermocouple is of nichrome–constantan have been calculated for $t_1 = 15°C$ from the data given in ref. 7 and are shown plotted in Fig. 9.7. If the instrument is calibrated at $t_1 = 15°C$, these curves give the value of G to be inserted in (6) in order to obtain the true velocity v_2 at any

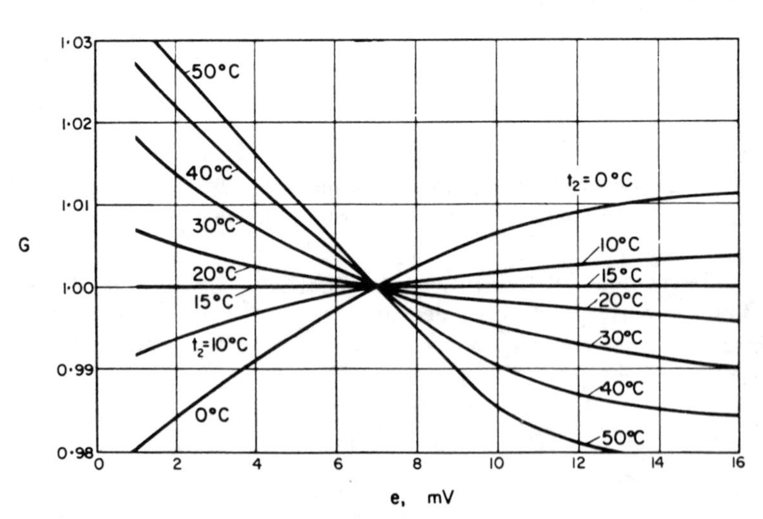

FIG. 9.7. Shielded hot-wire anemometer — corrections for variations in ambient temperature.

other ambient temperature t_1 from the calibration curve $v_1(\theta_1)$. Conversely, the curves and equation also serve to establish what the calibration curve would have been at 15°C ($=t_1$) from values obtained whilst calibrating at some other temperature t_2, or to establish directly the velocity corresponding to prevailing ambient temperature and pressure from a calibration at any other values. It is interesting to note that the variation of μ with t and the variation of θ with e for nichrome–constantan prove to be such that G lies between 0·98 and 1·03 over the entire range of t from 0° to 50°C and over the extreme range of e from 1 to 16 mV. When velocities are required no closer than to within ± 3 per cent, therefore, the value of v_2 under the new conditions of ambient temperature and pressure is given simply by

$$v_2 = \frac{\varrho_1}{\varrho_2} v_1.$$

Hot-film Anemometer

One advantage of the shielded hot-wire instrument over the unshielded wire is its greater mechanical strength. Another way of securing this is to replace the (unshielded) wire by an electrically conducting thin film deposited on an insulating support, usually at the nose of a wedge-shaped or conical probe: typically, the film is of platinum or nickel deposited on Pyrex glass. An appreciable quantity of heat is transferred to the air by conduction through the glass instead of by forced convection, and the extent to which this happens in time-variant flow depends on frequency.[9] Since compensation is difficult, the hot-film probe – unlike the hot wire – is not well-suited to turbulence measurements. It is valuable in steady flows, however, and has even been used in slurry liquids, since contamination does not affect the film as seriously as it does the wire.

Industrial Heated-element Anemometers

The constant-temperature bridge circuit shown in Fig. 9.3 is used in modified form in some robust heated-element anemometers suitable for industrial use.

The Anemotherm. This instrument, marketed by the Anemostat Products Division of the Dynamics Corporation of America, is self-compensated for changes in calibration due to changes in air temperature (see p. 242) by having the resistor X of Fig. 9.3 replaced by a duplicate of the heated element W, as shown in Fig. 9.8 in which the two elements W_1 and W_2 are identical. W_1 is heated separately by a heater element H placed close to it, and not by the bridge current. Both W_1 and W_2 are exposed to the air flow, so that changes in their

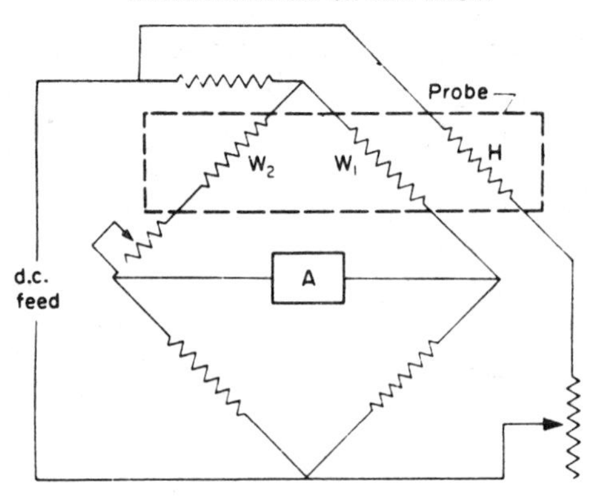

FIG. 9.8. Anemotherm measuring circuit.

resistance due to changes in air temperature cancel out, and measured changes in resistance of the heated element W_1 are due entirely to the cooling effect of the flow.

The probe is a stainless-steel cylindrical tube, the head of which (Fig. 9.9) encloses the two sensing elements W_1 and W_2. These consist of helical windings of Bako wire (70 per cent nickel and 30 per cent iron), set in epoxy-resin and encased in stainless-steel tubes. They are mounted side-by-side and exposed to the flow through a window as shown. The heater element H, which is not exposed to the flow, raises the temperature of W_1 in still air to about 10°C above that of W_2 by means of a heating current of 16–18 mA, obtained in a portable form of the instrument from dry batteries. In operation, this current is not varied to

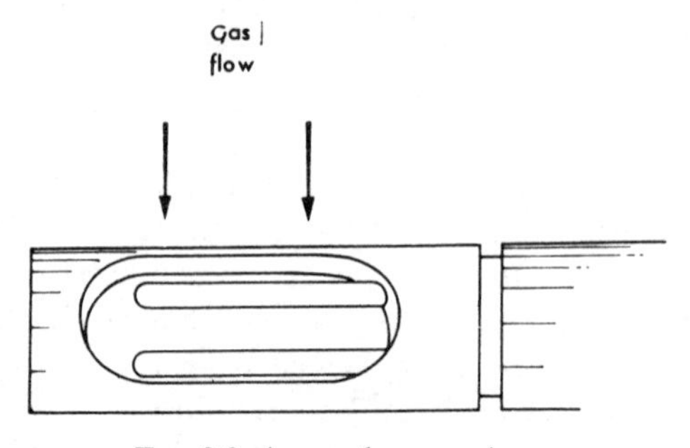

FIG. 9.9. Anemotherm probe.

restore the temperature of W_1 to its still-air value; instead, it is kept constant, and the out-of-balance current of the bridge passes through a micro-ammeter A graduated to read velocity directly. Alternatively, automatic balancing can be employed in conjunction with chart recording.

The diameter of the probe is 10·3 mm for a portable model and 21·3 for an automatically balanced type. Air speeds of between 0·05 and 40 m/sec can be measured; in addition, static head can be measured over the range ± 25 cm of water by fitting to the probe a cap which is, in effect, a calibrated orifice passing a quantity of air over the sensing element directly related to the inlet static pressure. The instrument, which may be either mains or battery operated, can also be used to measure temperature to within 0·3°C over the range $-30°$ to 125°C.

The ETA 3000 portable air velocity meter. This instrument, developed by Airflow Developments Ltd., of High Wycombe, is a temperature-compensated battery-operated, heated-element anemometer using an electrical system which includes a bridge circuit based partly on that of Fig. 9.3 with a thermistor (T_1) replacing the resistor W as the heated element. Another thermistor (T_2) is used for temperature compensation. The probe is a stainless-steel tube 170 mm long and 12 mm in diameter with an 8-mm diameter hole 10 mm from the tip through which the air flows past the heater thermistor T_1. Thermistor T_2 is close to T_1 but is shielded from the air flow.

The thermistor T_1 has a high resistance at ambient air temperature and the bridge is initially out of balance; the circuit is such that the resulting current causes a voltage to be applied to T_1 which then heats up until its resistance falls to a value at which the bridge is in balance. With the probe in still air, the instrument meter is then set to zero when the same voltage is automatically applied to thermistor T_2, which forms part of an auxiliary circuit used for zero setting. In operation, the bridge is out of balance by a current which depends only on the cooling effect of the flow and is read on the instrument meter, a milli-ammeter graduated directly in air speed within the range 0·1–15 m/sec.

Notes on the Use of Unshielded Wires

Measuring circuit for constant-resistance method. This is an ordinary Wheatstone-bridge circuit in which X (Fig. 9.3) is a resistor of very low temperature coefficient; its resistance is equal to that of the hot-wire W at its working temperature (see below). Y and Z are two equal resistors made of the same material. When the bridge is in balance, the temperature and resistance of the wire will be at their desired values, and the calibration of the wire will be the relation between air speed and the current in the external circuit or the wire itself necessary to maintain balance. These current changes can be observed by voltmeter or potentiometer readings as indicated in Fig. 9.3. In principle, this circuit can easily be modified for automatic control of the balance point by feedback, but for the measurement of steady speeds, as distinct from turbulence measurements, this complication will rarely be worth while in practice.

Wandering calibration. The changes in calibration that often occur with unshielded wires are generally accompanied by changes in I_0 (or P_0), the still-air values of current (or potentiometer reading). Wandering of the calibration curve can be considerably reduced if the relation between current and air speed is plotted as $(I - I_0)/I_0$ or $(P - P_0)/P_0$ against v. In this form the calibration curve will usually be found to remain constant for much longer periods. It seems that, for a small change of I_0, the reading I at a given speed changes in the same ratio as I_0.

Wire temperature. From the standpoint of sensitivity, the higher the temperature of the wire the better. This is illustrated by Fig. 9.10, which reproduces some of King's results.[3] The same diagram also shows the characteristic advantage that the hot wire possesses over pressure anemometers, namely increasing sensitivity with decreasing speed. On the other hand, too high a wire temperature may limit the low-velocity end of the useful range because of

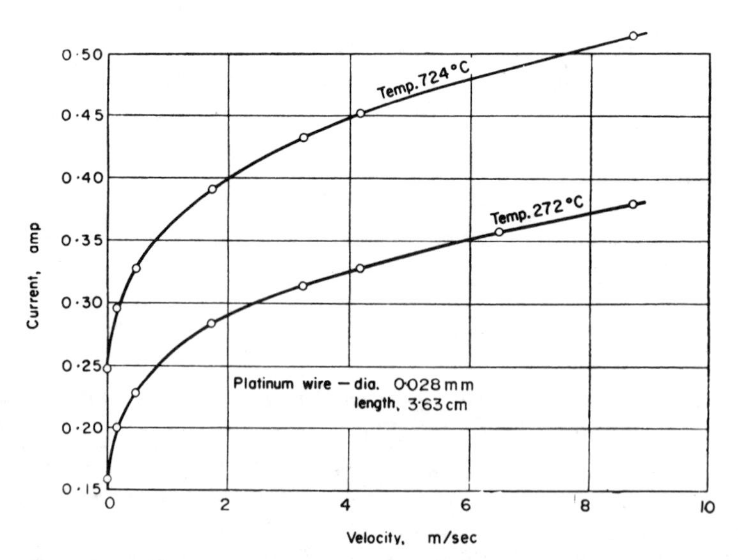

FIG. 9.10. Effect of wire temperature on calibration of unshielded hot-wire anemometer.

the natural convection currents due to the heated wire. Some calculations made by King indicate that the natural convection current is about 0·15 m/sec from a 0·08-mm diameter wire at 1000°C and 0·08 m/sec at 200°C; Collis and Williams[4] show that convection effects are small provided that the Reynolds number is greater than

$$\left(\frac{gd^3(T_w - T_a)}{T_a v^2}\right)^{1/3},$$

where T_w and T_a are the absolute temperatures of the wire (diameter d) and the ambient air respectively. Bradshaw[1] gives twice this expression as the criterion for neglecting convection effects altogether.

When the wire is at rest, the natural convection currents exercise a certain amount of cooling, part of which is reflected in the current I_0 necessary to maintain the wire at the desired temperature (resistance) when $v = 0$, see (4). At very low air speeds, the natural convection currents are disturbed, and the overall rate of cooling may actually be less than when the wire is at rest. This is illustrated by Fig. 9.11, which shows a calibration curve of the form $(I - I_0)/I_0$ against v obtained by Ower and Johansen.[10] At speeds below about 0·05 m/sec the cooling is less than it is in still air. The same effect was noted later by Cooper and Linton,[11] who also found that the range of such speeds was considerably less for a vertical than a horizontal wire. Hence a wire intended for use at very low speeds should always be used in the attitude in which it is calibrated; and its temperature should not be too high. Ower found that a horizontal platinum wire of 0·025-mm diameter, run at a temperature of 60–70°C, would satisfactorily measure air speeds down to about 0·03 m/sec. This agrees well with the

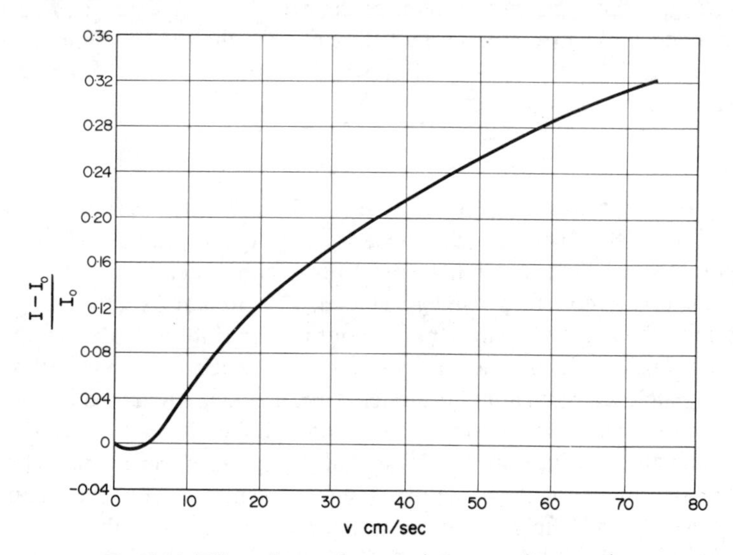

FIG. 9.11. Effect of natural convection at very low speeds.

criterion of Collis and Williams (see above) which, applied to these conditions, shows that natural convection effects are small at speeds above 0·03 m/sec.

Another factor that influences the choice of wire temperature T_w is the effect on the calibration of changes in the air temperature T_a, which cause changes in $\theta(=T_w - T_a)$ in (1) if T_w remains constant. A 1°C change in T_a may cause a change in calibration of the order possibly of 1 per cent. Errors due to this cause will be less the higher the value of T_w; together with the increased sensitivity (see above), this strengthens the case of a high wire-working temperature. Against this, however, we must note that calibration drift seems to be more acute at high T_w. A practical compromise is to run the wire at about 900°C for platinum (but not above 250°C for tungsten), except when very low speeds of the order of 0·03 m/sec have to be measured when, for reasons already given. T_w should not exceed about 70°C.

Bradshaw[1] points out that, if the change in T_a is small, a calibration curve of the linear form of (4) between I^2 and $v^{0\cdot45}$ can be obtained for the new T_a by measuring I at one known value of v at this T_a, and so obtaining a single point on the new calibration curve, which can then be assumed to be a straight line through this point parallel to the original calibration line. This follows from the fact that B in (3) is much less affected than A by changes in T_a.

The resistance of the wire for a given temperature t_w°C can be calculated sufficiently closely from the equation

$$R_{t(w)} = R_{t(a)} \frac{1 + \alpha t_w}{1 + \alpha t_a}, \tag{7}$$

where $R_{t(w)}$ is the resistance in ohms at the wire temperature t_w°C, $R_{t(a)}$ is the resistance at ambient temperature t_a°C, and α is the temperature-resistance coefficient per °C of the wire material, which is about 36×10^{-4} for platinum and 51×10^{-4} for tungsten.

Calibration of wires. Each hot wire has to be calibrated before use and the calibration must be checked for constancy at frequent intervals. This is most conveniently done in a small, low-speed wind tunnel such as that described in ref. 12. Alternatively, the wire may be moved in a circular path through still air at the end of a small whirling arm of about a metre radius. To shield the wire against draughts, which could easily upset the measurements, the path should be totally enclosed by an annular tunnel of about 30-cm square section, the hot-wire support and leads entering through an annular slit in the roof or floor. One disadvantage of this apparatus is that the speed of the anemometer wire through the air is not the same as its speed calculated from the radius and revolutions per minute of the arm, because the rotation of the wire and its supports causes a "swirl" velocity in the annular tunnel. However, this swirl takes some time to build up; and if the observations at each speed are completed during the first revolution or two of the arm, which is quite possible, and if the hot-wire probe is small, errors due to neglect of swirl will be negligible. If necessary, a correction can be obtained by using the wire as calibrated in this way to estimate the swirl speed: a dummy wire is rotated on the arm, and velocity measurements are made with the calibrated wire, which is now stationary and as close to the path of the rotating dummy as possible without fouling it.

Electric Flowmeters

Thomas[13] and Callendar[14] designed electric flowmeters which indicated the mass flow of air from measurements of the heat energy that must be supplied to the moving stream to maintain a specified temperature difference between two cross-sections, one upstream and one downstream of the heat source. If $\Delta\theta$ is the temperature difference and no heat is lost between the two sections, the mass of air M flowing per second is given by

$$M = \frac{H}{C_p\,\Delta\theta}, \tag{8}$$

where H is the heat supplied per second and C_p is the specific heat capacity of air at constant pressure.

If the heat is supplied by an electric current I flowing through a resistance R placed in the airstream, (8) becomes

$$M = \frac{I^2 R}{C_p\,\Delta\theta}. \tag{9}$$

Alternatively, if V is the voltage drop across the resistor, we may write

$$M = \frac{V^2}{R C_p\,\Delta\theta}. \tag{10}$$

In Callendar's instrument use was made of the relationship in the form of (10). Briefly, the meter comprised a resistance mat in the airstream, on either side of which were two electrical-resistance thermometers. The circuit was such that it was possible to measure the voltage drop necessary to maintain a specified temperature difference – in one case quoted in ref. 14 this was 2·5°C – between the two thermometers.

The Thomas meter used a similar arrangement of heating and thermometer coils; but the power supplied to the heating circuit was measured directly by a wattmeter. In a form of this meter that was marketed for industrial use, the heating current required to maintain the constant temperature difference was automatically adjusted at each flowrate so that no manual control was required.

In the instrument described in ref. 15, the temperature difference was registered by a thermocouple array with one junction upstream of the heat source and the other downstream, so that the thermocouple voltage for a given rate of heat input served as a measure of airspeed or rate of mass flow. (Alternatively, the airspeed or flowrate can be deduced from the heater current needed to maintain a given thermocouple output.) Airspeeds were measured down to about 0·3 cm/sec and flowrates down to about 14 cm^3/min.

This type of instrument, developed originally for measuring airflow in the blast freezing of meat, has also been used in ventilation systems.[(16)]

The Kata-Thermometer

Similar in principle to the hot-wire anemometer, but different in that the supply of heat whose rate of loss is measured is discontinuous and non-electrical, is the kata-thermometer. This instrument was first designed by Hill,[(17)] for use in his researches on the effect of atmospheric conditions on health and industrial efficiency, to measure the rate of loss of heat from a surface at approximately body temperature. The kata-thermometer is essentially an alcohol thermometer of special type; in its original form it had a bulb about 4 cm long and 2 cm in diameter, and a stem leading from the top of the bulb with the temperature range 100°F to 95°F† marked on it. The bore of the stem was enlarged at its upper end, so that the instrument could be heated considerably above 100°F without the risk of fracture.

This instrument was used by heating it until the top of the alcohol column was above the 100°F mark and observing the time taken for the column to fall‡ to the 95° mark. The amount of heat lost during this drop was always the same but the rate at which it was lost, and hence the time taken, depended on the atmospheric conditions. Although the instrument was not designed primarily as an anemometer, experience with it has shown that, provided that the bulb and stem are quite dry, the mean rate of cooling over the marked temperature range is related to the air speed v by the equation

$$H = (a + b\sqrt{v})\theta, \tag{11}$$

where a and b are constants, θ is the mean excess temperature of the kata-thermometer over that of the surrounding air, and H is the total heat lost in cooling from 100° to 95°F divided by the area of the cooling surface and by the cooling time.

The similarity of this law to that given by King for hot wires will be noted.

The total heat lost in cooling through a given temperature range, although constant for a given instrument, varies somewhat from one kata-thermometer

† No metric form of this instrument has been made.

‡ Hence the name given to the instrument: "kata" = Greek "down".

to another. Each instrument is calibrated in still air by the makers and is given a factor, F, which is the total heat lost per unit area of cooling surface in cooling through the specified range of temperature. Therefore, in order to calculate H in (11), it is necessary only to observe the time T in seconds taken by the alcohol column to fall from the higher to the lower temperature mark, and to divide the factor F by this time.

The first form of kata-thermometer was subject to two major objections: firstly, that it was liable to be affected by radiation from its surroundings, and, secondly, that the cooling times in hot atmospheres were very long.† Various modifications were introduced at different times,[18] without, however, any change in the main dimensions: modern kata-thermometers intended for measuring air speeds have brightly silvered bulbs in order to reflect incident heat radiation as much as possible; and the working temperature range is either from 130° to 125°F, or from 150° to 145°F in order to reduce the time of operation in hot spaces. The working forms of (11) for these instruments are as follows:

For the kata-thermometer working between 130° and 125°F:

$$\text{(a)}\quad F/T = (127{\cdot}5 - t)(0{\cdot}061 + 0{\cdot}0139\sqrt{v}).$$

$$\text{(b)}\quad F/T = (127{\cdot}5 - t)(0{\cdot}011 + 0{\cdot}017\sqrt{v}).$$

For the kata-thermometer working between 150° and 145°F:

$$\text{(a)}\quad F/T = (147{\cdot}5 - t)(0{\cdot}074 + 0{\cdot}0184\sqrt{v}).$$

$$\text{(b)}\quad F/T = (147{\cdot}5 - t)(0{\cdot}018 + 0{\cdot}0223\sqrt{v}).$$

In these formulae, t is the temperature of the air in degrees Fahrenheit and v is in feet per minute; formulae (a) are used for velocities below about 200 ft/min and formulae (b) for higher velocities.‡ The numerical values of the constants that occur in these formulae were obtained experimentally; variations from one instrument to another are covered by variations in the factor F as determined by the maker's calibration.

The use of the kata-thermometer as an anemometer is restricted mainly to the measurement of air movement in the study of ventilation problems in rooms and enclosed spaces. For that purpose it has the advantage of being largely non-directional. Provided that the angle of the flow direction is not vertical or nearly so, the instrument indicates the total air speed; and it is this total speed that is one of the factors associated with the cooling effect of air movement on human beings and with comfort, which are among the primary concerns of ventilation. Thus the kata-thermometer measures directly the velocity that one needs to know in such work, and there is no difficulty, as there would be with other types of anemometer, in aligning the instrument correctly

† Indeed, the instrument was useless at air temperatures above 95°F.

‡ These were the published formulae. If v is to be in metres per minute, the multiplying factors of $\sqrt{v}$ become 0·0252, 0·0308, 0·0333, and 0·0404 respectively.

with an air current whose local direction is probably unknown. Another advantage of the instrument is that it can measure low speeds.

Disadvantages of the kata-thermometer are, firstly, the length of time necessary to take readings and, secondly, the comparatively low accuracy. To take a reading, the instrument is first of all heated to well above the upper temperature, generally by immersion in hot water, then thoroughly dried and exposed to the air current with the stem vertical. The top of the alcohol column should then be well above the upper temperature mark in order that, by the time it reaches the latter, the rate of cooling shall have become steady. The time taken for the top of the column to fall from the upper to the lower mark is then observed by a stop-watch. The instrument must not be allowed to move during an observation, and it should be held at arm's length or, better, held in a clamp or suspended well clear of the observer. At least three readings should be taken for each velocity determination. The error of a determination may be about 5–8 per cent.

Further particulars and working instructions will be found in ref. 19.

References

1. P. Bradshaw, *An Introduction to Turbulence and its Measurement*, Pergamon Press, Oxford, 1971.
2. P. Bradshaw, Thermal methods of flow measurement, *J. Scient. Instrum.* (*J. Physics E*) Series 2, **1** (1968) 504.
3. L. V. King, On the convection of heat from small cylinders in a stream of fluid, *Phil. Trans. Roy. Soc.* A **214** (1914) 373.
4. D. C. Collis and M. J. Williams, Two-dimensional convection from heated wires at low Reynolds numbers, *J. Fluid Mechs* **6** (1959) 357.
5. L. F. G. Simmons and A. Bailey, A hot-wire instrument for measuring speed and direction of air flow, *Phil. Mag.* (7) **3** (1927) 81.
6. L. F. G. Simmons, A shielded hot-wire anemometer for low speeds, *J. Scient. Instrum.* **26** (1949) 407.
7. C. F. Cowdrey, Temperature and pressure corrections to be applied to the shielded hot-wire anemometer at speeds for which natural convective cooling is negligible, *British J. App. Phys.* **9** (1958) 112.
8. R. Hilpert, Wärmeabgänge von geheitzten Drähten und Rohren im Luftstrom, *Forsch. Ing. Wes* **4** (1933) 215.
9. B. J. Bellhouse and D. L. Schultz, The determination of fluctuating velocity in air with heated thin film, *J. Fluid Mech.* **29** (1967) 289.
10. E. Ower and F. C. Johansen, On a determination of the pitot-static tube factor at low Reynolds numbers, with special reference to the measurement of low air speeds, *R. & M.* 1437 (1931) Appendix IV.
11. Cooper and Linton, *Proc. Nova Scotian Inst. Science* **19** (Pt. I) (1934–5) 119.
12. C. Salter, *Low Speed Wind Tunnels for Special Purposes*. N.P.L. Rep. N.P.L./Aero/1218 (1966).
13. C. C. Thomas, The measurement of gases, *J. Franklin Inst.* **172** (1911) 411.
14. H. Moss and W. J. Stern, The Callendar electric air-flow meter, Aeronautical Research Committee, Technical Report II (1920–21) 868.
15. R. F. Benseman and H. R. Hart, A thermocouple anemometer, *J. Scient. Instrum.* **4** (1955) 145.

16. D. E. Sexton, A flowmeter for field tests of ventilation systems, *Heat. Vent. Engr* **38** (1964) 245.
17. L. Hill, O. W. Griffith, and M. Flack, The measurement of the rate of heat loss at body temperature by convection, radiation, and evaporation, *Phil. Trans. Roy. Soc.* B **207** (1916) 183; see also L. Hill, H. M. Vernon, and D. Hargood-Ash, The kata-thermometer as a measure of ventilation, *Proc. Roy. Soc.* B 93 (1922) 198.
18. T. C. Angus, L. Hill, and H. E. Soper, Calibration of the kata-thermometer for hot atmospheres and a simplified method for computing, *J. Industrial Hygiene* **12** (1930) 66. See also C. P. Yaglou and K. Dokoff, Calibration of the kata-thermometer over a wide range of air conditions, *J. Industrial Hygiene* **11** (1929) 278, and T. C. Angus, The kata-thermometer and its uses, *J. Instn Heating and Ventilating Engrs* **50** (1936) 4.
19. T. Bedford, *Environmental Warmth and its Measurement*, Medical Research Council War Memorandum No. 17 (1946).

CHAPTER X

MANOMETERS

PRACTICALLY always in incompressible flow, and very often also in compressible flow, the static, velocity, and total pressures to be measured are obtained by the use of differential pressure gauges, generally called manometers, in which the pressure difference is balanced by the weight of a column of liquid of known density. The height of this column is used as a measure of the pressure difference; and the instruments described in this chapter, with only one or two exceptions, employ this principle. They differ only in the systems adopted for increasing their sensitivity over that of the simplest instrument of the type – the plain U-tube manometer with vertical limbs.

We have already seen that, in incompressible flow, the velocity heads that have to be measured in practice are usually not large: for example, the velocity head, under ordinary atmospheric conditions of temperature and pressure, does not reach 2 cm of water until the air speed is roughly 18 m/sec. Further, since the head is proportional to the square of the speed, it falls off rapidly at the lower speeds; it is in fact less than 2 mm of water at 6 m/sec, a speed by no means outside the ordinary working range. It follows that, in order to determine air speeds with sufficient accuracy by means of pitot and static observations, sensitive manometers, capable of reading pressures of the order of 0·02 mm of water, will often be necessary.

Numerous and varied devices have been designed to supply this need. A number of these are, for different reasons, unsuitable for industrial use; and it is not proposed in this chapter to attempt a description of all the types of manometer that have been constructed or suggested. In the main we shall confine our attention to a few types of different degrees of sensitivity, which have been proved by experience to be suitable for engineering practice, as distinct from laboratory conditions. The chief difficulty in designing such manometers lies in combining high precision with the portability, robustness, and ease and rapidity of manipulation that are essential for their successful application to general practical use.

In compressible flow the pressure differences to be measured are much larger. At a Mach number of 0·5, for instance, the difference between the isentropic stagnation pressure p_0 and the static pressure p is 0·16 p_0 (see Table 4.1). If the stagnation pressure is atmospheric, therefore, $(p_0 - p)$ is about 12 cm of mercury, and becomes correspondingly greater at higher Mach numbers.

The Simple U-tube

A simple, vertical U-tube with water as the gauging liquid may provide adequate accuracy for pressure differences upwards of 5 cm head of water, and mercury may be substituted for water as the gauging liquid when the water column becomes inconveniently high. Unfortunately, there are few liquids with relative density intermediate between that of water and mercury (13·6). Ethylene bromide has a relative density of 2·2 and acetylene tetrabromide 3·0, but both are corrosive and their use is to be avoided wherever possible. Pressure differences slightly less than the lower useful limit of the vertical water-filled U-tube can be measured by using alcohol (relative density 0·8); but the main reason why alcohol is preferable is because its surface tension is much less, although its density varies with temperature much more rapidly and it tends to absorb water. Other liquids of about the same relative density as alcohol but giving better performance in other respects are available (see below).

The sensitivity of a U-tube can be increased, and the instrument used at lower pressure differences, by tilting its place so that it makes an angle θ with the horizontal. The vertical distance h between the liquid levels in the two limbs is then given by the equation

$$h = l \sin \theta, \tag{1}$$

where l is the difference between the lengths of the liquid columns measured in the plane of the instrument. The difference in head is therefore $l\sigma \sin \theta$, where σ is the relative density of the gauging liquid. In practice a lower limit to θ is set by non-uniformity of the tubes and by meniscus distortion. The latter depends on the tube bore and on certain properties of the gauging liquid discussed in more detail below. With alcohol in tubes of about 3-mm diameter, the lower limit of θ is about 4°, which corresponds to a fourteenfold increase in sensitivity. In the modern single-limb inclined-tube manometer described on p. 251, θ can be set to angles of 3° or less, giving magnifications of 20 or rather more.

At these low angles, $\sin \theta$ varies proportionally to θ: care must therefore be taken in the inclined U-tube manometer to ensure that the two limbs of the U-tube are coplanar, to determine θ accurately (a bubble protractor is a suitable means), and to provide the instrument with a rigid support.

As already indicated, choice of gauging liquid involves consideration of its density, thermal-expansion coefficient, proneness to water absorption, surface tension, and corrosive properties. To this list one must add wetting properties and viscosity (which affect response time) and volatility. Some of these data have been tabulated for various gauging liquids in ref. 1. Smith and Murphy,[2] while designing the micromanometer described on p. 264, made experiments on the uniformity of adhesion of selected gauging liquids to the walls of manometer tubing by determining how nearly the liquid level in inclined tubes of several diameters returned to its original position after being displaced by an applied pressure difference. On balance, the best liquid for their purpose proved to be a proprietary fluid manufactured by the Dow-Corning Corporation (U.S.A.) and known as DC-200 (Ics):† this has a relative density of 0·82, surface tension about a quarter that of distilled water, and viscosity coefficient 0·82 cP (distilled water, 1·00 cP). Its performance in the tests described was very much better than that of water with detergent added: in tubes of 2-mm or 3-mm bore inclined at 10°, the repeatability of the meniscus reading corresponded to a pressure difference of only 0·002 mm water. The liquid is colourless, but can be dyed red with Biebrich scarlet.

Single-limb Inclined-tube Manometer

As already remarked, the two limbs of an inclined U-tube manometer must be coplanar; they should also be parallel, and the axis of tilt must be perpendicular to both limbs and lie either in their plane or in a parallel plane. These requirements can be met by care in construction; but the increase in sensitivity that tilting confers can probably be obtained more easily in the type of inclined-tube manometer shown diagrammatically in Fig. 10.1, which is a generally convenient form of instrument for industrial use. One limb of the U-tube is replaced by a reservoir A containing most of the gauging liquid, and only the tube B forming the other limb can be tilted. A and B are connected by a flexible coupling and the two sources of pressure whose difference is to be measured are connected one to the top of the reservoir and the other to the open end of the tube B. The diameter of A is so large in relation to the bore of B that practically the entire movement of the liquid occurs in the tube, along which there is a linear scale. Adjustment of the liquid level to the zero of this scale can be effected either by raising or lowering the reservoir, or, more conveniently, by operation of a plunger or similar device which displaces the liquid level in the reservoir.

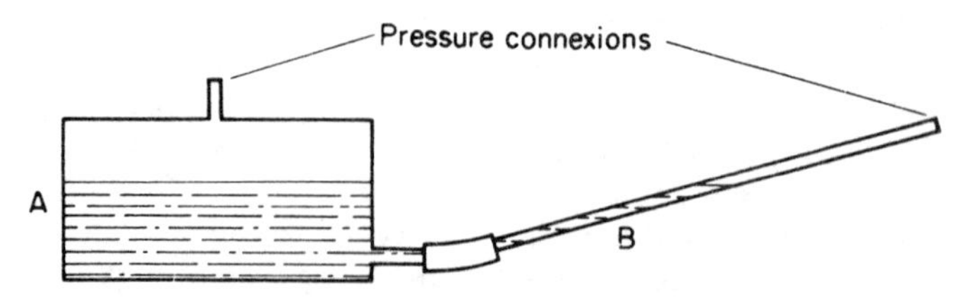

FIG. 10.1. Principle of inclined-tube manometer.

† This "Silicone Fluid" (polydimethylsiloxane) is available in England as Midland Silicones MS 200/Ics; it is supplied by Hopkin & Williams Ltd., Chadwell Heath, Essex.

In most designs of this type of manometer for general use, it is possible to set the inclined tube at more than one slope so that a wider pressure range can be covered than would be possible with a single setting. The advantage of this for measuring air speed will be realized when it is remembered that the velocity pressure is proportional to the square of the speed, and that it is desirable to allow for the instrument to be used over as wide a speed range as practicable. If, therefore, the tube is set to a slope such that the full-scale reading corresponds to a high air speed, the accuracy at low air speeds, for which the readings are small, will be poor.

For accurate work, this type of manometer requires calibration against a primary manometer, since various factors, such as lack of straightness of the inclined tube and variations in surface-tension effects due to want of uniformity in the bore, preclude the possibility of calculating the pressure differences sufficiently accurately from the observed readings and known inclination of the tube. However, in order to estimate initially the angle to which the tube should be set for any desired range, (1) may be used, with h and l now representing the maximum water column to be observed and the total length of the scale respectively; it is the formula that would be valid if the instrument were not subject to the errors mentioned above, and if the area of the reservoir were infinitely large compared with the bore of the inclined tube, so that the whole of the change of level takes place in the tube. It is easier to include the small correction for level changes in the reservoir in the overall calibration than to attempt to calculate it from accurate measurements of reservoir area and tube bore.

Manometers of this type are to be recommended for general use: they are consistent in their readings, accurate when calibrated, and inexpensive compared with other types of like sensitivity. Moreover, they are convenient and robust, and enable pressure observations to be made rapidly. The chief disadvantage is the need (already mentioned) for each instrument to be calibrated initially against a standard such as the tilting micromanometer. However, such a calibration can easily be made in the manner described at the end of this chapter.

Figure 10.2 shows a modern inclined-tube manometer having two independent adjustable precision-bored tubes, each with its own reservoir. The shorter limb, with a scale length of 320 mm, can be set vertical or to one of three fixed inclinations giving total full-scale readings equivalent to heads of 250, 50, 25, and 12·5 mm of water respectively; corresponding maximum heads for the longer limb (scale length 643 mm), for which there are the vertical and two fixed slopes, are 500,100, and 50 mm. Each reservoir contains an immersed bellows which can be expanded or contracted by operation of a knob on the supporting panel so as to adjust the liquid level in the inclined tube to the scale zero. An aneroid barometer is also included, and the whole instrument can be levelled with the aid of screw feet and the two spirit-levels shown.

Separate scales are provided to be clamped to each tube according to its slope, and graduated to read in millimetres of water column or directly in pressure units.†

The gauging liquid is a blend of kerosene of relative density 0·784 at 20°C, dyed deep orange red; this is found to give a well-defined and freely moving meniscus in the 5·5-mm bore inclined tubes.‡ The slopes of the tubes are calculated by the makers – Airflow Developments, Ltd., of High Wycombe – to take into account the tank/tube-bore ratios (about 118 to 1) as well as the relative density of the kerosene.

The approximate magnifications (compared with a vertical tube) for the three slopes of the shorter limit are 20, § 10, and 5; and for the two slopes of the longer limb 10 and 5. An experienced observer can read the position of the meniscus to 0·5 mm, so that the sensitivity at the highest magnification with the short limb is about 0·02 mm. At this slope, therefore, for which the full-scale reading (equivalent to 12·5 mm of water column) corresponds to an air speed of about 14 m/sec, the lower limit of air speed that can be measured by velocity-head readings with an error not exceeding 1 per cent is about 4 m/sec, and, with an error not exceeding 2 per cent, 3 m/sec. Higher air speeds can of course be measured with better accuracy, up to the limits of about 65 and 90 m/sec with the short and long limbs respectively vertical.

Multi-tube and Curved-tube Manometers

If it is desired to measure a number of pressures simultaneously, it is often convenient to use a multi-tube manometer in which a number of tubes are connected to a common reservoir. An inclined-tube form of such an instrument is described in ref. 4, and ref. 5 presents a simple method of tube adjustment which extends the lowest usable slope from 12° to 4°, giving (with alcohol) a pressure sensitivity of the order of 0·02 mm of water without special precautions. A two-way pressure switch[6] is useful for connecting a multi-tube manometer to several sets of pressure leads in turn.

In a modification of the inclined-tube manometer, due to J. L. Hodgson, the tube, instead of being of constant slope over its entire length, was curved in such a way that a uniform scale of velocity was obtained. The instrument was, in effect, an inclined-tube manometer having a tube of continuously varying slope, such that the inclination was least at the lowest speeds, i.e. when the velocity pressures were least. In this way the main advantage of the inclined-tube manometer – adequate sensitivity at low velocities – was retained without

† Scales graduated in British units are also available.

‡ This agrees with the bore of 5 mm recommended by Hunsaker with alcohol as the gauging liquid.[3]

§ A higher magnification could be obtained at a lower slope but only at the expense of practical difficulties, e.g. sluggishness and an ill-defined meniscus.

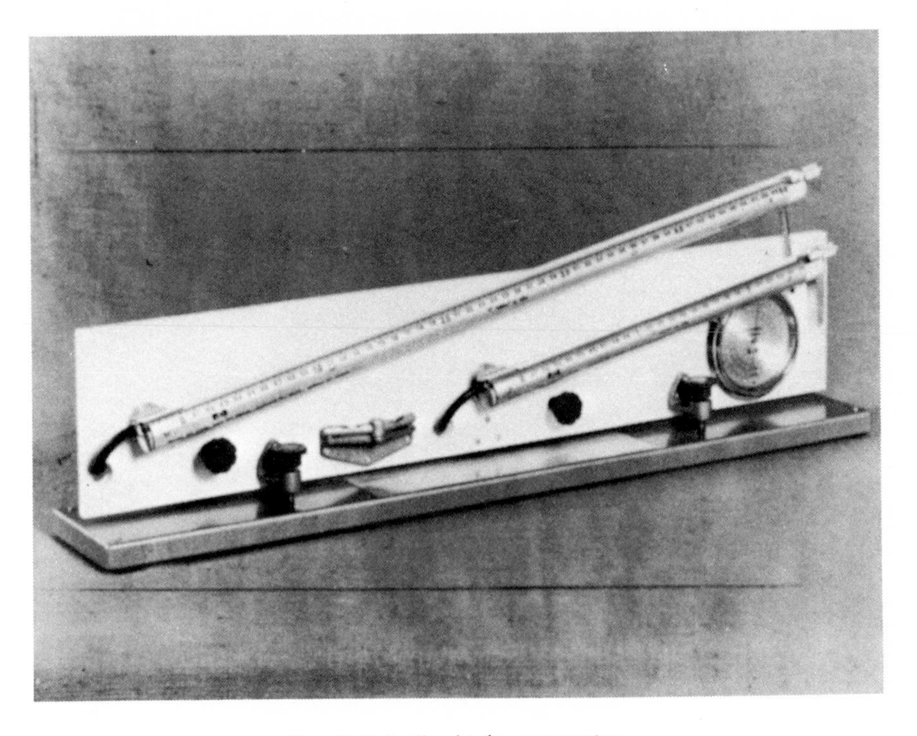

FIG. 10.2. Inclined-tube manometer.

either an unduly restricted velocity range, an impracticably long tube, or the inconvenience of having to adjust the inclination of the tube to suit the velocity range. Curved-tube manometers can be supplied either with pressure, velocity, or quantity scales.

Theory of the Curved-tube Manometer

Assume that the differential pressure p (Pa) applied to the manometer is the velocity pressure $\frac{1}{2}\varrho v^2$, and that the scale is to be graduated in metres per second. If the differential pressure p causes the manometer liquid (relative density σ) to rise a height of h cm, and if the reservoir is so large that this change in level takes place entirely in the curved tube, we can write

$$98{\cdot}07\, h\sigma = \tfrac{1}{2}\varrho v^2$$

or

$$v\sqrt{\varrho} = \sqrt{(196{\cdot}1\, h\sigma)}. \tag{2}$$

In Fig. 10.3, let O be the zero position of the liquid meniscus against the upper side of the curved tube, and take O as the origin of co-ordinates. The condition to be fulfilled is that, for a given increment in $v\sqrt{\varrho}$, the meniscus is to move a given distance along the tube to O'. Hence the equation of the curve of the tube is

$$s = Av\sqrt{\varrho}, \tag{3}$$

where s is the distance moved measured along the tube from O and A is a convenient constant.

Thus from (2), $$s = A\sqrt{(196{\cdot}1\, h\sigma)} = B\sqrt{h}.$$

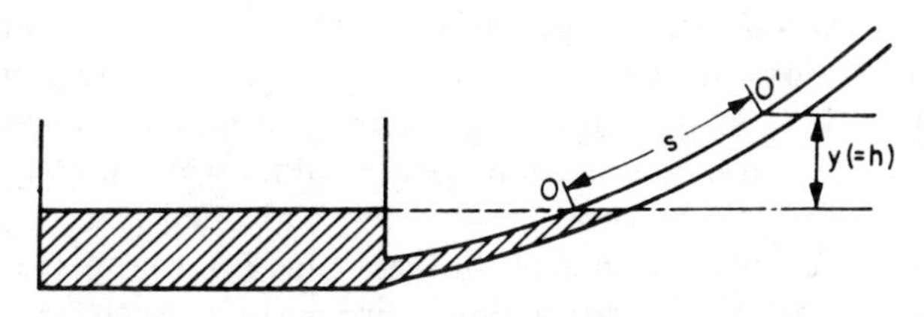

FIG. 10.3. Movement of meniscus in curved-tube manometer.

If x and y are the co-ordinates of O', measured from the origin O,

$$s = B\sqrt{h} = B\sqrt{y}.$$

Differentiating, we have

$$ds = \frac{B}{2\sqrt{y}}\, dy,$$

i.e.
$$\sqrt{(\mathrm{d}x^2 + \mathrm{d}y^2)} = \frac{B}{2\sqrt{y}}\,\mathrm{d}y$$

or
$$\frac{\mathrm{d}x^2}{\mathrm{d}y^2} = \frac{B^2}{4y} - 1.$$

Hence
$$\mathrm{d}x = \sqrt{\left(\frac{B^2}{4y} - 1\right)}\,\mathrm{d}y,$$

and the curve to which the tube is to be bent is obtained by integrating this equation. This gives

$$x = \frac{B^2}{4}\sin^{-1}\sqrt{\left(\frac{4y}{B^2}\right)} + \frac{B}{2}\sqrt{\left[y\left(1 - \frac{4y}{B^2}\right)\right]}. \tag{4}$$

If, therefore, the axis of the tube is bent to the form represented by (4), the gauge will show equal movements of the liquid meniscus along the tube for equal changes in $v\sqrt{\varrho}$. An instrument constructed on these lines requires calibration against a standard manometer, just as the ordinary inclined-tube gauge does.

It is clear from the above analysis that, if the instrument is furnished with a velocity scale, this can be accurate only at one value of the air density ϱ, which must be specified. For any other density ϱ', the readings must be multiplied by $\sqrt{(\varrho/\varrho')}$; thus if the scale is said to be accurate for air (assumed dry) at a temperature T absolute and at a pressure p then the readings observed at temperature T' and a pressure p' must be multiplied by $\sqrt{(T'p/Tp')}$ in order to obtain the true velocity.

Tilting Micromanometers

The instrument shown in Fig. 10·4, which has been used by one of the authors in factory conditions, is capable of giving all the accuracy required in most circumstances. It is in principle a sensitive U-tube water gauge, in which provision is made for microscopic observation of the water-level in one of the vertical limbs. The U-tube consists of a long horizontal glass tube A, communicating at each end with one of the glass cups BB, whose axes should be parallel and approximately vertical in the zero position. Each cup is provided with a removable glass cap, ground in to fit, to which rubber tubes leading to the static and pitot heads may be attached. Short vertical glass tubes with sealed ends project from the base of each cup and fit into corks which in turn fit firmly in holes in the upper movable steel frame C. The latter rests, by means of a three-point support, on the lower fixed frame D, and is held firmly thereon by the spring E. The three-point support is arranged with two points FF under the left-hand cup, the line joining these points being perpendicular to the axis of the tube A. These two supports are fixed; and the upper frame, with the U-tube, is tilted about the line joining them by means of the third supporting point *G*, which is formed by the hardened end of a vertical micrometer screw working in the lower frame. Of the two left-hand points, one engages a conical cup on the upper frame and the other a V-groove with its length horizontal and perpendicular to the axis of the tube *A*; the end of the micrometer screw bears against a small flat surface on the upper frame. In this way the tilting frame is given definite location without constraint. All the points and their seatings are of hardened steel. The position of the movable frame with respect to the fixed frame is observed by means of two indicators attached to the former, of which one projects above the graduated rim of the micrometer wheel and records fractions of a turn of the screw, and the other indicates complete turns on the fixed vertical scale on the lower frame. A microscope *K*, with cross-wires in the eyepiece, is carried by the upper frame for observation of the level of the liquid in the right-hand cup; and it is advisable for facilitating observation to have also a small adjustable mirror *L*, by means of which light from any convenient source can be reflected through the glass vessel into the microscope. The field of vision is then brighter above the water-level than below, and the line of demarcation can be sharply focused. The tap *M* is used to prevent spilling of the water whilst the manometer is being moved from place to place.

The instrument is often best used when mounted on a tripod. For this purpose it is desirable to provide the lower frame with a screwed hole to take a tripod, such as that ordinarily used for surveying instruments. It is as well to have also three levelling screws *NN* on the lower frame, to adjust the zero of the instrument if it should be found more convenient to rest it on a bench or table. A further convenient practical step is to bore eccentric holes to take the two cup supports in the two corks fitting into the holes provided in the tilting frame. In this manner some adjustment is provided for small variations from the standard dimensions in the glassware, since it is difficult to make this exactly to size in every case.

In using the manometer the tubes from the pitot and static tubes should be connected to the cups before the zero reading is taken, as the instrument may be displaced whilst the tubes are being fitted. The caps fitting into the necks of the cups should be greased with a small quantity of vaseline over the region of contact in order to prevent leaks at the joints. The glass tube *A* should initially be approximately horizontal; and the zero reading is obtained by observing the readings on the fixed scale and the micrometer wheel when the cross-wires of the microscope intersect the image of the water-level. The total pressure and static tubes are then inserted in the pipe in which the air speed is being measured; and the movable frame is tilted by means of the screw until the image of the water-level again coincides with the intersection of the cross-wires.

Let d be the difference between the initial and final readings expressed in turns and fractions of a turn of the wheel, l_1 be the distance between the cup centres, l_2 be the perpendicular distance of the axis of the micrometer screw

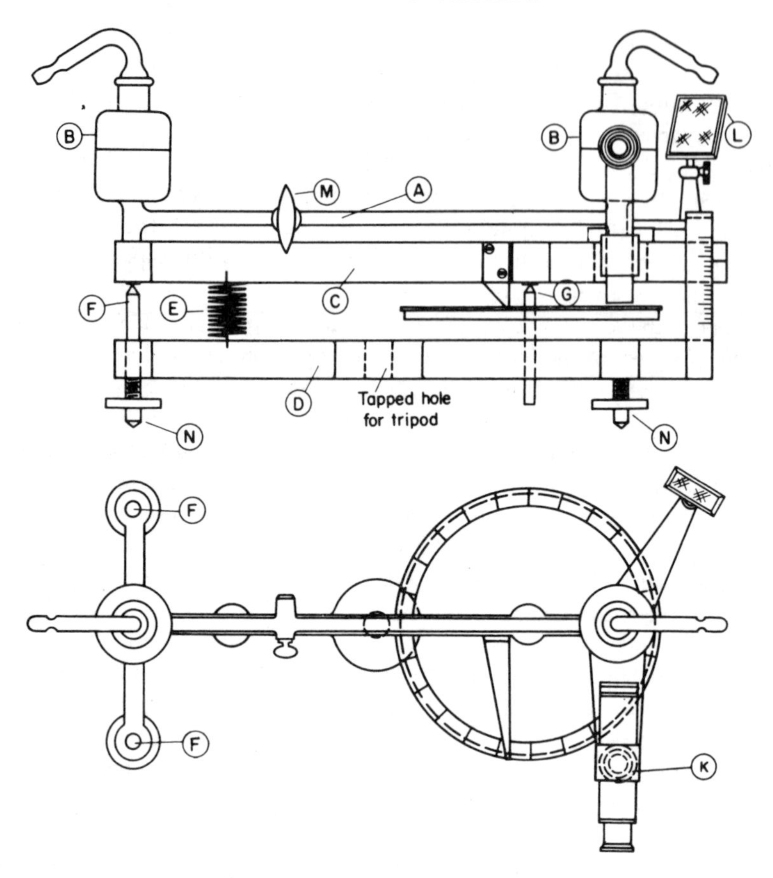

FIG. 10.4. Tilting micromanometer.

from the line joining the two fixed points, t be the pitch of the micrometer screw, and p be the pressure difference in head of water.

Then, if the angle of tilt be not too great,

$$p = td\frac{l_1}{l_2}. \tag{5}$$

Convenient practical values of l_1, l_2, and t are 33 cm, 25 cm, and 1·25 mm respectively. The available travel of the micrometer screw should not exceed 20 turns, giving the instrument a range of about 33 mm of water, and the diameter of the screw over the thread should be about 20 mm. Smaller sizes lead to "sloppiness" as the thread wears.

If a higher range is required, it should not be obtained by increasing the length of the screw, since the large angles of tilt that then occur at the higher end of the range introduce errors for which it is necessary to make allowance, and the above equation is no longer valid. It is preferable to increase the distance between the cup centres: a 66-cm U-tube used in place of a 33-cm tube would have double the range, and such a tube could quite conveniently be mounted on a frame designed for a 33-cm tube. In practice, however, when the velocity head exceeds that covered by the range of a 33-cm tilting manometer, an ordinary U-tube water gauge will probably give sufficient accuracy in most cases.

An important feature of the tilting manometer is that it is an absolute standard, which does not require calibration against another manometer. Of the dimensions that determine the reading for a given applied pressure difference, two, namely the pitch of the screw thread and the distance of its axis from the line joining the fixed points, can be made to size or measured sufficiently closely to limit errors in the readings to less than 0·1 per cent. The third important dimension, the distance between the cup centres, cannot be made to a given size with such precision, but can be measured, by calipers, to 0·2 mm without difficulty. With careful workmanship, the overall accuracy of the instrument should be within 0·1 per cent over the entire range.

If this degree of accuracy is required at the higher pressures, a small correction must be applied to the readings in this part of the range. In (5), the product td represents the height h through which the micrometer screw moves, that is the vertical distance by which a certain point on the movable frame is raised or lowered above another point, which does not move vertically. The assumption underlying (5) is that the length of the line joining these points is constant and equal to l_2, the distance between the axis of the micrometer screw and the fixed axis of rotation of the tilting frame. Actually, of course, the length of this line on the moving frame, i.e. the leverage, is not constant, and has its value assumed in (5) only in the initial position. As the micrometer screw moves, its point slides along the hardened steel surface on the moving frame with which it is in contact, so that the leverage changes. The true leverage for the reading in (5) is not l_2 but $\sqrt{(l^2 + t^2d^2)}$, so that the true difference in head is obtained from the equation

$$p = td\frac{l_1}{\sqrt{(l_2^2 + t^2d^2)}}. \tag{6}$$

The error† incurred by using (5) instead of (6) is less than 1 per cent for the full-range reading (33 mm) of the 33-cm instrument, and decreases as td becomes smaller relatively to l_2.

The necessity for applying this correction can be avoided if the upper end of the micrometer screw terminates in a horizontal, hardened, flat surface instead of in a point; and a hardened steel ball, fixed to the moving frame and resting

† Sometimes called the cosine error because the leverage increases by $1/\cos\alpha$, where α is the tilt of the moving frame, assumed horizontal to start with.

on the flat end of the micrometer screw, is substituted for the plane bearing surface. The leverage will then be constant and equal to l_2, the distance between the line joining the two fixed left-hand points (*FF* in Fig. 10.4) and the centre of the steel ball: the latter will now simply slide on the fixed end of the micrometer screw.

With a microscope having a magnification of 20 or 30 diameters, the tilting manometer as described will be sensitive to pressures of less than 0·02 mm of water; this means that, when this instrument is used with an ordinary pitot-static tube, a wind speed of 3 m/sec can be observed with an accuracy of about 98 per cent. By a modification of the glass-work, the sensitivity can be greatly increased, unfortunately, however, at the expense of ease and rapidity of working. The alteration consists in cutting the horizontal limb of the U-tube and connecting the two parts to a central vessel, one part being connected to the outside of this vessel near its lower end and the other being bent vertically upwards and passing up into the centre of the vessel. The modified arrangement is shown in Fig. 10.5, which illustrates the glass-work of a manometer constructed on these lines. It is mounted on a metal frame similar to that previously described.

This design of glass-work is an elaboration of that first suggested by Chattock.[7] The vertical tube in the central vessel extends upwards for some distance as shown, and is cut off square and ground. Distilled water is poured into the lower parts of the two cups and the central vessel; and into the upper portion of the latter, completely filling it, is admitted a quantity of medicinal paraffin, by the aid of the filling reservoir and tap shown in the diagram. The paraffin does not mix with water, and two well-marked surfaces of separation are formed, one outside the vertical tube in the interior of the central vessel and one

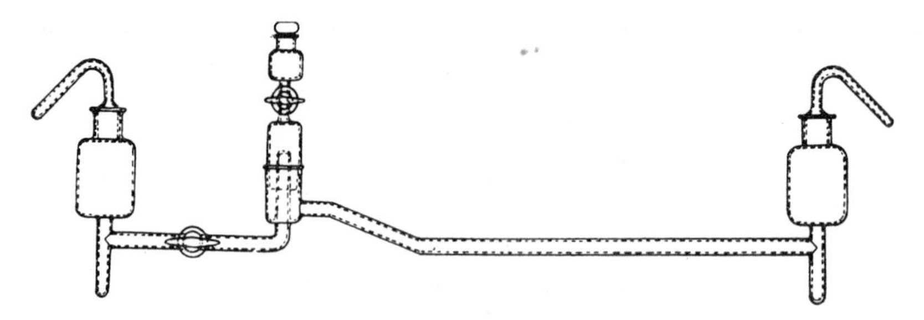

FIG. 10.5. Glass-work for 66-cm Chattock manometer.

within this tube. The liquid levels are adjusted by regulating the quantities admitted, so that one surface of separation is located on the open end of the central vertical tube, and has the appearance of a bubble resting on the mouth of the tube. If now a pressure difference is applied to the two end-cups, this bubble tends to become larger or smaller; and its movement, which can be followed by suitable illumination from behind and microscopic observation, is arrested by giving an appropriate tilt to the frame by means of the micrometer screw.

Instruments of this type, having cups whose centres are 33 cm apart, are sensitive to pressure differences of the order of 0·002 mm of water, the increased sensitivity over the type illustrated in Fig. 10.4 being attributable to the fact that the motion of the liquid is observed in a narrow tube instead of in the relatively wide end-cups. Apart from surface-tension effects, the motion of the bubble for a given pressure difference will be greater than the motion of the surface in one of the end-cups, in the ratio, approximately, of the area of the water surface in the cup to the internal area of the central tube. With a microscope of given magnification, therefore, the sensitivity of a Chattock type of manometer will be greater in this ratio than that of an ordinary tilting micromanometer. Actually, surface-tension effects are by no means negligible, and reduce the sensitivity considerably. Nevertheless, as will be seen from a comparison of the figures quoted, this instrument is considerably more sensitive then the plain tilting micromanometer. Its chief disadvantage from an engineering standpoint is that, particularly with fluctuating pressures, the bubble is easily broken, in which case measurements have to be interrupted whilst a fresh zero setting is taken. Further, small zero changes may also occur as the temperature alters, because of the different coefficients of thermal expansion of the water and the paraffin. A modification to eliminate this zero creep has been suggested by W. J. Duncan;[8] and for further details of the Chattock gauge the reader is referred to his paper, and to a full description that appeared in *Engineering*.[9]

Another useful form of null-reading tilting U-tube micromanometer is described on p. 264.

Long-range Micromanometers

A useful and robust type of manometer consists in effect of a flexible U-tube, the lower ends of the two vertical limbs being connected by a length of rubber tubing. One of the limbs remains stationary, the other being raised or lowered, by an amount corresponding to the applied pressure difference, by means of a micrometer screw to which it is attached. Figure 10.6 illustrates a manometer that utilizes this principle and incorporates some useful features introduced at the University of Toronto.[10] Rotation of the nut supporting the movable limb of the U-tube is prevented by means of a keyway, which engages with a vertical key cut on the right of the vertical supporting standard carrying the head of the micrometer screw. In place of a cup for the moving limb, an inclined glass tube of variable slope is employed, having a horizontal mark etched on it. The zero reading is first established with the liquid meniscus in this tube in coincidence with the fixed mark; when a pressure difference is applied, the inclined tube is raised on its nut until the liquid is again in equilibrium at the mark. Complete turns of the micrometer screw are indicated by the pointer moving over the vertical scale, and fractions of a turn are observed on the graduated head. The fixed limb of the U-tube consists of a reservoir of large cross-sectional area in comparison with that of the inclined tube, so that the change in level in the reservoir can be neglected. A convenient feature of this manometer is the vertical glass tube fixed alongside the scale as shown: this tube is connected to the reservoir in parallel with the inclined tube by means of a T-piece; and if its open end is connected to the same source of pressure as the open end of the inclined tube, the level of the liquid in it indicates the approximate height to which the nut should be raised when the order of magnitude of the applied pressure difference is unknown beforehand. The final adjustment is made by reference to the meniscus in the inclined tube.

With alcohol as the gauging liquid, the best sensitivity for general use was found(10) to be obtained by the use of an inclined tube of about 5-mm bore set at a slope of 3° or so, corresponding to a magnification of roughly 20 to 1. Reducing the slope below this makes the instrument too sluggish; tubes of smaller bore reduce the sluggishness to some extent, but the meniscus is not so good. Instrument lag can be reduced in some experiments by applying the pressure to be measured to the inclined tube (and, through a T-piece, to the vertical tube) instead of to the reservoir; if either pitot or velocity pressure is being measured, the reservoir should then be located at the top of the instrument instead of at the base as shown in Fig. 10.6.

This instrument, though not so simple or rapid in use as the plain inclined-tube manometer, is certainly more accurate, and has the advantage that it does not require calibration: null-reading avoids errors due to non-uniformity of tube bore (since the meniscus is always returned to the same position), and the readings need correction only for the relative density of the gauging liquid to convert them to head of water. Once the accuracy of the micrometer screw thread has been checked, the manometer can therefore be used as an absolute standard. It was stated(10) that the movement of the liquid column can be read to within about 0·005 mm on the average with an extreme possible error of 0·015 mm. A further important advantage of this instrument is the fact that the range is not as restricted as it is in other sensitive manometers, since the micrometer screw may be made some 25 cm long, or more.

A manometer similar in principle and construction to that just described has been designed for use at the National Physical Laboratory,(11) but the position of equilibrium is determined in a different manner. The two limbs of the U-tube in this case consist of two glass cups, one of which is similar to those used in the tilting micromanometer (Fig. 10.4) and can be raised or lowered by means of a micrometer screw to balance applied pressure differences as in the instrument illustrated in Fig. 10.6. The other cup is fixed, and is constructed as shown in Fig. 10.7; it will be seen that the arrangement is precisely similar to that of the

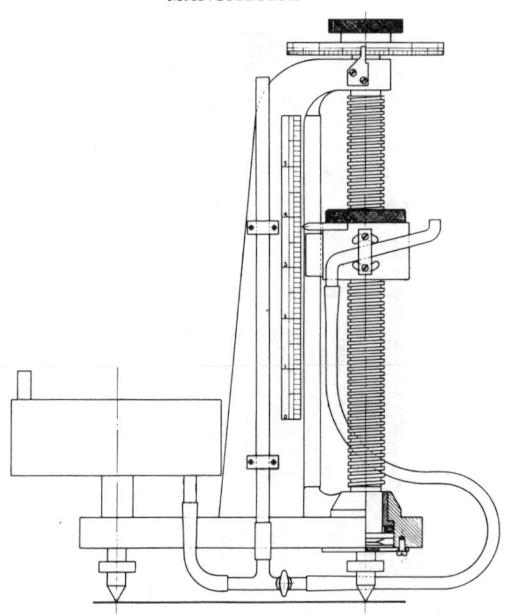

FIG. 10.6. Large-range micromanometer.

central vessel in the Chattock manometer and, as in that instrument, use is made of a water–paraffin meniscus on the mouth of the central tube to indicate equilibrium. The meniscus is suitably illuminated from behind and is viewed through a microscope. In this instrument the cross-sectional area of the rubber connecting tube is not small compared with that of the cups. Hence, in order to avoid appreciable errors due to changes in volume of the rubber tubing, it is advisable to have a long length of tubing, so that it hangs approximately vertically from each cup over the entire range of the movable cup. The changes of shape that then occur will produce negligibly small volume changes unless the internal pressure is large (see also p. 264).

One further example of this class of manometer may be mentioned, not only because of the ingenious method used for indicating balance, but also because it can be used in industrial conditions to measure heads up to 20 cm of water to an accuracy of $\pm 0{\cdot}02$ mm and at absolute pressures up to 7 kg/cm^2. The instrument is illustrated in Fig. 10.8. The reservoir on the left can be clamped in any convenient position on the cylindrical rod, and the measuring chamber connected to it, with the vernier, can move over the complete range of the square column which carries the scale. To accelerate setting, the measuring chamber can be rapidly clamped to the column, the final fine adjustment being obtained by means of the vertical screwed rod. The flexible connexion between the two limbs is made of plastic tubing with a terylene-braided insert to prevent appreciable change of shape due to movement or high internal pressure.

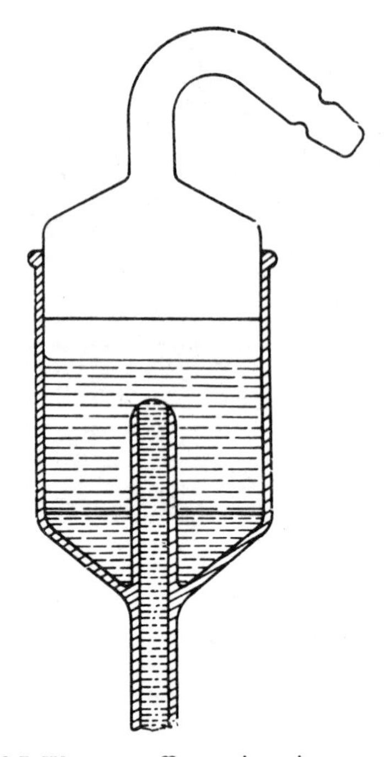

FIG. 10.7. Water-paraffin meniscus in manometer cup.

Details of the observation chamber are shown in Fig. 10.9, which is reproduced from the *Engineer*.[12] The height of the chamber is adjusted until the water-level in it is flush with the point fixed in its centre, when, with the illumination and viewing arrangements shown in Fig. 10.9, the point appears just to touch its reflected image. This system of adjustment is said to be capable of an accuracy within 0·01 mm of water.

Micromanometers with Small Time-lag

When the dimensions of the pressure tapping or pressure probe are small, as in boundary-layer explorations, the response of a sensitive micromanometer such as the Chattock gauge may become unacceptably slow. In any case, each test run tends to be long in experiments of this type, and it is therefore important to reduce zero drift as far as possible. Both considerations have been catered for in an instrument described by MacMillan.[13] This is a null-reading U-tube, basically identical with that shown in Fig. 10.6, except that MacMillan chose to adjust the height of the reservoir instead of the inclined sight-tube: the meniscus is observed through a microscope, and it is a matter of practical convenience to have the optical system fixed. The principal novel features of

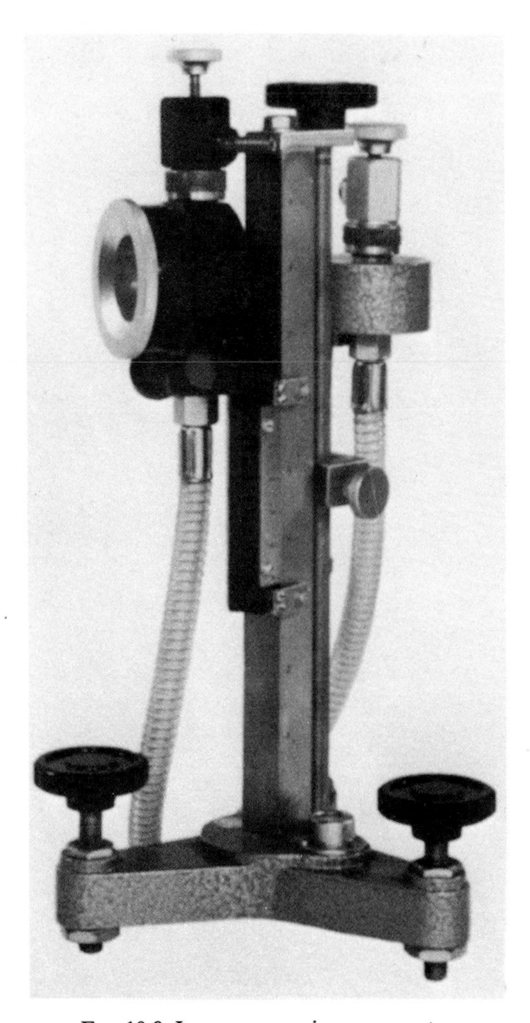

FIG. 10.8. Large-range micromanometer.

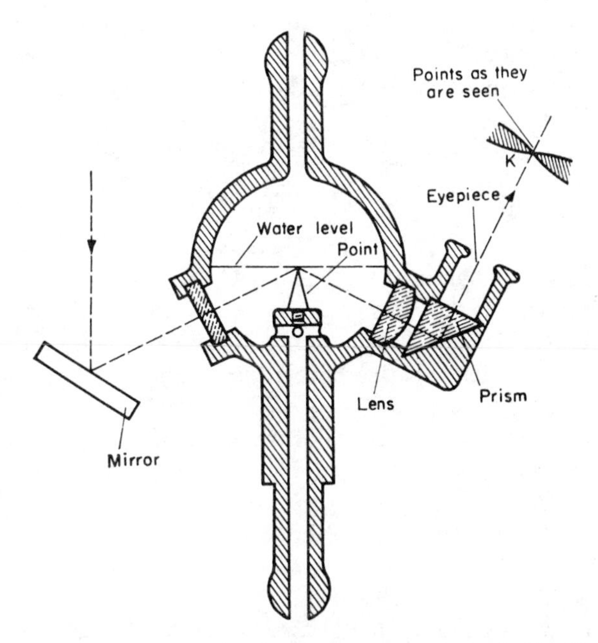

FIG. 10.9. Observation chamber for instrument shown in Fig. 10.8.

MacMillan's contribution are the attention paid to the reduction of instrument lag, and the analysis of the cause of zero drift due to temperature changes, and its elimination. The null-indicating meniscus is formed in the small-bore tube of 1–3 mm internal diameter, inclined at a small angle θ to the horizonal: variable sensitivity can be obtained by making θ adjustable. The other arm of the U-tube is the liquid reservoir, whose height is adjusted by means of a micrometer screw: a standard micrometer screw can be used if the range of the instrument does not need to exceed 25 mm. The reservoir volume is kept small in order to minimize lag; even so, it is preferable always to connect the pressure probe to the inclined tube if possible, as the volume in this arm of the instrument is smaller still. The sensitivity increases with the ratio of the cross-sectional area (A) of the reservoir to that (a) of the inclined small-bore tube, although increasing the reservoir area increases instrument lag at the same time. The sensitivity also increases as θ is reduced; negative angles may be used provided that the tube bore is sufficiently small. In theory the sensitivity becomes infinite when $\sin \theta$ is equal to $-a/A$, but the position of the meniscus then becomes unstable.

Since the effects of temperature changes on surface tension oppose those of thermal expansion as far as change in manometer zero reading is concerned, the instrument can be designed so that zero drift due to temperature changes is eliminated.

It is shown in refs. 13 and 14 that, if one neglects the effects of temperature changes on the volume of the manometer itself (as distinct from changes in volume of the manometric liquid), the thermal-expansion and surface-tension effects cancel out if

$$\frac{Vr}{A} = -\frac{2}{g\beta}\frac{\mathrm{d}(\gamma/\varrho)}{\mathrm{d}t}, \tag{7}$$

where V is the total volume of the manometric liquid, r is the radius of the inclined sight tube, A is the cross-sectional area of the liquid in the reservoir, t is the temperature, γ is the surface tension ($\mathrm{d}\gamma/\mathrm{d}t$ is negative), ϱ is the density of the liquid, β is its coefficient of thermal expansion, and g is the acceleration due to gravity.

Thus if the dimensions V, r, and A of the manometer are matched with the properties γ, ϱ, and β of the manometric liquid in accordance with (7) the zero drift is very small.

Errors due to inadvertent tilting of the base of the instrument are reduced by decreasing the distance between the reservoir centre and the zero position of the meniscus. The accuracy attainable with this type of instrument is typically 0·005 mm of gauging liquid. Similar conclusions were drawn by Smith and Murphy,[2] who developed a similar type of instrument. As sight tube they employed a tube of 3-mm internal diameter inclined at 10°, with the silicone fluid DC–200 (Ics) (see p. 250) as the gauging liquid. For differences of head up to 20 mm the reservoir height was measured by means of a dial gauge; a commercial vernier scale was employed to extend the range to 60 cm. The accuracy attainable in these two ranges was estimated to be $\pm$0·005 and $\pm$0·02 mm of water respectively.

A tilting U-tube instrument with rapid response has been described by Bradshaw.[14] In this instrument, shown diagrammatically in Fig. 10.10, the reservoir is raised by means of a tilting table whose axis of rotation passes through the zero position of the null indicator: this is the meniscus in a sight tube inclined at a small angle to the axis of rotation, so that its inclination scarcely changes as the reservoir height is adjusted. The liquid container, including the tubing between the reservoir and the sight tube, is entirely rigid, so that no errors arise from volume changes in the tubing: such errors can be appreciable in precise measurements of small pressure differences. To eliminate cosine errors at large angles of tilt, the foot of the reservoir is shaped as part of a sphere whose centre is on the liquid level, or as part of a cylinder whose axis passes through that point and is parallel to the axis of tilt. With polydimethylsiloxane (see p. 250) as the gauge liquid, the sensitivity is about 0·002 mm of that liquid. The instrument, which is made by Combustion Instruments Ltd. of Staines in two models of ranges 3 and 4 cm of gauge liquid (relative density 0·822), is compact and robust. Its response is much faster than that of a

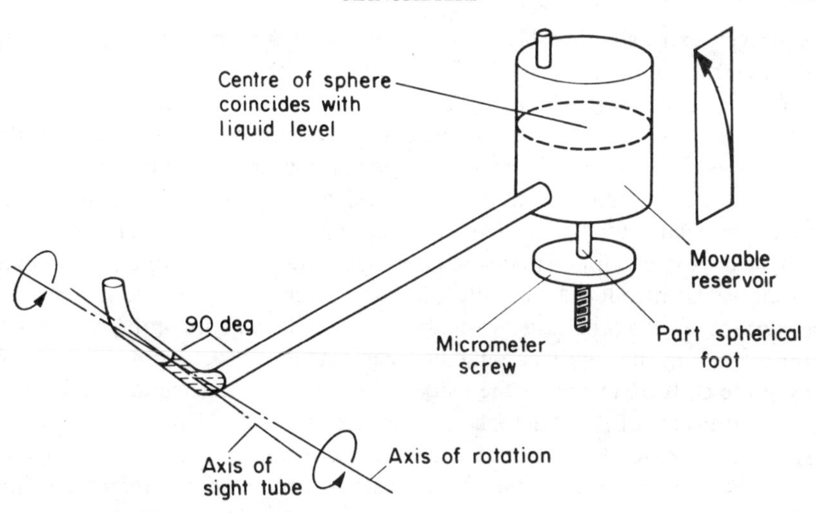

FIG. 10.10. Bradshaw's micromanometer.

Chattock gauge, its time constant being about half a second. Zero drift can be made very small by satisfying the conditions expressed by (7) for balancing zero changes due to variations of surface tension with temperature against those due to thermal expansion.

Very Sensitive Manometers

Various types of manometer having a sensitivity ten or more times that of the most sensitive of those already described have been designed for special purposes. Such instruments are more suitable for laboratory than for general industrial use, since they are precision instruments that require careful handling, setting up, and manipulation, and have not the speed in use of the less-sensitive types. Brief descriptions follow of a few of these extra-sensitive manometers.

The first, which is due to Hodgson,[15] is an inclined-tube gauge having a large reservoir fitted with a cylindrical displacer, partly submerged and partly above the liquid in the reservoir, actuated by means of a vertical micrometer screw. The liquid meniscus in the inclined tube is maintained at a fixed mark, any departure therefrom, due to an applied pressure difference, being restored by an appropriate vertical motion of the displacer in the reservoir. The meniscus is specially illuminated, and is viewed through a microscope having a magnification of 60. The calibration is determined by the relation between the internal diameter of the reservoir and the external diameter of the displacer, so that the instrument may be used as a fundamental standard. The manometer is relatively expensive, but it is robust and portable, and the sensitivity is said to be about 2×10^{-4} mm of water. This high precision is attained by the use of a displacer having a small diameter in relation to that of the reservoir.

Another type of sensitive manometer is that designed by Ower for the calibration of the standard pitot–static tube at low speeds.[16] Essentially, this is a Chattock manometer with the central cup arrangement replaced by a piece of glass capillary tubing. The two end-cups and the connecting tube are filled with the manometer liquid (xylol was found to be the most suitable), and a small air-bubble is introduced into the capillary. When the pressure in one cup exceeds that in the other, the air-bubble, which fills the bore of the tube, tends to travel along the capillary, but the operator prevents it from doing so by tilting the cups by means of the usual micrometer arrangement. The applied tilt is a measure of the difference of pressure. For full details of the various expedients adopted to overcome the serious difficulties that occur when very low differential pressures have to be measured, the reader is referred to the original paper. This instrument can be made to respond to a pressure difference equivalent to a head of about 10^{-4} mm of water.

Falkner,[17] by swinging the glassware of a Chattock gauge through a large angle, so that it was at an angle of almost 90° to the length of the tilting frame, i.e. almost parallel to the axis of tilt, obtained a sensitivity about equal to that of the type last mentioned. A further ingenious innovation introduced by Falkner was to fill the cups, the horizontal limbs, and the lower part of the central vessel (see Fig. 10.5) with a lighter liquid than that used in the upper part of the vessel, so that the controlling force on the surface of separation, or the bubble, which forms the indicator in this type of gauge, was considerably less than in the conventional Chattock instrument. In this way a much more mobile bubble was obtained, which responded with greater movements to very small pressure differences. The liquids used by Falkner were xylol for the main filling, and water.

Various sensitive manometers based on other principles have also been proposed, including the application of the torsion balance,[18] and the direct weighing device described in ref. (19). The latter is worth mentioning because of its basic simplicity in using a top-loading balance, with a range of 0–1 kg and sensitive to 10 mg, to "weigh" the applied pressure difference directly. As described in ref. 19, its sensitivity is about 2½ times that of the Chattock gauge. This could theoretically be increased, but probably only by incurring serious practical difficulties, particularly sluggishness.

Betz Gauge

The Betz Gauge (see Fig. 10.11) is one of the best available long-range micromanometers. It is essentially a U-tube in which one limb is a reservoir of

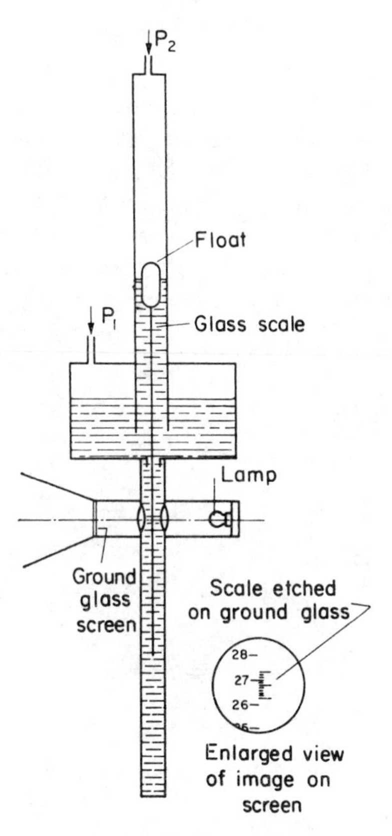

FIG. 10.11. The Betz micromanometer.

large cross-section, as in the type shown in Fig. 10.1. The liquid level in the other, much narrower, vertical limb is indicated by means of a translucent graduated scale carried by a float. An illuminated and magnified image of the graduations is projected on to a ground-glass screen provided with a scale for obtaining fractional parts of a graduation. A zero adjustment (not shown in Fig. 10.11) is provided.

An excellent form of Betz gauge is manufactured by Instrumentenfabriek van Essen, NV of Delft, and is obtainable in the United Kingdom from R. E. Swan, Ltd., of Royston, Herts. Two types are available, one with a range of 250 mm of water and the other of 500 mm, graduated in Newtons per square metre (1 mm H_2O at 4°C = 9·81 N/m^2). The sensitivity is about 0·01 mm.

This instrument combines long range with robustness and accuracy, and is very convenient to use; but its contained air volume is large enough to cause unacceptably long lag if used with pressure probes of less than about 2 mm internal diameter.

Ring-balance Manometer

The ring-balance manometer consists essentially of an axially pivoted hollow ring partly filled with liquid as sketched in Fig. 10.12. The space above the liquid is divided into two compartments by the partition *S*, and the two pressures p_1, p_2 whose difference is to be measured are applied to the two compartments through leads L_1, L_2. The liquid levels are thereby displaced and cause the ring to rotate about its axis until equilibrium is restored by the weight of the mass *M* (attached to the ring) in its displaced position. The amount of the angular displacement serves as a measure of $(p_1 - p_2)$.

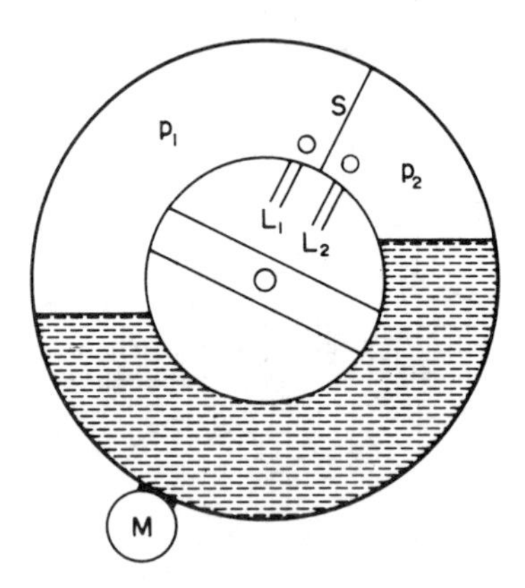

FIG. 10.12. Ring-balance manometer.

The instrument is robust, and is well suited for incorporation into direct-reading or recording velocity and flowrate meters (see p. 299). In this form it is supplied for a variety of uses, mainly in Germany. A number of industrial types are made by the West German firm of Rixen and Co. KG of Bochum-Gerthe, covering pressure ranges of 4–1700 mm of H_2O, using different gauging liquids with densities ranging from 0·87 kg/dm^3 (oil) to 13·6 kg/dm^3 (mercury) according to the pressure range to be covered. The accuracy is quoted as 1 per cent of the full-scale reading, subject to an overriding figure of 0.1 mm H_2O.

The Two-Liquid Differential Manometer

A simple modification of the plain U-tube may sometimes be found useful for measuring pressures below about 5 cm of water when high accuracy is not required, but a sensitivity considerably higher than that of the U-tube is desirable and circumstances do not permit the use of an inclined U-tube. As shown in Fig. 10.13, the upper part of each limb of the vertical U-tube is enlarged to form a reservoir of much greater cross-sectional area than the tube

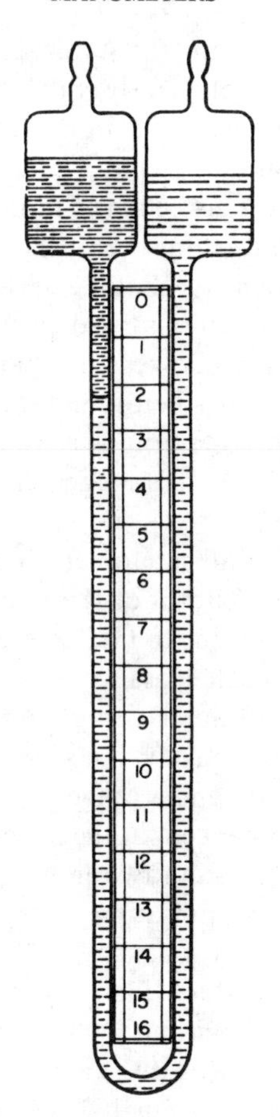

FIG. 10.13. Two-liquid manometer.

itself. The gauge is filled with two liquids of nearly equal density, which do not mix and are practically insoluble in each other; the meniscus (surface of separation) is arranged to be within the tube below one of the reservoirs. Assume that when the pressure difference is applied the meniscus moves a vertical distance h down the tube. If A_1 and A_2 are the areas of the reservoirs, a is the area of the tube, and σ_1 and σ_2 are the relative densities of the lighter and heavier liquid respectively, the relation between the applied pressure difference p (in head of water) and the movement h is

$$p = h\left[(\sigma_2 - \sigma_1) + \frac{a(A_1\sigma_2 + A_2\sigma_1)}{A_1 A_2}\right]. \tag{8}$$

If A_1 and A_2 are very large in relation to a, and σ_1 and σ_2 are of the order of 1, the second term is unimportant, and the system has a magnification of approximately $1/(\sigma_2 - \sigma_1)$. Thus the more nearly equal the relative densities of the two liquids, the greater will be the magnification. But the extent to which this fact can be used in practice is limited: if σ_1 and σ_2 are too nearly equal, the movement is sluggish and the meniscus becomes ill-defined and tends to break up. In a manometer of this type used by one of the authors, the two liquids were benzyl alcohol of relative density about 1·05 (coloured with aniline black) and a calcium-chloride solution of relative density about 1·15. The liquids were well shaken together and then allowed to separate out and filtered, so that each was saturated with the other before use. Provided that the glass U-tube was chemically cleaned before the liquids were introduced, this combination gave satisfactory performance, with a magnification of about 10, in a U-tube of 5-mm bore.

Other forms of the two-liquid manometer have been used. Reference 20 describes one in which the two limbs of the U-tube are concentric instead of side-by-side. This avoids errors due to tilt. Reference 21 gives particulars of an instrument with a more elaborate glass-work system in which a magnification of 35 was obtained, the two liquids being kerosene and alcohol of relative densities 0·797 and 0·777 respectively. This instrument was stated to have a sensitivity of 0·02 mm of water, and a range of head of 0·02 to 1·0 mm. Errors were thought not to exceed 4 per cent at the lower end of the scale.

We may note also the interesting two-liquid manometer used by Kastner and Williams[(22)] for measuring pulsating pressures (see Chapter XII). It was filled with water and oil of relative density 0·89, and the general construction was the same as that used by MacMillan (see p. 262). One of the surfaces of separation always remained within an inclined sight tube fixed in height but variable in inclination; it was viewed through a fixed microscope and always brought to the zero position by adjusting the height of the other meniscus, which was located in a second, vertical, glass tube carried by a slide whose height could be determined by a scale and vernier. A noteworthy feature of this manometer was that, to reduce pulsation errors in the leads, as explained in Chapter XII, the oil fillings above the water in both limbs were continued to within 12 mm of the sources of pressure to be measured. Polyvinyl-chloride tubing was used instead of rubber because it was found to be cleaner and to maintain a more constant cross-section when bent.

Calibration of Manometers

In calibrating any type of manometer against a standard, it is convenient to take simultaneous readings of the two instruments when one side of each is connected to the same source of pressure, which can be varied at will, the other sides being open to atmosphere or to any constant pressure. For the standard it is necessary to use an instrument, such as the tilting micromanometer, from whose indications, which depend solely upon its mechanical construction and dimensions, the applied pressure differences can be accurately calculated.

The simplest method of conducting the calibration is to open one side of each gauge to atmosphere, and to connect the other sides together through a T-piece and tap, by means of which the pressure can be applied to both gauges simultaneously. This method is subject to three serious disadvantages. In the first place, when the tap on the variable-pressure side is shut, the connecting tubes will enclose a small volume of air; and the small local atmospheric temperature fluctuations which are continually occurring will produce serious changes in the readings of both gauges, so that simultaneous observation of both under steady conditions will not be possible. A similar effect arises from the alteration in the enclosed volume while the standard instrument is being adjusted: this volume should be sufficiently large to render negligible the pressure changes to which it is subjected by the movement of either gauge. Finally, there are almost certain to be local variations of pressure in different parts of a room due to draughts and similar causes, which are by no means negligible when very small pressure differences are being considered. Hence it cannot be assumed that the pressures acting on the sides of the two gauges open to atmosphere are constant, or, more important, the same for both at any given instant.

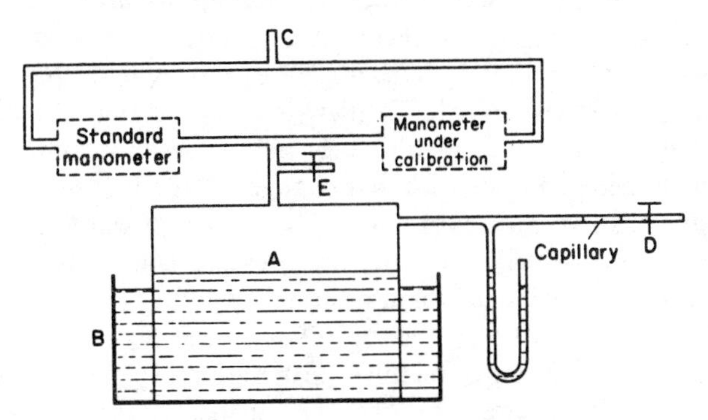

FIG. 10.14. Calibration of manometers.

Thus much more refined methods are necessary for the calibration of sensitive manometers. Preston[23] has described a method by which he obtained highly satisfactory results although no precautions were taken to avoid the effects of draughts mentioned at the end of the preceding paragraph. It seems probable that his work was done in an unusually still room, or in unusually calm weather conditions.

Figure 10.14 shows a calibration system based on Preston's, but with provision against draughts. A large, bottomless reservoir *A* stands in water in the tank *B*, and water flows between them when the pressure in *A* changes. One side of each manometer – the standard and the one under calibration – is connected to the air space above the water-level in the reservoir; the other sides of the two manometers are connected to the atmosphere through the T-piece *C*. The calibration pressure can be varied through the connexions at the right-hand side of the reservoir *A*, which include a small U-tube water manometer to provide an indication of the applied pressure, and a length of about 10 cm of capillary tubing to exercise some control over the rate of application of pressure and keep it within the ranges of the two manometers. A convenient method of applying pressure differences is to suck (rather than to blow, which causes a temperature rise in the reservoir) on the end of the tube beyond the tap *D*, after which *D* is closed. Another tap *E* and bypass enable one side of both manometers to be connected to atmosphere for obtaining zero readings; the other sides of the manometers are permanently open to atmosphere through the T-piece *C*. In very disturbed atmospheric conditions it may be necessary to shield the open end of *C*, e.g. by surrounding it with cotton-wool in a small open container, but this will rarely be necessary, particularly if the pressure leads to the two manometers are kept as short as possible and are of reasonably large-bore tubing – say not less than 5 mm.

As compared with a completely closed reservoir, the water seal reduces considerably the pressure changes in the air space that result from temperature variations. If the reservoir were completely closed, a temperature change of 1°C would, at an air temperature of 15°C, produce a pressure change of $p_0/288$ in the reservoir ,where p_0 is the absolute pressure in the reservoir before the temperature changes from 15°C. If p_0 is approximately the atmospheric pressure (10 m of water) – as it will be for most cases of incompressible flow for which manometers will be used – a change of temperature of 1°C will be equivalent to a change in a closed reservoir of 3.5 cm of water. Preston shows that, with the water-sealed reservoir, the pressure change is $1/(1 + p_0/H_0)$ times the pressure change in a closed reservoir, where H_0 is the initial level of the water in the reservoir below the top. Thus even if H_0 is as much as 30 cm the pressure rise per °C is only one thirty-fifth of that in a closed reservoir, and by making H_0 smaller the pressure rise can be further reduced. But by doing so, we obviously reduce the volume in the reservoir and thereby increase the tendency for the adjustment of each manometer to affect the reading of the other if null-reading instruments are being used. The reservoir used by Preston was made from a 20-dm^3 drum of 30-cm diameter and 20-cm height; the height of the air space was at first 30–38 cm, but better results were obtained when this was reduced to 5 cm.

Very often, particularly if two observers are available to take simultaneous readings of the two manometers, satisfactory calibrations can be made with a simpler system, which can be built up entirely of equipment found in any chemical laboratory – glass-ware, rubber or plastic tubing, and corks and bungs. The open-ended reservoir is replaced by a large glass flask – minimum size about a Winchester quart – which is completely immersed in a water-bath. Otherwise the connexions are as in Fig. 10.14, the leads to the reservoir entering through a cork or bung in the neck, which can be made leak-tight by having molten paraffin wax poured round the joints. With the water-bath, there should be no difficulty in limiting temperature changes within the reservoir to 0·1°C over sufficiently long periods of time. Although a change of 0·1°C will produce a change of pressure of 3·5 mm of water in the reservoir, such changes will occur only slowly and, which is more important, will affect both manometers simultaneously. Hence if the two manometers are read simultaneously, the disturbing effects of temperature changes will be very small. One of the authors used this simplified set-up with success before Preston published the account of his more refined method.

Manometer Connexions

In all air-flow measurements that include observations of pressure, care should be taken to ensure the absence of leaks in the various instruments, connexions, and connecting tubing. Glass joints, such as taps and the caps of the two vessels of the tilting manometer described on pp. 254–8, should be greased with a thin coating of vaseline; and, if the internal pressure exceeds atmospheric, tubes should be secured to nipples with copper wire.

All parts of a pressure circuit should be tested for leaks. Particular attention should be paid to the concentric type of pitot–static tube, which may develop internal leaks from one tube to the other, and to all rubber tubing, which is apt to perish and crack, especially at the ends where it is stretched over metal or glass connecting pieces.† Even new tubing should not be regarded as above suspicion: appreciable porosity is sometimes found in rubber tubing taken from stock, and it is therefore advisable to test separately each length of tubing used.

A simple form of leak tester is shown in Fig. 10.15. This consists of a small U-tube water gauge with a tap and a T-piece incorporated in one of its limbs. One end of the tube or connexion to be tested is connected to the tube A by means of a piece of rubber or plastic tubing which has previously been ascertained to be itself free from leaks; the other end is closed, and a small pressure (about 5-cm head of water) is applied by blowing gently through the tube B and closing the tap before the pressure is released. Any small leaks will now be indicated by a gradual creep of the water level in the two limbs.‡ In order to observe the motion better, it is convenient to attach a scale to the board to which the U-tube is fixed. The function of the two bulbs in the leak tester is to

† This tendency can be greatly reduced if a touch of glycerine is applied to the outer surface of the glass or metal before the joint is made.

‡ With any sealed system, care must be taken to distinguish between this creep (due to leaks) and an inherent wandering tendency. It may be necessary to resort to a water-sealed gasometer-type system as recommended for manometer calibration (see p. 271).

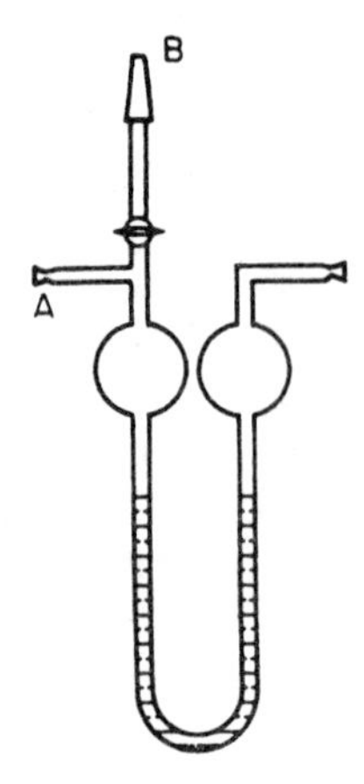

FIG. 10.15. Leak tester.

reduce the risk of ejecting water from the limbs if excessive pressure is applied inadvertently.

Testing for leaks can often be done by using the manometer itself as the means of observing the response of the system to pressure or suction applied to a T-piece (with tap) inserted at some convenient point in a connecting tube; with a liquid manometer such as that shown in Fig. 10.6, the pressure or suction can be applied by raising or lowering the reservoir.

In tests of the concentric type of pitot–static tube for leaks, each tube should be dealt with separately. The pitot orifice can be sealed by inserting a small rubber bung or covering it with a short length of tubing plugged at the other end. The static-pressure holes are best sealed by covering them with a short length of soft rubber tubing which has been rolled into position: the rubber impresses itself into any small surface scratches that may be present, thus avoiding the air leaks that are liable to occur along the scratches if an attempt is made to seal the orifices with self-adhesive tape.

As regards the choice of connecting tubing, it is found that for general use fairly thick-walled tube of red rubber, with a bore of 5 mm and an external diameter of about 10 mm, is suitable. Such tubing is usually found to be practically non-porous† and considerably more lasting than thin black tubing, which is not to be recommended for this type of work. Thick-walled rubber pressure tubing is sometimes troublesome when making connexions, but this difficulty does not occur with correctly designed nipples. Rubber tubing which may come into contact with mercury should be sulphur-free in order not to contaminate the mercury.

Plastic tubing is a useful alternative to rubber. Besides not perishing or

† Both rubber and plastic tubing are very slightly porous over a period of hours, but this is seldom of practical importance.

cracking, it is resistant to most chemicals and can be obtained in a great variety of sizes, grades of flexibility, and colour. Being translucent, it can easily be inspected for blockages, such as those caused by escape of gauging liquid. It is also available in multi-tube strip form, which is convenient when a number of pressure points have to be connected to the recording system. Polyvinyl chloride (PVC), however, suffers from the disadvantage that its hardness varies too much with temperature: connexions are easily dislodged in cold weather whereas in hot weather the tubing kinks easily. Polythene is much better in these respects, and is therefore to be preferred despite its greater cost.

With any type of flexible tubing, care should be taken to avoid kinks. One advantage of reinforced plastic tubing is that it does not kink. To prevent flexible tubing from collapsing under reduced internal pressure, a wire helix may be inserted. Tubing can be prevented from kinking, and sometimes also from collapsing under reduced pressure, by winding soft copper wire round the *outside*.

For connecting two flexible tubes together, it is best to use a short length of metal tubing. Such connexions are especially convenient if their ends are slightly tapered with a total angle of about 7°, as the taper facilitates the connexion and accommodates a range of size of tube bore. PVC tubing, however, is easily dislodged from a tapered connector when cold: a connector with corrugations superimposed on the taper ensures a joint that is less easily pulled apart, although liable to leak if loosened.

For permanent installations, copper or soft "compo" tubing is sometimes preferred (with flexible-tubing connexions at the ends, if necessary). Copper tubing work-hardens with vibration, and is easily cracked, however, whilst soldered joints deteriorate and very easily become broken; and compo tubing is so soft that it is very vulnerable to damage. Plastic reinforced with wire or nylon is highly suitable for permanent installations: well-designed end connexions are commercially available.

Whenever small pressure differentials are being measured, care must be taken to avoid errors due to volume changes in flexible leads, difference in temperature between the leads to either side of the manometer, and temperature changes in the system as a whole. The manometer itself, if of the liquid type, must be provided with a firm support, and the meniscus or bubble indicator must not be heated appreciably by the illuminating system.

Lag in Leads

Whenever the pressure being measured changes, a finite time lapses before the new value is registered by the indicating or recording system, because time is needed for the air and liquid movements that have to take place. With a pitot tube, for instance, an increase in wind speed increases the pressure presented to the orifice, and air enters the tube until the new equilibrium condition has been established. If the tube is connected to a liquid manometer, air has also to pass through the pitot orifice into the connecting lead and the air volume of the instrument, both to establish the increased pressure and – except in a null-reading instrument – to fill the additional air volume created by liquid displacement, as when the liquid level in the limb of a U-tube is depressed.

All these air and liquid movements introduce time-lag into the system. That associated with the manometer itself can only be reduced in the instrument design by minimizing the internal air volume and volume changes due to movement of the liquid. A null-reading instrument is ideal in the latter respect: examples are provided by the micromanometers described on pp. 262–5. For the same reason, leads connecting orifices to a manometer should be kept as short as possible. Tubing of small bore, however, although reducing the air volume, offers a greatly increased resistance to the air flow through it: the pressure drop along a tube for the laminar flow of a given air volume is inversely proportional to the fourth power of the internal diameter (p. 91, eqn. (15)). In addition, a greater velocity in the tube would be needed if the air-volume change in the manometer were to be supplied in the same interval of time; this too, would increase the resistance. In practice, therefore, runs of small bore are kept to a minimum.

Pressure Transducers

A pressure transducer converts the response of a pressure-sensing element into an electrical output, with advantages over a liquid manometer for remote control, automatic recording, signal processing, and miniaturization if required. Moreover, since it usually has an inherently faster response and a smaller internal volume than the liquid manometer, it reduces the time-lag in the measurement of steady pressures, particularly by obviating long pressure leads; while for the measurement of rapidly fluctuating pressures some form of pressure transducer is essential.

Many forms are available commercially, in which the movement of a pressure-actuated diaphragm, bellows, or bourdon tube effects a corresponding change in voltage, resistance, reluctance, inductance, or capacitance which can be measured by standard electrical methods. These types include some highly accurate instruments of secondary-standard accuracy (errors not exceeding $\pm 0{\cdot}02$ per cent of reading. Such devices in portable form have made it possible to extend standards-laboratory accuracy to every-day applications. The response times of these accurate instruments, however, are generally of the order of several seconds.

Pressure transducers require calibration. Although the maker usually supplies calibration data, for accurate measurements these data should be checked by the user. The re-calibration should determine not only the sensitivity in terms of electrical output (voltage per unit of applied pressure), but also the extent to which this varies over the pressure range to be covered (i.e. non-linearity of response). Zero offset (drift) and hysteresis effects must also be established, and careful attention given to temperature conditions, both in calibration and use. These variations cause the chief practical difficulties in using transducers; they can be allowed for if the temperature is recorded and the necessary corrections, which may be of the order of 0·004 to 0·04 per cent of the full-scale reading per °C, are applied. Alternatively provision can be made for re-zeroing the transducer and applying a calibration pressure at intervals during use.

Provided that it is calibrated, a transducer can be used outside the manufacturer's quoted range, if this was restricted to that over which the sensitivity is practically constant.

Pressure transducers may often be more convenient than liquid manometers for measuring large pressure differences; but their usefulness for low-pressure differences is limited not only by their calibration drift due to the thermal expansion already mentioned, but also by mechanical friction in some types. As a general guide it may be taken that the lower limit of usefulness of a pressure transducer for a steady applied pressure difference is about 0·2 mm of water. Hence if errors in air speed are not to exceed 1 per cent, the lowest velocity head that a pressure transducer should be used to measure is that corresponding to an air speed of about 13 m/sec.

When measuring a number of pressures by transducer, it is usual practice to employ only one transducer, and to connect it in turn to the various pressures by means of a scanning valve. This avoids the expense of a separate calibrated transducer at each pressure station.

A valuable practical account of pressure-transducer technique, although orientated towards wind-tunnel requirements, is to be found in ref. 24.

References

1. D. W. Bryer and R. C. Pankhurst, *Pressure-probe methods for determining wind speed and flow direction*, H.M.S.O., London, 1971.
2. A. M. O. Smith and J. S. Murphy, Micromanometer for measuring boundary-layer profiles, *Rev. Scient. Instrum.* **26** (1955) 775.
3. J. C. Hunsaker *et al.*, Reports on wind tunnel experiments in aerodynamics, *Smithsonian Miscellaneous Collection*, Publication No. 2368, **62** (1916) 27.
4. R. Warden, An improved multitube tilting manometer, *R. & M.* 1572 (1933).
5. D. W. Bryer, A simple method of tube adjustment for multitube tilting manometer, N.P.L./Aero/345 (1958); *AGARD Rep.* **163**, 77.
6. D. W. Bryer, Two-way switch and isolating valve for multi-tube manometer, N.P.L./Aero/347 (1958); *AGARD Rep.* **163**, 171.
7. A. Chattock, W. E. Walker, and E. H. Dixon, Specific velocities of ions in the discharge from points: note on a sensitive pressure gauge, *Phil. Mag.* (6), **1** (1901) 96.

8. W. J. Duncan, On a modification of the Chattock tilting pressure gauge designed to eliminate the change of the zero with temperature, *J. Scient. Instrum.* **4** (1927) 376.
9. J. R. Pannell, Experiments with a tilting manometer for measurements of small pressure differences, *Engng* **96** (1913) 343.
10. J. H. Parkin, University of Toronto, School of Engineering Research, Bulletin No. 2, Paper No. 1 (1921).
11. N.P.L. Report for 1921, 170.
12. Micrometer water level and pressure gauge, *Engr* **151** (1931) 248.
13. F. A. MacMillan, Liquid manometers with high sensitivity and small time lag, *J. Scient. Instrum.* **31** (1954) 17.
14. P. Bradshaw, A compact, null-reading, tilting-U-tube micromanometer with an entirely rigid liquid container, *J. Scient. Instrum.* **42** (1965) 677.
15. J. L. Hodgson, A sensitive micromanometer, *J. Scient. Instrum.* **6** (1929) 153.
16. E. Ower, A micromanometer of high sensitivity, *Phil. Mag.* (7) **10** (1930) 544; see also *R. & M.* 1308 (1930).
17. V. M. Falkner, A modified Chattock gauge of high sensitivity, *R. & M.* 1589 (1934).
18. P. Reichardt, Die Torsionswaage als Mikromanometer, *Z. f. Instrumentenkunde* **55** (1935) 23.
19. M. R. Head, Weighing pressures – a simple micromanometer, *Aero. J.* **76** (742) (1972) 615.
20. F. Short and J. S. Dines, Gust research in the Royal Aircraft Factory's Tetranemograph, *R. & M.* 144 (1913).
21. J. E. Brow and F. A. Schwertz, A simple micromanometer, *Rev. Scient. Instrum.* **18** (1947) 183.
22. L. J. Kastner and T. J. Williams, Pulsating flow measurements by viscous meters, *Proc. Instn Mech. Engrs* **169** (1955) 419.
23. J. H. Preston, Simple method comparing manometers, *Engng* **172** (1951) 645.
24. D. S. Bynum, R. L. Ledford, and W. E. Smotherman, Wind-tunnel pressure measuring techniques, *AGARDograph* No. 145 (1970).

CHAPTER XI

MISCELLANEOUS METHODS OF FLOW MEASUREMENT

Mean Rate of Flow in a Pipe from a Single Observation

WHEN the distributions of velocity and static pressure at the section of measurement are fairly symmetrical and conditions there are reasonably steady, a rapid and simple method of determining the mean rate of flow is to establish, by a preliminary exploration, the relation between the mean and the axial velocity, and thereafter to take single readings of the velocity pressure at the axis of the pipe. If conditions at the section of measurement are such that fully developed turbulent smooth-pipe flow is established, the preliminary exploration would be unnecessary were it not for the unexplained discrepancy between the results of Stanton and Panell and of Nikuradse to which we have already referred in Chapter V. This discrepancy prevents us from using a standard curve of the form given in Fig. 5.6 for the relation between mean and axial velocity in smooth-pipe flow; and a preliminary exploration must therefore always be made if this method is used, unless a possible error of perhaps 2 or 3 per cent can be tolerated.

If fully developed flow exists, the difficulty due to the uncertainty about the precise value of the ratio of the mean to the axial velocity can be largely overcome by locating the measuring point about $0{\cdot}75a$ (a = pipe radius) off-centre instead of axially. On the basis of Nikuradse's results (see Chapter V), Aichelen(1) recommended a position of the measuring point at $0{\cdot}762a$ from the axis of the pipe, where, over a Reynolds-number range of 4×10^3 to 3×10^6, he found that the measured velocity was equal to the mean velocity within $\pm 0{\cdot}7$ per cent. Over the same range of Reynolds number the corresponding deviation from the mean velocity in rough pipes was $\pm 2{\cdot}5$ per cent for a range of relative roughness a/ε of 15–100, where ε is the average height of the excrescences forming the roughness.

Preston, Gregory, and Norbury,(2, 3) using Stanton's results (Chapter V (ref. 11)), adopted $0{\cdot}75a$ as the location of the measuring point in an instrument they termed the Three-quarter Radius Flowmeter.† They gave values(3) of the amounts by which the velocity measured at that point differed from the mean

† This consisted basically of four total-pressure tubes equally spaced around the pipe section, with four static-pressure side holes (see p. 136) between them.

velocity, which were -1 per cent at a Reynolds number of 4×10^4, $-0{\cdot}4$ per cent at 10^5, $-0{\cdot}1$ per cent at 10^6, and zero at 10^7 and over. Winternitz,[4] however, using more recent data than those of Stanton, concluded that the Aichelen position of $0{\cdot}762a$ gives slightly more accurate results than the three-quarter radius position in fully developed flow between Reynolds numbers of 2×10^6 and 10×10^6. He also showed that the three-quarter position can give rise to large errors when the flow differs markedly from the fully developed type. The same is presumably true for the $0{\cdot}762a$ position.

It appears therefore that, when the flow approaches the fully developed type, a velocity measurement at $0{\cdot}762a$ (rather than $0{\cdot}75a$) will as a rule give adequate accuracy without the need for initial calibration. In practice, however, the occasions on which we can take advantage of this fact will be rare. In the first place, fully developed swirl-free flow will occur comparatively rarely outside the laboratory; but, more important, even if these conditions are believed to exist it will generally be necessary to verify that they do by an initial velocity traverse along more than one diameter. Thus from a practical standpoint, there seems to be little to choose between these special measuring positions and any other single point conveniently located, preferably where the velocity-profile curve is flattest.

Whatever position is chosen for the single measuring point, the velocity there can be determined either by means of a pitot–static tube of suitable size or a pitot tube in conjunction with a static-pressure hole in the wall of the pipe. The initial calibration that will nearly always be necessary in practice to establish the relation between the velocity at the measuring point and the mean velocity can be obtained by a velocity traverse across the section followed by an integration, using one of the methods described in Chapter VI and observing the precautions there mentioned to avoid errors due to changes in the rate of flow that may occur during a traverse. If the rates of flow that it is subsequently desired to measure extend over a considerable range, enough traverses at different rates of flow should be made to cover the range adequately. If the measuring point is on the axis of the pipe, Fig. 5.6 provides an indication of the variation that is likely to occur with Reynolds number in the ratio of mean to axial velocity; and the extent of the preliminary explorations will be governed by this variation and by the degree of accuracy required. Sometimes a single exploration at a rate of flow mid-way between the minimum and maximum likely to be encountered in practice will suffice.

General Remarks on the Single-observation Method

The method may be used for large as well as for small pipes, provided that a favourable velocity distribution exists. It is because a condition making for such a distribution, namely an upstream length of straight pipe of some 20 diameters or so, more usually exists with pipes of small diameter that the method finds its largest field of application for such pipes. This condition of an adequate length of pipe upstream of the section of measurement is essential for a symmetrical distribution of velocity and static pressure in a pipe system on the blowing side of a fan, since fans usually deliver the air in a disturbed turbulent condition, and the flow close to the fan outlet often does not fill the pipe symmetrically. On the suction side, however, 3 or 4 diameters of pipe upstream of the section may suffice, with a further 6 diameters or so between it and the fan inlet. Although the distribution of velocity at such a section will naturally not be of the fully developed type, it will nevertheless usually be symmetrical and have a steady value for the ratio of mean velocity to that at the selected measuring point, which can be determined by an initial exploration as already described.

A word of caution is necessary for this case. The open end of the pipe will probably act to some extent as a sharp-edged orifice, and the airstream will contract and not fill the pipe completely for a short distance from the end. This

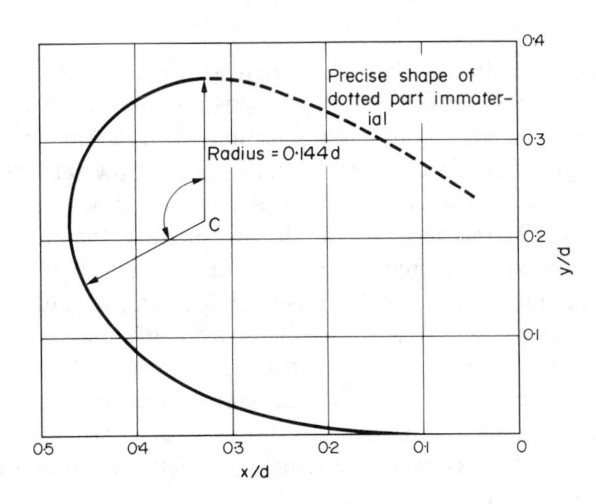

FIG. 11.1. Shaped inlet for pipe.

d = internal diameter of pipe; co-ordinates of C: (0·325, 0·219)

Co-ordinates for flared inlet

x/d	0	0·094	0·109	0·125	0·141	0·156	0·172	0·188	0·203
y/d	0	0	0·001	0·001	0·002	0·003	0·004	0·006	0·008
x/d	0·218	0·234	0·250	0·266	0·281	0·297	0·312	0·328	0·344
y/d	0·010	0·013	0·016	0·019	0·024	0·029	0·034	0·041	0·048
x/d	0·359	0·375	0·391	0·406	0·422	0·438	0·453	0·469	0·453
y/d	0·057	0·067	0·078	0·091	0·107	0·127	0·154	0·219	0·284
x/d	0·438	0·422	0·406	0·391	0·375	0·359	0·344	0·328	
y/d	0·308	0·325	0·338	0·347	0·353	0·358	0·361	0·362	

effect can be avoided by fitting the pipe with a shaped inlet having an easy entry, as shown in Fig. 11.1[5] The simpler, cheaper type illustrated in Fig. 11.2 will also effect a large improvement.

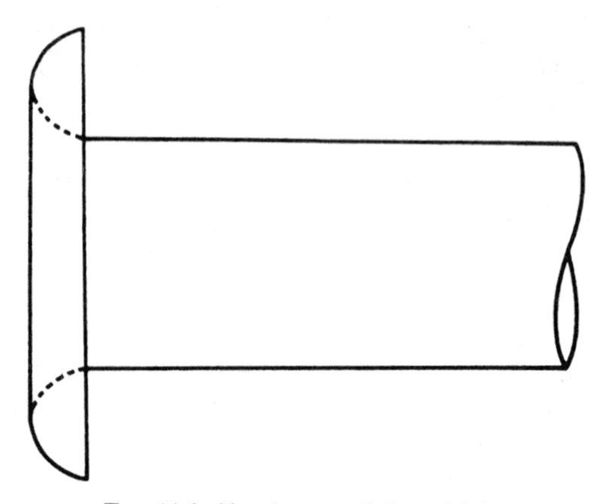

FIG. 11.2. Simple type of shaped inlet.

A modification of this method may often be found convenient. Instead of determining the mean rate of flow from velocity measurements, we may use the total pressure at some fixed point in the section, or the static pressure at a side hole according to circumstances. With steady parallel flow at the section, and if the air density is constant, each of these quantities will be a function only of the rate of flow. The relation between the pressure and the mean rate of flow must be established by a preliminary calibration. Thus if P is the pressure (either total or static), we have to measure P for various values of the mean velocity v_m covering the range that will subsequently be required, v_m being measured by one of the methods described in Chapter VI. We then draw a calibration curve of v_m against P, which is afterwards used to determine v_m for any measured value of P. Usually, as shown below, P will vary nearly as the square of the speed for constant air density, so that the relation between the two will be of the form

$$P = K\varrho v_m^2 \text{ (approx.)}, \tag{1}$$

where K is a coefficient determined from the initial calibration.

It will be useful to examine this method in greater detail. Let us confine our attention to a section X in a given length of pipe AB having a fan at B, which may be made either to blow from B to A or to exhaust from A to B (Fig. 11.3). For simplicity, imagine that in the latter case there is a shaped entry at A. We shall assume initially that the velocity across the section is uniform and equal to v_m. Let O be a point in the undisturbed atmosphere where the air is at rest.

If we neglect the beginning of the motion when the fan is starting up, and consider only the steady conditions when the flow is established and proceeding at a constant rate, we see that the only work done is that against friction between the moving air current and the pipe walls, plus the energy equivalent of the internal losses in the current itself due to turbulence. These may be grouped with the frictional losses, since both are inseparable from the flow. We may regard the flow as a complete circuit, which begins and ends at the point O in the external atmosphere: the air starts from rest, acquires a velocity v_m in the pipe, and is returned to a state of rest. (We neglect losses in the fan itself.) The change of total head between any two sections of the circuit is equal to the energy loss per unit volume between those two sections (see p. 78).

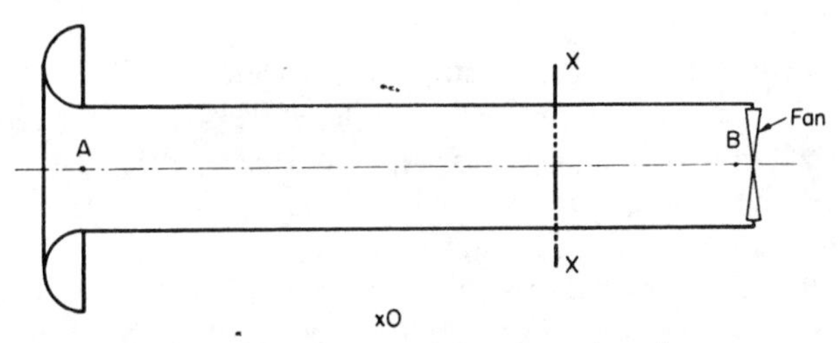

FIG. 11.3. Fan-actuated flow through a pipe.

(*a*) *Fan sucking air from A to B*. Assume that at A, just inside the pipe inlet, the air has acquired its velocity v_m. At O, the total pressure, velocity pressure, and static pressure are all zero (the atmospheric pressure is taken as the datum according to the usual convention). Between O and A, there is a loss of pressure f_1 due to friction between moving and still air, and to inlet losses. With a shaped entry, however, f_1 is very small. At A, the air has acquired its velocity v_m; and since f_1 is the only loss of pressure between O and A, the total pressure at A differs from that at O by this amount only. Thus the total pressure at A is $-f_1$. Now the static pressure is equal to the total pressure minus the velocity pressure and since the velocity pressure at A is $\frac{1}{2}\varrho v_m^2$, it follows that the static pressure at A is $-\frac{1}{2}\varrho v_m^2 - f_1$. Between A and X there is an additional pressure loss f_2 due to the resistance of the length AX of the pipe. The total pressure at X is therefore less than that at A by an amount f_2; and the static pressure is again the total pressure less the velocity pressure, the latter being $\frac{1}{2}\varrho v_m^2$ as before since we are assuming that the pipe is parallel. Thus we have

$$\text{total pressure at } X = -(f_1 + f_2),$$

$$\text{static pressure at } X = -(f_1 + f_2 + \tfrac{1}{2}\varrho v_m^2),$$

the negative signs indicating that both these pressures are below atmospheric.

We have already remarked that f_1 is small and, on the assumption that it can be neglected in relation to the other pressures (which seems to be generally true), we may say that, approximately,

$$\left.\begin{aligned}&\text{total pressure at } X = -f_2,\\ \text{and}\quad &\text{static pressure at } X = -(f_2 + \tfrac{1}{2}\varrho v_m^2).\end{aligned}\right\} \quad (2)$$

The quantity f_2 is the pressure loss due to the frictional resistance of the length AX of the pipe, and is related to the mean velocity v_m in the manner described in Chapter V. Thus for smooth pipes the relation between f_2 and v_m is given by the curve of Fig. 5.3, or by eqn. (5) of Chapter V; and for rough pipes f_2 varies as the square of v_m. The relation for smooth pipes is not far from a square law, and hence both the total pressure and the static pressure vary approximately as the square of the speed. The approximate eqn. (1) is therefore established for the inlet side of the fan.

Thus on the inlet side of a fan the static pressure in a pipe always has a greater negative value than the total pressure, and therefore gives higher readings on a manometer with one side open to atmosphere. Hence, if this method of measurement is adopted, it is preferable, on the inlet side, to utilize the static pressure as a measure of the speed.

The conclusions to which the reasoning has led will not be invalidated if there is no shaped entry at A; the effect of this will simply be to alter the value of K in (1).

(*b*) *Fan blowing air from B to A*. We are again concerned only with the length XA of the pipe. Assume that the mean rate of flow is v_m as before.

At A, the air will have a velocity pressure $\frac{1}{2}\varrho v_m^2$ and the static pressure referred to the atmospheric pressure will be zero, so that the total pressure at A will also be $\frac{1}{2}\varrho v_m^2$. The resistance of the length XA will produce a pressure loss f_2 as before, so that the total pressure and static pressure at X will both be higher by f_2 than they are at A. Thus,

$$\left.\begin{aligned}&\text{static pressure at } X = f_2,\\ &\text{total pressure at } X = f_2 + \tfrac{1}{2}\varrho v_m^2.\end{aligned}\right\} \quad (3)$$

Following a similar line of argument to that previously adopted, we arrive at the conclusion that in this case, i.e. when X is on the delivery side of the fan, the total pressure, being greater than the static pressure, should be taken as a measure of the speed, and that it varies very nearly as the square of the speed.

In both cases (*a*) and (*b*) the assumption has been made that the velocity v_m is uniform across the section. Actually, of course, it is not, but the effect of this will merely be to alter the value of K in (1). Even if f_2 does not vary as the square of the speed (so that (1) is no longer true) the method may still be used to measure the flow; all that is necessary is to draw up, by actual test, a calibration curve between the pressure and the air speed in the duct. The choice of the pressure – static or total – to be used as a measure of the speed will depend on which of the two is numerically greater.

The Conical Inlet Nozzle

If an open inlet end of the pipe system is accessible, an adaptation of the single-observation method will often prove both useful and reasonably accurate. It is among the methods specified for the testing of fans both by the British Standards Institution[6] and by the Fan Manufacturers' Association.[7] A conical inlet nozzle of the form shown in Fig. 11·4 is fitted to the pipe inlet and measurements are made of the average value of the pressures obtained from the

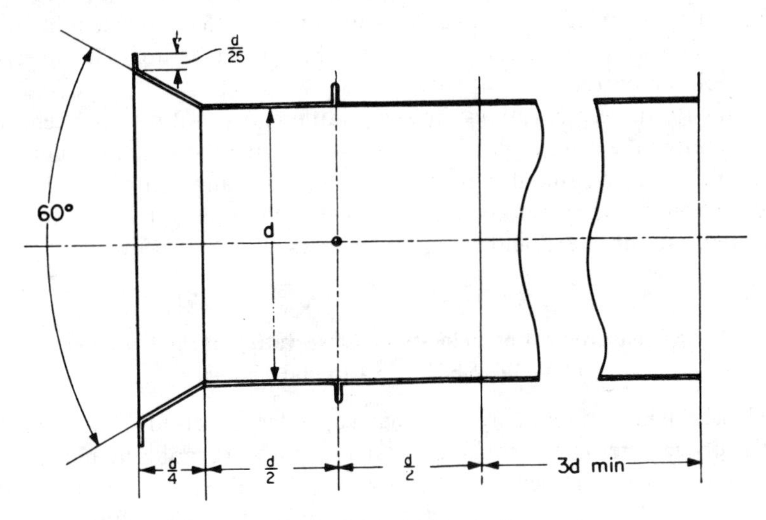

FIG. 11.4. Conical inlet nozzle.

four wall-pressure taps located half a pipe diameter downstream of the end of the inlet cone itself. The arrangement is in fact equivalent to an orifice in which $m = 1$ (see Chapter VII) and in which the upstream pressure p_1 is assumed to be the atmospheric pressure, and the downstream pressure p_2 is the average pressure reduction at the four wall taps. For the rates of flow for which this device will generally be used, incompressible conditions can be assumed, and the standard orifice equation then applies with $m = 1$ and $p_1 = 0$, i.e.

$$Q = \alpha A \sqrt{(2 p_2 \varrho)},$$

where Q is the mass rate of flow, A is the area of the pipe, and α is the nozzle coefficient which takes account of the differences between the assumed and actual conditions.

Reference 6 gives values for α as shown in Table 11.1†

TABLE 11.1. COEFFICIENTS OF CONICAL INLET NOZZLE

Reynolds number	2×10^4	4×10^4	6×10^4	10^5	2×10^5	3×10^5	4×10^5
α	0·930	0·940	0·945	0·953	0·967	0·973	0·975

Some care is necessary in the manufacture and installation of this device. The B.S.I. specify that the bore shall be smooth, parallel, and truly circular within a tolerance of $d/200$ (see Fig. 11.4) for a length d, and continue at the same nominal diameter for a further length of at least $3d$ before any fittings or obstructions are introduced into the pipe. Moreover, within a cylindrical space surrounding the inlet to the cone, of radius at least $1{\cdot}5d$ and extending for at least $1{\cdot}5d$ forward and back of the inlet, there must be no solid objects or extraneous air currents.

Even with these precautions, however, differences of 2 or 3 per cent have been found in the rates of flow obtained by applying the standard values of α to the pressures measured with different nozzles manufactured to the same specification.(8) If, therefore, errors greater than ± 2 per cent cannot be tolerated, each nozzle should be calibrated *in situ*.

The Measurement of Velocity by Observation of the Frequency of Eddies Shed by a Circular Cylinder

It has been known for many years that a cylinder immersed in a moving fluid sheds vortices alternately from the two sides. At low Reynolds numbers (based on cylinder diameter) between about 40 and 150, these vortices are shed in stable, well-defined patterns, which persist for long distances downstream as what are generally termed "vortex streets". Periodic shedding occurs also at values of the Reynolds number up to about 10^5 or more, but the vortices then quickly lose their identity in a generally turbulent wake. However, close behind the cylinder it is still possible to detect the frequency.

The properties of these vortex streets have been studied by many investigators; and it has been established that the dimensionless parameter nd/v, where n is the frequency of vortex shedding from a cylinder of diameter d in a stream of speed v, is a function of R, the Reynolds number vd/ν. Strouhal(9) first showed that the shedding frequency depends on the speed, and the parameter nd/v is generally known after him as the Strouhal number, denoted by S. It had long been known that S tends to a constant value at values of R

† It should be noted that these values differ appreciably from those given in earlier B.S.I. publications.

above about 600; but in 1954 Roshko[10] established the functional relationship between S and R within certain ranges of R with sufficient accuracy to enable it to be used as a means of measuring air speed.

Roshko found that the results of a large number of observations that he made could be well represented by the equation

$$S = 0{\cdot}212\left(1 - \frac{C}{R}\right), \tag{4}$$

where C has the values 21·2 and 12·7 for the Reynolds number ranges 50–150 and 300–2000 respectively.

Equation (4) can be reduced to a form which expresses the direct relationship between v and n, viz.

$$v = 4{\cdot}72nd + \frac{C\nu}{d}. \tag{5}$$

These results may be used to measure air speeds as follows: a suitable cylinder is inserted into the airstream and the vortex-shedding frequency n is determined.† Then a first approximation to the value of v can be obtained from (5) by inserting the known value of d and neglecting the second term $C\nu/d$. The ratio of the second term to the first ($4{\cdot}72nd$) can be shown to be equal to $C/(R - C)$, the maximum value of which for the minimum Reynolds number (50) for which Roshko's results can be used is, from (4), about 0·7. At higher Reynolds numbers the ratio of the two terms in (5) is less; and it will generally be found that neglecting the second term in (5) in calculating v will give a sufficiently close approximation to the Reynolds number R to show which of the two values of C (12·7 or 21·2) should be inserted in (5) to recalculate v using the complete equation. The value of ν appropriate to the air temperature and pressure can be obtained from Table 5.1.

It is important that the Reynolds number should be within one of the ranges specified by Roshko, i.e. 50–150 or 300–2000, and particularly that it should not lie between 150 and 300 since the observations in that intermediate range showed a much wider scatter than elsewhere. Whether or not the Reynolds number for a given wind speed lies within one of the desired ranges will depend on the diameter of the cylinder selected. If the first choice gives a value of R not in either range, another size should be substituted. Very little experience is required to enable a suitable size to be chosen at the first attempt. Table 11.2 may help in making the first choice, based on a guess as to the value of the speed v; the table shows, for various cylinder diameters, approximate values of n in cycles per second and corresponding values of R within the ranges specified by Roshko. The figures assume a value of $1{\cdot}461 \times 10^{-5}$ m²/sec for the kinematic viscosity ν.

† See below.

TABLE 11.2. RELATION BETWEEN CYLINDER DIAMETER AND VORTEX FREQUENCY WITHIN THE SPECIFIED RANGES OF R

v (m/sec)	d = 10 mm		d = 5 mm		d = 2 mm		d = 1 mm	
	R	n (Hz)	R	n (Hz)	R	n (Hz)	R	n (Hz)
0·5	342	10	—	—	68	37	—	—
1·0	684	21	342	41	137	90	68	146
1·5	1026	31	513	62	—	—	103	252
2·0	1368	42	684	83	—	—	137	358
2·5	1710	53	855	104	342	255	—	—
5	—	—	1710	210	684	520	342	1021
10	—	—	—	—	1368	1050	684	2080
15	—	—	—	—	—	—	1026	3142

Determination of the Vortex Frequency

An accurate method of determining the vortex frequency is to use a hot wire (see Chapter IX), with its rapid response to velocity fluctuations, in an electronic circuit containing an oscilloscope. This was the method used by Roshko. The hot wire should be not more than 5 or 6 cylinder diameters behind the cylinder, because Roshko's records show that, although for Reynolds numbers up to 150 satisfactory observations can be made some 50 diameters downstream, this is not true for Reynolds numbers above 200.

It is possible that a simpler method of observing n, e.g. by means of smoke trails used in conjunction with a stroboscope, could be employed; there are no published records of the direct use of such a method, but there is evidence from other low-speed work on the visualization of flow that it could be used with success unless the stream is naturally highly turbulent.

Accuracy of the Method

The main advantage of this method lies in its application to the measurement of low air speeds, for which the velocity pressures are very small and therefore difficult to determine accurately. Roshko stated that at normal velocities the accuracy is as good as that obtainable with a "conventional" manometer, while at velocities below 4 m/sec it is much better. Roshko used a manometer sensitive to about 0·02 mm of water, which is equivalent to about 2 per cent of the velocity pressure at 4 m/sec. Roshko's statement therefore implies that the accuracy of the method for the determination of velocity is better than 99 per cent at speeds down to 4 m/sec.

But his statement that it is much better at lower speeds implies not that it is better than 99 per cent at these speeds also, but only that the method gives better accuracy than that with which velocity pressures can be measured at these low speeds with conventional manometers. Inspection of Roshko's records indicates that errors will in general be within ± 2 per cent.

The Kent Vortex Meter

Using the eddy-shedding principle, Kent Instruments Ltd. have developed a flowmeter suitable for industrial use for measuring the flow of liquids or gases in pipelines from 2 to 30 cm in diameter. In this instrument,[11] the eddies are generated by a sharp-edged cylinder of rectangular cross-section, which produces a better-defined eddy pattern than a circular cylinder. The periodic pressure fluctuations induced on the surface of the cylinder by the eddy-shedding are detected by two sensitive pressure transducers, inset one on each side of the cylinder, which convert them into analogue or wave-frequency signals showing the shedding frequency in terms of which the meter is calibrated; the flowrate is directly proportional to the shedding frequency. The length-to-breadth ratio l/b of the cylinder is chosen to correspond to a peak that is found to exist in the variation of signal strength with l/b: a high signal-to-noise ratio is thus obtained.

The minimum rate of flow that can be measured with the full accuracy of which the meter is capable (errors not exceeding $\pm 0{\cdot}5$ per cent) corresponds to a pipe Reynolds number R_D of 3×10^4. At lower R_D the accuracy deteriorates, the possible errors at $R_D = 10^4$ being within the range $+$ 6 per cent to -1 per cent. A useful feature of this instrument is that it can be used to measure flows in which the static pressure is as high as 6900 kPa. The pressure loss across the meter is two velocity heads.

The Fluid-jet Anemometer

The lower limit to which flow speed can be measured with a pitot–static tube is set by the difficulty of measuring small pitot–static pressure differences with sufficient accuracy. The problem can be circumvented by use of the device sketched in Fig. 11.5. In this instrument,[12] a jet is directed across the flow and is deflected to an extent which depends on (among other variables) the ratio of the flow velocity v to the jet velocity. The pressure difference Δp registered between two "receiver" tubes R_1 and R_2 therefore provides a means for determining v. Suitable design of the geometry of the probe, together with appropriate choice of jet supply pressure p_s, creates pressure differences Δp which are many times the dynamic pressure $\frac{1}{2}\varrho v^2$. (Hence the term *fluidic amplification* used in describing such devices, by analogy with electronics.)

The instrument studied in ref. 12 had a jet exit diameter d equal to $0{\cdot}7$ mm; the jet-to-receiver distance h was $30d$, and the distance of R_2 downstream of the

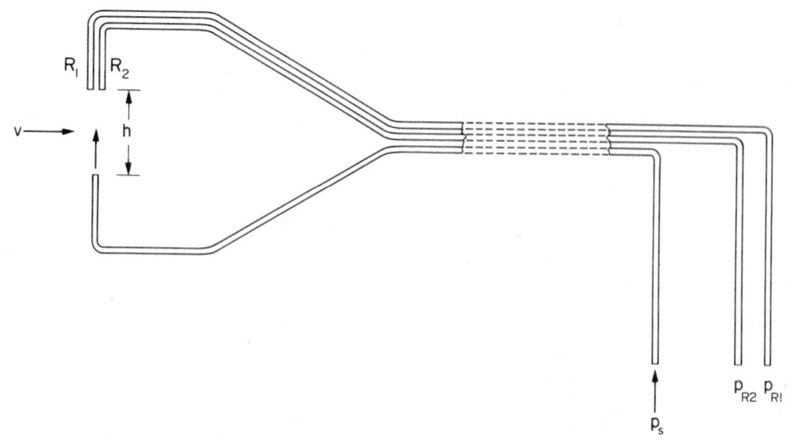

FIG. 11.5. Principle of fluid-jet anemometer.

axis of the undeflected jet was 1·62*d*. With a jet supply pressure of approximately 100 kPa (gauge), the pressure difference generated (in air) amounted to about 6 cm H_2O when v was 1 m/sec, whereas $\frac{1}{2}\varrho v^2$ at this speed is only 0·06 mm H_2O. Moreover, Δp varied linearly with flow speed (up to nearly 2 m/sec). The pressure "amplification" at lower speeds than 1 m/sec was therefore correspondingly greater. The limiting errors quoted were 5 per cent at 1·4 m/sec and 12 per cent at 0·3 m/sec.

Rugged forms of the device are available commercially from FluiDynamic Devices Ltd. (Toronto) and Lear Siegler Inc (Englewood, Colorado, U.S.A.) with velocity ranges up to 35 m/sec. Such instruments have been used successfully in various industrial processing plants, and in hostile environments.

The Laser Doppler Anemometer

When light (or sound, for that matter) reaches an observer from a moving body, it does so with a radiation frequency that is different from that at which it was emitted. This is the well-known *Doppler effect*. The frequency change (or Doppler shift) depends on the velocity of the body relative to the observer. In the optical Doppler effect (also sometimes referred to as the Fizeau effect), the moving body may be either a luminous source or a particle or larger body that reflects or scatters light incident upon it from a stationary source. It is the latter alternative that has been used for measuring flow velocity, the moving body consisting of particles which are present in the flow (or which have been introduced artificially) and which are assumed to be moving at the velocity of the fluid. This application of the Doppler effect has become practicable with the advent of the laser, a light source with the requisite properties of a narrow, coherent beam of monochromatic light of very high intensity.

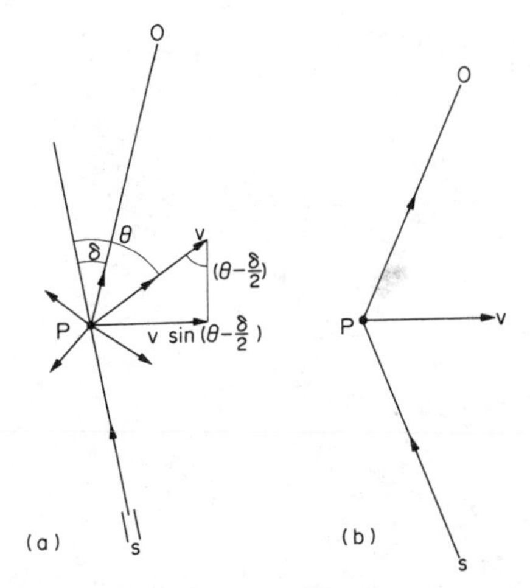

FIG. 11.6. Principle of laser Doppler anemometer.

Figure 11.6(a) illustrates the scattering of light emitted from a stationary source S by a particle at P moving with velocity v relative to an observer at O. The incident light is scattered from P in all directions; the observed change in frequency depends both on the vector velocity v and on the directional location of the observer (i.e. on the angle δ shown in Fig. 11.6(a). Provided the magnitude of the velocity is much smaller than the speed of light, as is the case in practice, the frequency change may be shown to be

$$\Delta f = \frac{2nv}{\lambda_0} \sin\frac{\delta}{2} \sin\left(\theta - \frac{\delta}{2}\right), \tag{6}$$

where n is the refractive index of the fluid, λ_0 is the vacuum wavelength of the incident light, δ is the angle between the direction of the incident light and the line joining P to the observer, and θ the angle between the incident direction and the direction of the velocity of the scattering particle at P. It can be seen from (6) that, for given positions of S, P, and O the frequency shift is proportional to the component of the vector v along the bisector of the angle SPO. Hence a suitable arrangement of source direction and observer location makes possible the measurement of any chosen component of the velocity. For example, if source and observer are symmetrically disposed with respect to a known flow direction as in Fig. 11.6(b), so that the bisector of the angle SPO lies along this direction, $[\theta - (\delta/2)]$ is equal to 90°, and (6) reduces to

$$\Delta f = \frac{\delta n}{\lambda_0} v \sin\frac{\delta}{2}.$$

In practice, various optical configurations and signal-processing schemes have been used. In that shown in Fig. 11.7, the scattered beam NPO reaches a

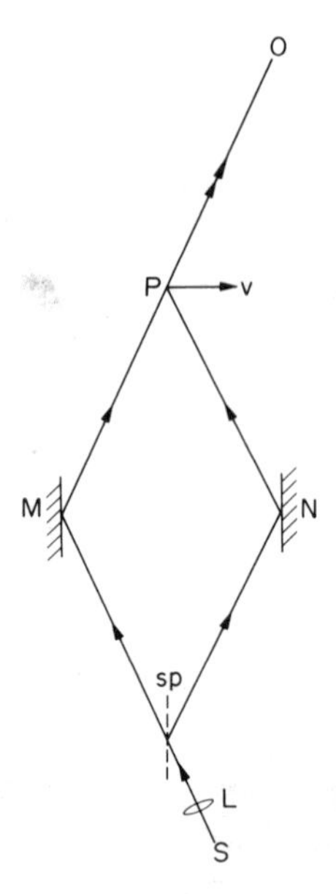

FIG. 11.7. A practical optical configuration ("reference-beam mode") for laser Doppler anemometry.

photo-detector at O, together with a reference beam MPO (derived from the same source S (using a splitter plate sp and mirrors N and H) which has passed directly through the flow field and retains the frequency of the source. The two beams are focused at P by the lens L. The output from the photo-detector results from the combination of the two beams of different frequency, and thus has a "beat" frequency equal to the Doppler shift, since this is the frequency difference between the two beams.

The development of laser Doppler anemometry has been rapid during the past decade, and is still continuing; it has now reached the stage of practical application at which it can fairly be regarded as a technique that is complementary to the pitot–static tube and the hot-wire anemometer. Its outstanding advantage is that the method is entirely free from disturbances due to a probe, since all the instrumentation is external to the flow. In addition, a single component can be measured, as well as temporal variations (including, with an appropriate optical configuration, reversals of flow direction); and the method can be used as a calibration standard. On the other hand, the laser source and the electrical measuring equipment are far more complex than simple pressure-probe systems, and it is a disadvantage if solid particles have to be introduced into the flow. A valuable recent review and bibliography is provided in ref. 13. Eye protection against direct or reflected laser beams is essential.

The Soap-film Method

For measuring very small rates of flow, which are perhaps more common in the chemical laboratory than in engineering practice, a method originally suggested by Barr(14) and afterwards elaborated by Gooderham(15) will often be found useful. The gas or air whose flow is to be measured is passed upwards through a vertical glass tube (Fig. 11.8). When the rubber bulb *B* is squeezed, the level of the soap solution in the reservoir *R* can be made to reach the bottom of the vertical tube; when the bulb is released, a soap film is formed (stretching across the tube), rises at the same speed as the gas, and is timed between two fixed points *P*, *Q*, or along a given length of a graduated scale etched on the

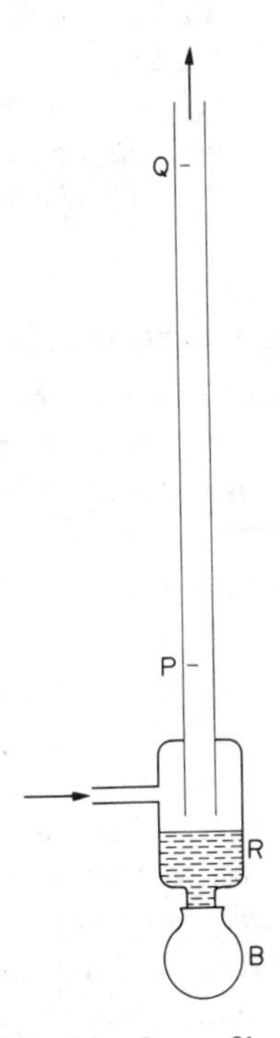

FIG. 11.8. Principle of soap-film flowmeter.

tube. The volume between the marks, or the calibration of the scale in terms of volume, is previously determined by direct measurement with liquids; and therefore the rate of flow of the air or gas can easily be deduced from the time measurement.

The accuracy with which the flow can be measured has been analysed in detail by Levy.(16) The method has been developed by the National Engineering Laboratory(17) for the calibration of gas flowmeters at low rates of flow up to about 600 cm^3/sec. Timing accuracy has been improved by photoelectric detectors which trigger an electric timer as the soap film passes. Errors are well within $\pm 0 \cdot 5$ per cent.

The Quartz-fibre Anemometer

Another type of meter for measuring low airflow rates (10–60 cm/sec in a 2·5-cm diameter tube) has been reported by Griffiths and Nicol.[18] A quartz fibre, of the order of 0·02 mm diameter, is fixed at one end of a diameter of the tube and the deflexion of the free end (typically of the order of 0·2 mm) is measured by means of a travelling microscope. The flow in the tube is laminar and the deflexion of the fibre can be calculated theoretically. In practice, however, the instrument is calibrated because the value of Young's modulus for quartz depends slightly on fibre diameter. Provided that the tube is transparent at the measurement station, this device is a simple and inexpensive alternative to a hot-wire or other type of low-flowrate meter.

Direct-reading Velocity and Quantity Meters

Many firms make instruments that give direct readings of velocity, quantity, or flowrate. They are useful not only because interpretation of their readings requires no calculation except (in most cases) for corrections for departures of fluid density from the calibration conditions, but also because they are well suited for application in process control.

(*a*) *The Rotameter*

This is a development of the Ewing ball-and-tube flowmeter which, in its original form,[19] consisted of a vertical glass tube containing a steel ball. The tube was slightly tapered in bore, with the diameter decreasing downwards; and the fluid flow to be measured passed upwards through the conical tube, which was inserted in the flow circuit. The ball, of diameter slightly greater than the minimum bore of the conical tube, was carried upwards by the passage of the fluid until it reached a position in the tube where its weight was balanced by the force due to the fluid flowing past it. The position of the ball in the tube depended on the rate of flow, and was used to measure it.

The main objection to the Ewing flowmeter was unsteadiness of the ball; this could be partly overcome by inclining the tube. However, a sphere is not an ideal shape for the float because, as is well known, the type of flow past it is liable to change suddenly in a narrow Reynolds-number range, with a consequent sudden change in resistance, and therefore in the rate of flow necessary to balance its weight. Moreover, if it acquires a rotation about a horizontal axis, which is not impossible, it will experience a lateral force tending to move it away from the axis of the vertical tube towards the side wall, and "chatter" may result.

Most modern versions of this instrument, of which a number exist, have specially shaped floats which avoid the disadvantages of spheres. In the rotameter one type of float is shaped as shown in Fig. 11.9(a). Some of these floats have spiral grooves in their upper periphery which keep them in rotation about a vertical axis centrally in the tube as the fluid streams past. The tapered glass tube has a quantity or speed scale marked on it (Fig. 11.10), and the position of the float in relation to the scale gives an immediate direct indication of the mean rate of flow, no calculation being required except for corrections to allow for deviations from standard conditions (see below).

As an indicator in a system in which the flow has to be constantly checked and maintained at a specified rate, this instrument serves a particularly useful purpose; if it is installed where it can easily be seen by the operator responsible for controlling the rate of flow, the necessary adjustments to the control valve can readily be made as required. Types are also available for use in automatic-control systems.

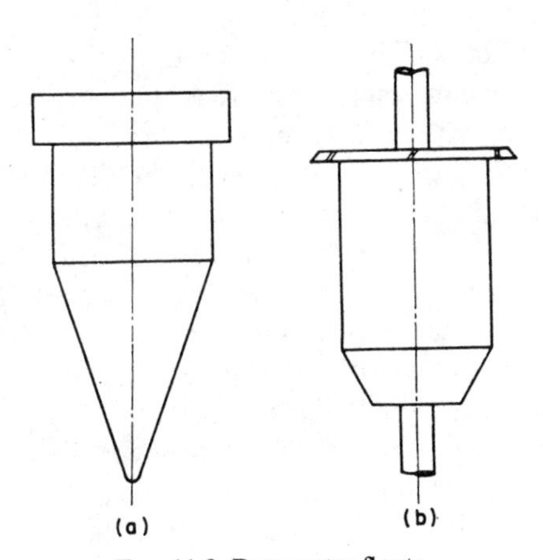

FIG. 11.9. Rotameter floats.

Rotameters can be used to measure both liquid and gas flows. For the latter purpose they are made in a number of sizes, ranging from those measuring rates of flow as low as 2 cc/min to those rated at 4000 m^3/h, which can be inserted in pipelines about 20 cm in diameter. In the larger sizes, the tapered tube is made of metal and the float carries a vertical sighting rod which is visible, together with the scale, above or below the instrument in a glass-fronted sighting compartment. The standard range of one of these meters is 10 to 1 and the normal accuracy is within about ± 2 per cent of the indicated flow. This can be improved in special instruments.

The theory of the rotameter has been discussed by a number of workers.[19–26]

If the float is in a stream of fluid of density ϱ and speed v, the upward force on the float, i.e. its resistance, can be expressed in the form

$$F = C\varrho A v^2,$$

where C is a coefficient, usually called the resistance coefficient, and A is the maximum cross-sectional area of the float at right angles to the flow.

When the float is in equilibrium in the tapered tube its weight W less the buoyancy balances the resistance, i.e.

$$C\varrho A v^2 = V_{ft}(\varrho_{ft} - \varrho)g, \tag{7}$$

where V_{ft} and ϱ_{ft} are the volume and density of the float respectively.

In the special case of air and similar gases, the density of the float is so large compared with that of the gas that the latter can be neglected as far as the buoyancy is concerned, and the equation can be correspondingly simplified to

$$C\varrho A v^2 = K, \tag{8}$$

where K is a constant for any given rotameter.

In a tapered tube, v will vary with the annular area between the float and the walls of the tube, and the height of the float in the tube will adjust itself until v is such that the above equations are satisfied. In general C for any particular design of float will vary with the Reynolds number, but manufacturers of these meters have discovered shapes of float for which C changes very little, if at all, over a wide range of Reynolds number. Figure 11.9(b) is a sketch of one such float whose resistance, according to the makers, GEC–Elliott Process Instruments Ltd., is virtually unaffected by changes in viscosity over a wide range. This is presumably because the resistance is largely due to the sharp edges of the upper disk, and very little to surface friction. This type of float is guided by a central rod, fixed in the tapered tube, and passing through the float.

When the resistance of the float is independent of Reynolds number, C in (8) is constant, and for any particular combination of float and tapered tube, i.e. for any given rotameter, this equation can be reduced to the simple form

$$Q\sqrt{\varrho} = \text{constant}, \tag{9}$$

where Q is the volumetric flowrate, in terms of which the scale is usually graduated rather than in terms of velocity. This is the relation that enables the user of an instrument installed for measuring air flow, whose scale will be accurate for specified temperature and pressure conditions, to correct the readings for air at different temperatures and pressures or for other gases. Thus if the calibration relates to air density ϱ_1, and the instrument indicates a volumetric flowrate Q when the density is ϱ_2, the true value of Q at this density (ϱ_2) is $Q\sqrt{(\varrho_1/\varrho_2)}$.

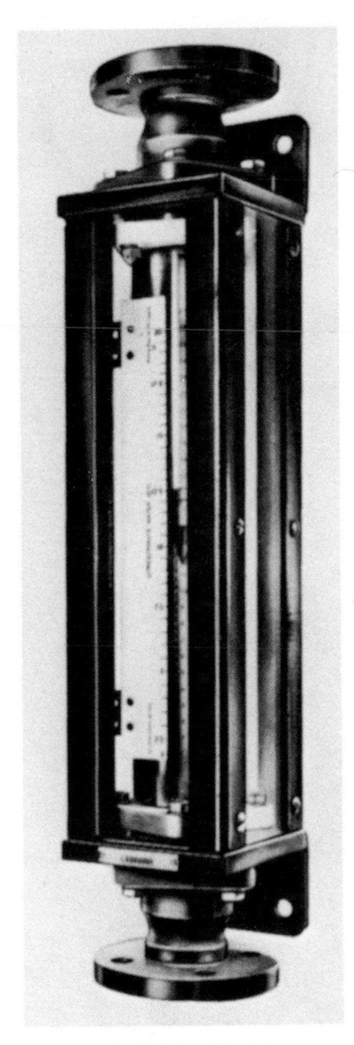

FIG. 11.10. Rotameter

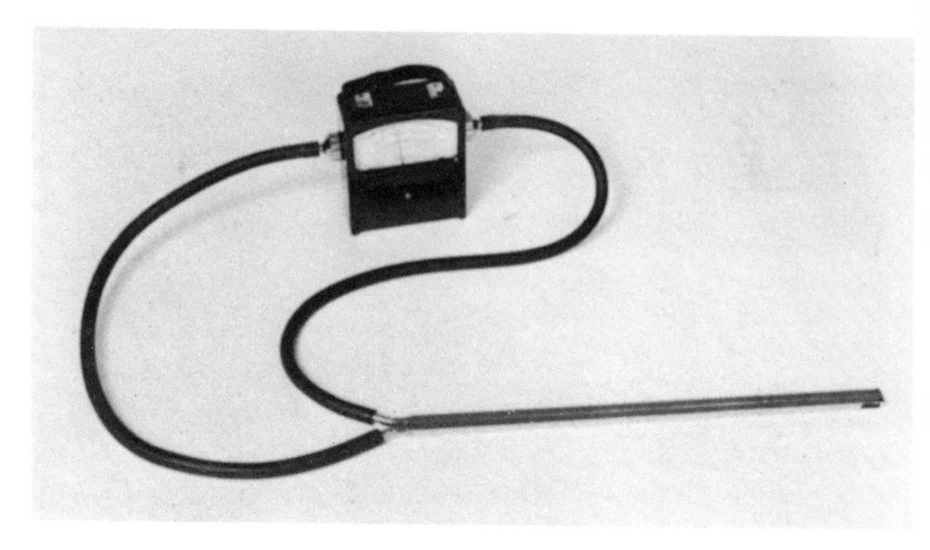

FIG. 11.11. Velometer.

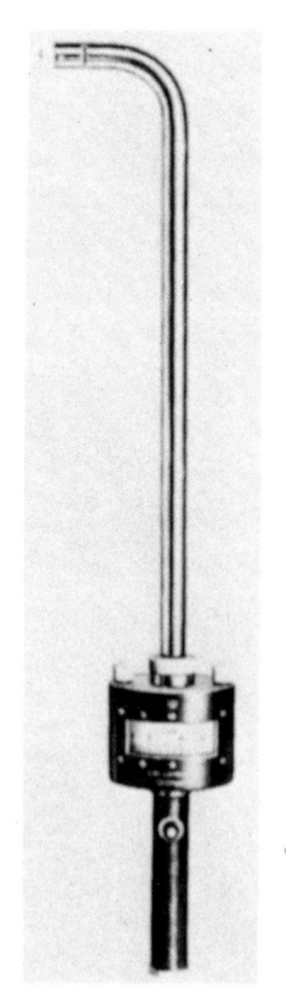

FIG. 11.12. Direct-reading meter for pitot–static tube.

If the resistance coefficient C of the float is not constant, but varies with Reynolds number, the equation corresponding to (9) can be written (still on the assumption that the density of the gas is negligibly small in relation to that of the float)

$$\frac{Q\varrho}{\mu} = f\left(\frac{\varrho}{\mu^2}\right),$$

where μ is the coefficient of viscosity and f represents a functional relationship.

This equation is a modified form of that derived by Martin;[23] it can be used as follows to draw a series of calibration curves that will enable any particular rotameter to be used for measuring the flow of gases of different ϱ and μ. For any one gas at a specified temperature and pressure, ϱ and μ will be known (or can be measured); each scale reading on the rotameter indicated by the height of the float when gas is passing will correspond to a known value of $Q\varrho/\mu$, which can be plotted against ϱ/μ^2. This gives one point on the $(Q\varrho/\mu,\ \varrho/\mu^2)$ curve for that scale reading. One such point can be obtained for each of a series of scale readings; and if this process is repeated for different values of μ and ϱ, either by changing the gas or the temperature and pressure (or both), a series of curves is obtained, one for each chosen scale reading, covering a range of gas density and viscosity. These curves can be used to determine the flowrate Q corresponding to any scale reading obtained for any gas for which μ and ϱ fall within the range covered by this calibration. Martin[23] describes experiments for obtaining the data from which these curves can be plotted when they are not supplied by the makers of the instrument.

(b) *Velocity Meters*

An example of this type of instrument is the velometer,[27] originally developed by Boyle in the United States, and now manufactured in England by Salford Electrical Instruments Ltd. The indicating component of this instrument consists of a balanced, damped, spring-controlled, pivoted vane, which can be deflected by a very light current of air or gas and carries a pointer moving over a graduated scale. The movement is enclosed in a case having two orifices through which the air enters and emerges. The vane moves in a chamber so shaped that the scale deflexion is proportional to the velocity. In large ducts or passages – mine ventilation passages, for example – this meter can be exposed directly to the air current with the inlet orifice facing upstream; it indicates air velocities as low as 6 m/min, the maximum error of any reading being within 3 per cent of the full-scale reading. A more sensitive type can be obtained scaled 0–18, 0–30, or 0–45 m/min.

The velometer can also be used to measure flow in pipes in which the meter is too large to be inserted, by introducing a special probe into the pipe (see

Fig. 11.11). This probe comprises a pair of pressure tubes, each with a pressure orifice, one of which, when the probe is in position, faces upstream and the other downstream; hence, when each is connected to one of the meter orifices, a current of air flows through and deflects the vane.

Other forms of probe are supplied for such purposes as measuring air velocities from ventilation grilles, discharge openings, or ventilating diffusers both in walls and ceilings. The meter has a number of scales, each for use with different probes and their rubber connecting leads. It is important to bear in mind that the calibration allows for the resistance of these leads, and is accurate only when they, or other lengths of rubber tube of identical length and bore, are used. Standard scales correspond to 0–60, 0–600, and 0–1800 m/min and there is in addition a static-pressure scale graduated in head of water; metric scales can be provided. All scales relate to air at 18·3°C and 762 mmHg, and when the temperature and barometric height differ from these values the indicated velocities must be multiplied by the ratio of the square root of the air density at standard conditions to that at the time of measurement.

Similar spring-balanced pivoting-vane types of meter are made by Paul Gothe of Bochum, West Germany, and Wilh. Lambrecht K.G., Göttingen. Like the velometer, the former can be used without probes in large ducts or passages, or with various types of probes, including a standard Prandtl-type pitot–static tube. The meter itself can be scaled to cover velocity ranges between 0–0·6 m/sec and 0–100 m/sec, with an error not exceeding 1·5 per cent of the full-scale reading.

In one variant the Prandtl tube is connected directly to the meter, as shown in Fig. 11.12 which illustrates the similar instrument made by Lambrecht. In this arrangement the air enters through the total-head orifice, passes through the inner tube of the pitot–static combination, then through the meter and back to the outer tube, emerging at the static slit. The Lambrecht instrument, which is designed only for direct coupling to the Prandtl tube as in Fig. 11.12 has four interchangeable scales, on the lowest of which (0–3 m/sec approximately) a velocity of about 0·2 m/sec can be observed; the high-velocity scale is graduated up to about 50 m/sec. The errors are said to be within ± 2 per cent of the full-scale reading.

A different type of direct-reading velocity meter, called the Florite, is made by the Bacharach Industrial Instrument Company of Pittsburgh, Pa. This is essentially a vane anemometer in which the rotor, instead of turning freely, is restrained by a spring whose deflexion is a measure of the air speed; this is indicated by a pointer moving over a scale marked on the casing. The use of a stop-watch which is necessary with the conventional vane anemometer is thus avoided.

Like the velometer, and the instruments by Gothe and Lambrecht described above, the Florite, which can be obtained in the United Kingdom from Shandon Southern Instruments Ltd., of Camberley, has useful applications in ventilation work, e.g. measuring the flow from grilles and inlet diffusers. In many such cases high accuracy is not essential; a more important requirement is to compare or equalize the quantities of air being delivered at various discharge points in a ventilation system, and for this purpose comparative readings are all that are necessary.

In the vane type of meter described by Head and Surrey[28] the vane is carried on a counterweighted and shielded arm attached to the movement of a moving-coil ammeter. A current is passed through the coil to balance the aerodynamic load on the vane, and thus provides a measure of the air speed. The range of the instrument is from about 6–140 cm/sec. For use in fluctuating flow, the instrument incorporates an electric circuit which makes it automatically self-balancing, and the reading is displayed continuously.[29]

(c) *Ring-balance Meters*

Useful types of velocity or flowrate measuring or recording instruments, developed in Germany, are based on the *Ringwaage* (Ring-balance) manometer (see p. 268). Briefly, these devices depend on the angular rotation under an applied pressure difference of a ring mounted on ball bearings. The movement of the ring is transmitted by mechanical linkage to an indicating pointer moving over a scale graduated in millimetres of water.

For velocity measurements, the pressure difference applied to the ring-balance is the dynamic pressure† from a pitot–static tube. For flowrate indicating or recording, the applied pressures are the upstream and downstream pressures from a calibrated orifice inserted in the pipeline. In this case, of course, the meter scale relates only to one particular installation for which the meter is supplied.

The maximum pressure difference for which the Rixen range of ring-balance manometers described on p. 268 is designed is 1700 mm H_2O. This is well above the velocity pressure corresponding, in normal atmospheric conditions, to the upper limit of air speed (60 m/sec – see p. 15) for which it can be assumed that, subject to errors not exceeding 0·5 per cent, the air speed can be calculated from the velocity pressure without allowance for compressibility effects.

(d) *Gas Meters*

Gas meters are almost exclusively used in the gas industry to measure the volume of gas delivered to gas-holders or to consumers. Their indications are proportional to the volume passing through them, so that the rate of flow cannot be determined unless the time interval between observations is also recorded.

† For pressure definitions, see p. 6.

Two main types are in general use, the dry meter and the wet meter; the former, which is found in most private premises to measure domestic gas consumption, is not a precision instrument; but the latter, which is generally used in the works of gas-supply undertakings, is capable of a high degree of accuracy – errors not exceeding 0·1 or 0·2 per cent – when accurately calibrated.

The operation of a wet meter is illustrated diagrammatically in Fig. 11.13.†

The drum *A*, revolving on its axis *B*, is divided into a number of compartments *C* by means of partitions *D*, which extend the full length of the drum. Gas is introduced into the centre of the drum from the inlet pipe *E*, which passes through one end of the drum in the form of a hollow spindle, and flows into the compartment marked "Filling" via port *F*. In this compartment it becomes a fluid wedge between the surface of the water and the partition *D*, pushing the partition upwards and turning the drum. When the compartment is filled, the port *F* is sealed by the water; and soon afterwards the discharge port *G* rises above the surface of the water and discharge begins. The drum continues to turn owing to the filling of the next compartment, so that water flows into

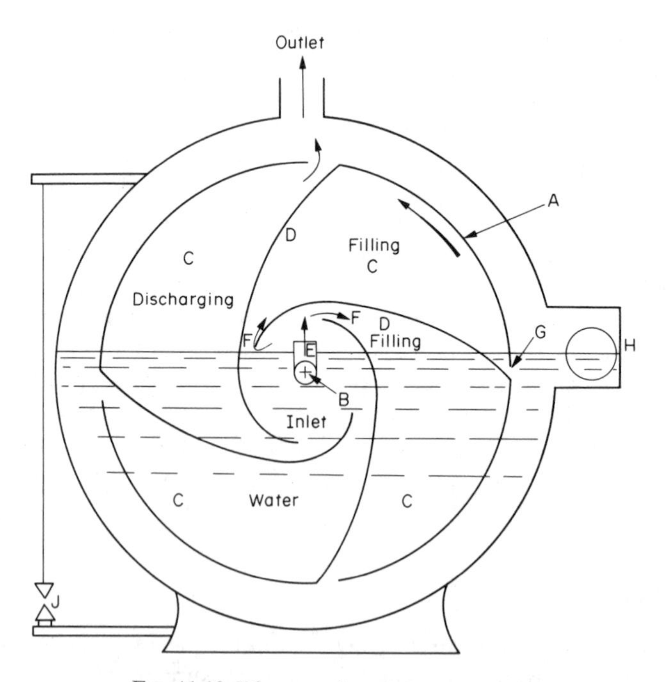

FIG. 11.13. Diagram of wet gas meter.

† For this diagram and for the description that follows of the method of operation of the wet gas meter, the authors are indebted to the Gas Council Watson House Laboratory.

port *F*, expelling gas from port *G* into the meter case and hence to the outlet. In turning, the drum rotates the index pointers.

As the drum revolves, each compartment fills alternately with gas and with water, so that it is completely filled with gas and emptied once per revolution. Discharge of gas from a compartment cannot begin before the charging is complete, so that, for every revolution of the drum, an accurately defined volume of gas is delivered.

The water-level, which controls the effective volume of each compartment, and therefore also the calibration of the meter, can be observed through the sight box *H*, and must be accurately maintained at the level mark there indicated. For the same reason the meter must always be levelled accurately by means of the plumb-bob *J*, or by a spirit level.

The drums used in practice contain spiral partitions, arranged in a manner similar to the rifling of a gun barrel, the ends of the partitions corresponding with radial ports cut in the end covers of the drum. Gas passes axially along the drum, from back to front; but otherwise the operation is as described above.

It will be clear from the foregoing description that a wet gas meter needs calibration, since one cannot determine by any convenient means the effective total volume of the compartments. The usual process of fundamental calibration makes use of a standardized apparatus, resembling in principle a cylindrical gasholder, from which a known volume of gas can be passed through the meter. Various difficulties complicate the process; in particular, the problems of measuring accurately the temperature of the gas, of maintaining it constant during the calibration, and of ensuring that the water used to seal the apparatus is completely saturated with gas before calibration begins.

Although admirably suited for the purpose for which it was primarily designed, the wet gas meter has several disadvantages for general use in engineering practice. Apart from the difficulties of calibration, it is very large in relation to its rating. Thus a meter rated at a maximum delivery of 35 m^3/h – not a high flowrate for the ventilating engineer – will occupy a space rather over one metre in length, breadth, and height and, when filled with water to the datum level, is extremely heavy. The rating cannot be increased appreciably by allowing the drum to rotate faster, since this results in a banking-up of the water-level to one side, with a consequent change in effective volume of the compartments. Thus if we exclude as impracticable for general purposes the very large station meters used in gasworks to meter large quantities of gas, the use of the gas meter will be restricted to the measurement of low rates of flow. The further difficulty then arises that a head of 5–10 mm of water is needed to operate even a small meter; and static heads of this order are not common with slow rates of flow.

Thus although in suitable conditions the wet gas meter may be one of the most convenient and accurate instruments for measuring low rates of flow, it is seldom suitable for general engineering work.

(*d*) *Turbine Meters*

During the last fifteen years or so, a type of gas meter known as the turbine meter (or sometimes the propeller meter) has shown itself to be an extremely useful instrument for the measurement of flowrate in pipelines. Hitherto it seems to have been mainly used for metering liquids; it can, however, also be used for gas flow, and types designed for this purpose have been described.[30, 31]

In principle, the turbine meter resembles the vane anemometer except that the rotor is totally enclosed, apart from the inlet and outlet ports, and that the whole of the flow passes through the rotor; in fact the meter may be regarded as a vane anemometer co-axial with the pipe and of overall diameter equal to the internal pipe diameter. The rotor rotates at a speed such that the total driving torque balances the frictional resisting torque of the bearings; and, as follows from the theory of the vane anemometer given in Chapter VIII, the rotor speed is a measure of the product $v\sqrt{\varrho}$. This has been confirmed experimentally by Gehre and Smits,[30] who describe a meter used in the German gas industry with a capacity of 60,000 m^3/h. The diameter of the meter was 1·2 m and that of a similar one of 1100 m^3/h capacity was 25 cm.[32] It is clear, therefore, that the turbine meter is greatly superior to the wet gas meter on the score of size. Other advantages are high accuracy (errors probably within ± 1 per cent in very unfavourable approach-velocity distributions), low-pressure loss, and a wide operating range, normally about 10 to 1. In addition, the type is well suited to electronic means of measuring and instantaneously indicating rotor speed, with the advantages this offers as compared with the older method of using gear trains.

For further information on this type of meter refs. 33 and 34 may be consulted, in addition to those already quoted.

(*e*) *The Mass Flowmeter*

There is an increasing demand in many industries for a meter that indicates the mass (as distinct from the volume) of fluid flowing per unit time or within a specified period. A mass flowmeter based on the rate of cooling of a heated element is described in Chapter IX, p. 243. Other types are being used, particularly in the United States, mainly for metering liquids, but also for air and gases. The true mass flowmeter indicates in units of $\varrho v A$ (where A is the cross-sectional area of the pipe or duct), instead of vA like a volume meter. Hence, once this instrument is calibrated, no measurements of temperature or pressure are required because it is unnecessary to calculate ϱ.

The principle of the Massa Strom Meter, or M.S. Meter[35, 36] illustrated in Fig. 11.14 is based on a suggestion by Schultz-Grunow[37] in 1941. The fluid enters the meter at A and divides into two paths around the cylinder B which is

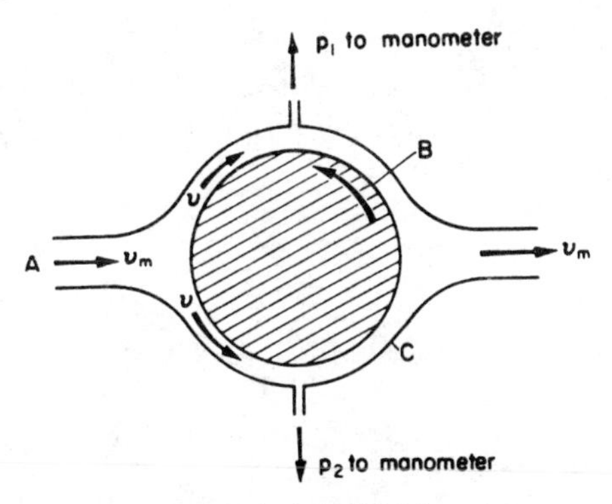

FIG. 11.14. Principle of M.S. meter.

driven at constant rotational speed inside the meter casing C. The gaps above and below the cylinder are equal, so that the mean speeds v in the two are equal when the cylinder is at rest. When rotating, the cylinder sets up a circulating velocity v_c, and, if this does not affect the main-stream velocity in either gap, the total velocities in the two gaps are $v - v_c$ in the upper and $v + v_c$ in the lower. As a first approximation, we may apply Bernoulli's equation to the flow through the meter in the two gaps; then if p_1 and p_2 are the pressure in the upper and lower gaps respectively

$$p_1 + \tfrac{1}{2}\varrho(v - v_c)^2 = p_2 + \tfrac{1}{2}\varrho(v + v_c)^2,$$

so that

$$p_1 - p_2 = 2v_c\varrho v = v_c\varrho v_m,$$

where v_m is the mean velocity of the main flow entering the meter ($=2v$ by assumption).

Thus since v_c is a meter characteristic which is assumed to depend only on the rotational speed at which the cylinder is driven and can be determined by calibration, the differential pressure $p_1 - p_2$ is proportional to the mass flow.

Although the assumptions underlying this elementary theory can obviously not be fully satisfied in practice, satisfactory results were obtained in the laboratory with a meter of this type,[36] particularly in the measurement of pulsating flow (see Chapter XII) for which it was primarily designed; but the type was not developed commercially. The same action, namely the addition of a velocity component to one branch of the main flow and the subtraction of an equal component from the other branch, is better obtained by using a positive-action circulating pump rather than by a rotating cylinder: and developments have been reported[38] on these lines, but so far only for metering liquids.

Most commercial mass flowmeters now available depend on the principle of momentum transfer. Meters of this type usually include some form of impeller which imparts angular momentum to the fluid, and in doing so, experiences a torque proportional to $M\omega$, where M is the mass flowing per second and ω is the rotational speed of the impeller. The various existing types of meter differ in the methods adopted to measure this torque, either directly or by transferring it to a separate measuring element.

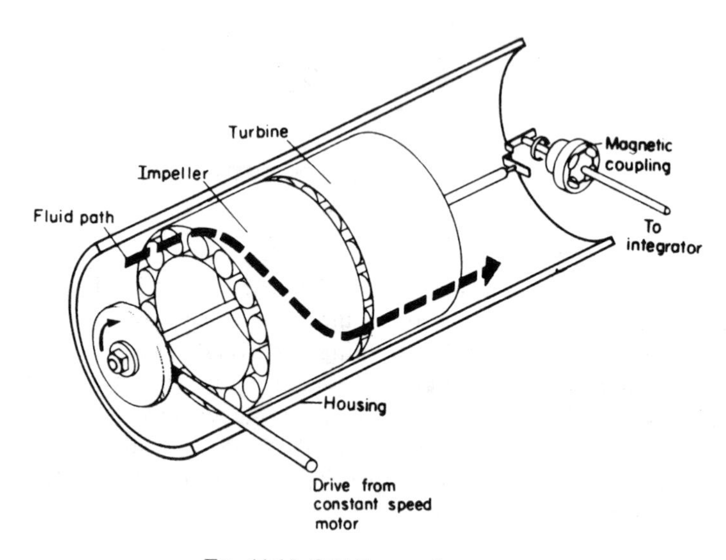

FIG. 11.15. G.E.C. mass flowmeter.

An example of the last-mentioned type is that jointly developed in the United States by the firm of Black, Swalls & Bryson and the General Electric Co. for metering both liquids and gases. It was first described by Orlando and Jennings.[39] The principle on which it operates is shown in Fig. 11.15. Angular momentum is imparted to the entering fluid by the left-hand component of the meter – the impeller – which is driven by a constant-speed motor. The right-hand component – the turbine – is similar to the impeller in construction; it incorporates a number of peripheral guide vanes or passages which remove the angular momentum and straighten the flow again. The torque is thus transferred to the turbine. In the original instrument the turbine was prevented from rotating by means of a helical restraining spring which deflected a pointer moving over a graduated scale; and the scale reading in degrees was related to the mass flow by calibration. In the commercial development of the meter the torque acting on the turbine, which again is prevented from rotating, is transmitted through a magnetic coupling to an integrating system in which a gyroscope is caused to precess at a rate proportional to the torque, i.e. to the mass flow. The standard type indicates total mass flux, but additional equipment can be supplied to indicate flowrate, i.e. mass per unit time.

In its original form this meter was used to measure the flow of mixtures of hydrogen and carbon monoxide at rates as low as 8 m^3/h with an error of less than 1 per cent.(40) The small torque produced by such low-density gases, however, prevents the standard commercial instrument from being operable at comparably low rates of flow. The makers state that the commercial type has a range of about 300 to 2000 kg/h (approximately 250 to 1700 m^3/h) for air at ordinary atmospheric temperatures and pressures. Errors are said not to exceed ± 1 per cent.

Further information on mass flowmeters will be found in refs 33, 41, and 42. At present these meters are very little, if at all, used in the United Kingdom for metering air or gases; possibly their high cost is a deterrent. But, especially in view of the possibilities they offer for measuring pulsating flow, it seems likely that they will be more widely used in future.

(*f*) *Applications of Ultrasonics*

The speed of propagation of acoustic disturbances in a moving fluid is equal to the sum of the speed of sound a and the component v_s of the stream velocity along the acoustic path. In a uniform flow field, therefore, the fluid velocity can be determined from the time of transit t of an acoustic pulse originating at a point A (Fig. 11.16) and arriving at a point B, since

$$t = s/(a + v_s),$$

where s is the distance AB. Thus

$$v_s = (s/t) - a.$$

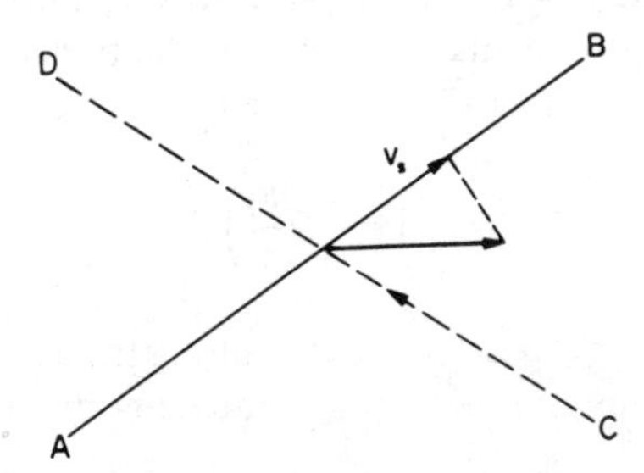

FIG. 11.16. Principle of ultrasonic method of measuring flow velocity.

At first sight this method of determining v_s appears to be highly attractive, particularly as the apparatus is located at the flow boundaries, so that no flow disturbances or pressure losses are introduced. Further, the results are not subject to the effects of viscosity, and hence are independent of Reynolds number.

Closer consideration, however, reveals substantial difficulties. In the first place the method depends basically on the measurement of the small difference between the transit time t in the moving fluid and the corresponding time t_0 with the fluid at rest, since $a = s/t_0$. Both t and t_0 are small (t/s and t_0/s are about 3×10^{-3} sec/m of acoustic path in air at standard sea-level conditions), and $(t_0 - t)/s$ is of an order of 3×10^{-4} sec/m when $v_s = 30$ m/sec. Large errors arise in the determination of $(t_0 - t)$ if the speed of sound a is not identical during the measurements of t and t_0: for instance, a change in a between the measurement of t and that of t_0 of 0·25 per cent (which would arise from a temperature change of 1·5°C in the case of air at 15°C) would cause a percentage change in $(t_0 - t)$, when v_s is one-tenth of a, of ten times the percentage difference in a; the error in the derived value of v_s would therefore amount to 2·5 per cent. Similar considerations apply if the time difference is determined using symmetrically disposed upstream and downstream paths as indicated in Fig. 11.16 (for which the transit times t_1 and t_2 are $s/(a \pm v_s)$), except that the resultant error of 2·5 per cent is reduced to 1·25 per cent because v_s is now given by $\frac{1}{2}(s/t_1 - s/t_2)$ (provided the path lengths AB and CD are accurately the same). These difficulties can be overcome, however, if a single acoustic path is employed and the pulse transmission and reception functions are provided alternately by similar apparatus at each end of the path.

When the velocity of the fluid varies along the acoustic path, the transit times become

$$t = \int \frac{ds}{a \pm v_s}, \tag{10}$$

where v_s is a function of s. If the form of this function is known or can be assumed, as is often the case in pipe flow, the above equation enables v_s to the determined from measurements of t. The evaluation is simplified if $v_s \ll a$ (as in the flow of a liquid), since (10) can then be replaced by the approximation

$$t \doteqdot \frac{1}{a} \int \left(1 \mp \frac{v_s}{a}\right) ds \, . \tag{11}$$

Equation (11) assumes that a does not vary along the acoustic path; and if a is identical in the determination of t_0 (or in the determinations of t_1 and t_2) we obtain simply

$$t_0 - t \doteqdot \frac{1}{a^2} \int v_s \, ds$$

and

$$t_2 - t_1 \doteqdot \frac{2}{a^2} \int v_s \, ds$$

respectively.

An alternative approach in the application of acoustic methods to the determination of the rate of flow in a pipe or duct is to measure, not transit times, but the distance x by which a transverse acoustic beam is displaced in the flow direction (Fig. 11.17). If the speed of sound is constant across the pipe or duct,

$$x = \int u \frac{ds}{a}$$

$$\div \int u \frac{dy}{a}$$

where u is the axial component of the flow velocity.

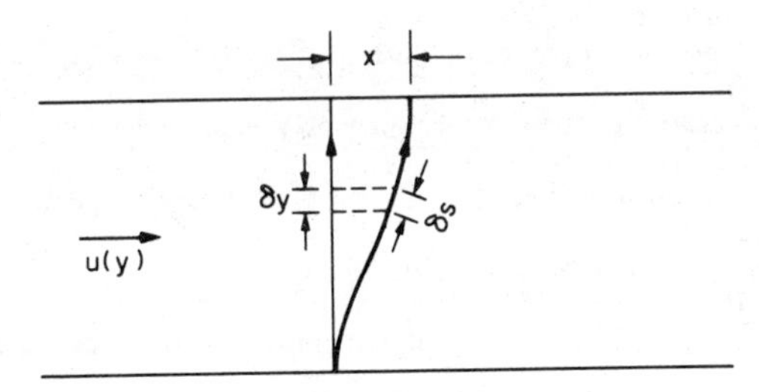

FIG. 11.17. Principle of ultrasonic-beam deflexion method of measuring flowrate.

Measurement of x thus determines $\int u\, dy$ directly; and if a number of transverse paths are employed in a plane yz perpendicular to the axis of the pipe or duct, the flowrate Q then follows at once (assuming the speed of sound is everywhere the same) since

$$Q = \int u \, dy \, dz \, .$$

An investigation into the possibilities of this type of air flow meter has been described in ref. 43, which draws attention to the fast response of such a meter. At the same time it is acknowledged that difficulties arise because of the rapid attenuation of signal strength at the high (ultrasonic) frequencies employed in order to exploit the highly directional properties of ultrasonic waves. The attenuation is greater in gases than in liquids.

Several meters for the measurement of flowrate have been developed along the lines described above, particularly for use in liquids.[44] Because of the practical problems indicated, however, together with considerations of cost and electronic complexity, such ultrasonic methods are not at present widely used in airflow measurement and seem likely to find only limited application despite their basic attractions.

References

1. W. Aichelen, Der geometrische Ort für die mittlere Geschwindigkeit bei turbulenter Strömung in glatten und rauhen Rohren, *Z. f. Naturforschung* **2** (1947) 108.
2. J. H. Preston and N. Gregory, The three-quarter radius pitot tube flow meter, *Engr* **190** (1950) 400.
3. J. H. Preston and J. F. Norbury, The three-quarter radius flowmeter – a reassessment, *Flow Measurement in Closed Conduits*, paper A-5, Vol. I, 46, H.M.S.O., Edinburgh, 1962.
4. F. A. L. Winternitz, Rapporteur's review of ref. 3, *ibid.* 88.
5. C. Salter, *Low Speed Wind Tunnels for Special Purposes*, N.P.L. Aero Rep., 1218, 1966.
6. B.S. 848: Part 1: 1963, *Methods of Testing Fans for General Purposes*, British Standards Institution, London, 1963.
7. F.M.A. Code 3: 1952, *Fan Performance Tests*, Fan Manufacturers' Association Limited, London, 1952.
8. I. S. Pearsall, *Calibration of Three Conical Inlet Nozzles*, N.E.L. Rep., 39, 1962.
9. V. Strouhal, Über eine besondere Art der Tonerregung, *Annalen der Physik u. Chemie* (neue Folge) **5** (1878) 216.
10. A. Roshko, On the Development of Turbulent Wakes from Vortex Streets, *N.A.C.A. Rep.* 1191 (1954).
11. D. J. Lomas, Vortex flowmetering challenges the accepted techniques, *Control and Instrumentation*, July/August (1975) 36.
12. J. W. Tanney, An anemometer for very low velocities, Canadian National Research Council Aeronautical Report LR-472 (1967).
13. R. J. Goldstein, Measurement of fluid velocity by laser-Doppler techniques, *Applied Mechanics Reviews* **27** (1974) 753.
14. G. Barr, Two designs of flowmeter, and a method of calibration, *J. Scient. Instrum.* **11** (1934) 321.
15. W. J. Gooderham, Soap film calibrators, *J. Soc. Chem. Industry* (*Trans*) **63** (1944) 351.
16. A. Levy, The accuracy of the bubble meter method for gas flow measurements, *J. Scient. Instrum.* **41** (1961) 449.
17. P. Harrison and I. F. Darroch, Air flow measurements by the soap film method, N.E.L. Report 302 (1967).
18. R. T. Griffiths and A. A. Nicol, A fibre flowmeter for very low flow rates, *J. Scient. Instrum.* **42** (1965) 797.
19. J. A. Ewing, A ball-and-tube flowmeter, *Proc. Roy. Soc. Edinburgh* **45** (1924–5) 308.
20. J. H. Awberry and E. Griffiths, Further experiments with the Ewing ball-and-tube flowmeter, *Proc. Roy. Soc. Edinburgh* **47** (1926–7) 1.
21. E. M. Schoenborn and A. P. Colburn, The flow mechanism and performance of the rotameter, *Trans. Amer. Instn. Chem. Engrs* **35** (1939) 359.
22. V. P. Head, An extension of rotameter theory and its application in new practical fields, *Instrument Practice*, Dec. 1946, 64 and Feb. 1947, 135.
23. J. J. Martin, Calibration of rotameters, *Engng Progress* **45** (1949) 338.
24. J. D. Hougen, Prediction of gas-flow performance of rotameters, *Instruments* **26** (1935) 1716.
25. M. C. Coleman, Variable area flow meters, *Trans. Instn Chem. Engrs* **34** (1956) 339.
26. H. H. Dijstelbergen, The dynamic behaviour of rotameters, Doctorial Thesis, Delft University, 1963; see also Rotameter dynamics, *Chem. Engng Science* **19** (1964) 853.
27. R. Poole and A. W. Leadbeater, Development in the measurement of air-flow in mines, *Trans. Instn Mining Engrs* **97** (1939) 225.
28. M. R. Head and N. B. Surrey, Low speed anemometer, *J. Scient. Instrum.* **42** (1965) 349.
29. M. R. Head and R. R. Thorpe, Direct reading low speed anemometer, *J. Scient. Instrum.* **42** (1965) 811.
30. H. Gehre and J. M. A. Smits, A turbine-type gas flowmeter, *Flow Measurement in Closed Conduits*, paper G–3, Vol. II, 701, H.M.S.O., Edinburgh, 1962.
31. J. W. Powell, Propeller meter with a gas bearing, *Engng* **192** (1961) 566.

32. F. L. FREELAND, (Discussion of ref. 30), *ibid.* 744.
33. E. A. SPENCER, Developments in industrial flowmetering, *Chem. and Process Engng* **44** (1963) 297, 305.
34. D. J. MYLES and P. HARRISON, *A Survey of Turbine Flowmeters*, N.E.L. Rep. 91, 1963.
35. F. C. L. VAN VUGT [Method of measuring the rate of mass flow of streams of liquids or gases in closed systems], *de Ingenieur* **62** (1950) 05.
36. D. BRAND and L. A. GINSEL. The mass flow meter. A method for measuring pulsating flow, *Instruments* **24** (1951) 331.
37. F. SCHULTZ-GRUNOW, Durchflussmessverfahren für pulsierende Strömungen, *Forsch. IngWes* **12** (1941) 117.
38. B. FISHMAN and G. BLOOM, Orifice meter that measures time mass flow, *Proc. Instrum. Soc. Amer.* **17** (1962) paper 16.4.62.
39. V. A. ORLANDO and F. B. JENNINGS, Momentum principle measures mass rate of flow, *Trans. A.S.M.E.* **76** (1954) 961.
40. D. W. TIMS and P. L. PALMER, Gas flow measurement: a guide to current methods and their limitations, *Research and Development* **3** (1961) 100, 105.
41. Y. T. LI and S.-Y. LEE, A fast response true-mass-rate flowmeter, *Trans. A.S.M.E.* **75** (1953) 835.
42. T. P. FLANAGAN and D. E. COLMAN, Mass flowmeters, *Control* **7** (1963) 242.
43. H. A. FITZHUGH, G. G. TWIDLE, and J. R. RICHARDSON, Proposals for the measurement of volume flow by ultrasonic methods, N.P.L. Mar. Sci. Report 3–71 (1971).
44. K. I. JESPERSON, A review of the use of ultrasonics in flow measurement, N.E.L. Report 552 (1973).

CHAPTER XII

THE MEASUREMENT OF PULSATING FLOW

It is often necessary to measure air speed or quantity when the flowrate is not steady, but pulsates with constant or approximately constant frequency and amplitude on either side of an average value; and the problem is to measure the time average of the flowrate. We must distinguish between this pulsating flow and the turbulent flow that always occurs in pipes and ducts when the critical Reynolds number is exceeded (see Chapter V). As explained in Chapter III, we can regard this turbulent flow as comprising a main flow of speed v along the pipe, on which is superposed a random turbulent velocity having a component $\pm v_x$ in the same direction. In "steady" turbulent flow v does not change with time, and the time average of v_x (taken over a sufficiently long interval) is zero; in pulsating flow as considered in this chapter, in which the pulsation frequency is assumed to be low compared with the turbulence frequency, the time average of v_x is still zero, but v itself changes with time. It is the time average of v that we wish to measure.

If we determine this average v accurately, we shall obtain the same value whatever the interval over which it is measured, provided that this interval extends over one or more complete cycles. There is, however, another type of flow, which we may term fluctuating flow, in which the variations of the flowrate are not entirely random, yet have neither definite periodicity nor constant amplitude; hence the quantity of fluid passing a given section in a given time will not be constant, and it is meaningless to talk about an average flowrate in such cases. If the fluctuations are not large, an approximate "average" rate of flow can probably be obtained by using one or other of the methods to be described in this chapter; but it is not possible to give more precise advice because of the variations in frequency and in excursions from the "average" that may occur in fluctuating flow, and we shall not further discuss this type of flow.

On the other hand, although the technique of measuring pulsating (as distinct from fluctuating) flow, despite the great amount of effort that has been devoted to it, is not yet entirely satisfactory, some progress has been made. If the period of the pulsations is long compared with the time needed to make an observation, and the amplitude is not too great, there is no inherent difficulty in making a sufficient number of measurements of the flowrate during a cycle to enable the time average of the flow to be established with reasonable accuracy. But when the frequency of the pulsations is high, difficulties may arise from two entirely separate causes, between which it is important to distinguish.

As was pointed out in Chapter I, we have to rely in air-flow measurements on observations of some effect – e.g. a pressure difference – produced by the motion and characteristic of it. The two possible causes of error in the measurement of pulsating flow are, firstly, that the time average of the effect we observe may not be related to the flowrate in the same way as it is in steady flow; and, secondly, that even if the relation is the same, certain properties of the instrument system we use to measure or record that effect – e.g. the inertia of a vane anemometer or the lag, damping, and inertia in a manometer and its connexions – may prevent its average reading from being the true time average of the effect we wish to measure.

To illustrate the problem, assume that the velocity fluctuates about an average velocity $\bar{v}$ according to the sine law $v = \bar{v}(1 + \lambda \sin \omega t)$; also let E represent the physical quantity we wish to measure to deduce $\bar{v}$. Suppose, firstly, that $E = a + bv$, where a and b are constants and v is the velocity at time t.

Then
$$E = a + b\bar{v}(1 + \lambda \sin \omega t), \tag{1}$$

and the time average of E over a cycle is

$$\bar{E} = \frac{\omega}{2\pi}\int_0^{2\pi/\omega} [a + b\bar{v}(1 + \lambda \sin \omega t)]\, \mathrm{d}t = a + b\bar{v}. \tag{2}$$

Thus when E bears a linear relationship to $\bar{v}$, i.e. in a linear-output meter, we can use the steady-flow equations to calculate the average rate of flow from the time average of E. But we still have to satisfy ourselves that the instrument or measuring system does record the time average of E.

The great majority of air-flow measurements depend on measurements of a differential pressure p, which, in incompressible flow, varies not as the first power of the speed but as its square. In these cases

$$p = av^2 = a\bar{v}^2(1 + \lambda \sin \omega t)^2;$$

and the time average of p is given by

$$\bar{p} = \frac{\omega}{2\pi} a\bar{v}^2 \int_0^{2\pi/\omega} (1 + \lambda \sin \omega t)^2\, \mathrm{d}t = a\bar{v}^2\left(1 + \frac{\lambda^2}{2}\right). \tag{3}$$

Thus, for this sinusoidal velocity variation, the time average of the pressure is higher by the factor $[(1 + \lambda^2/2)]$ than it would be if the motion were steady. In other words, if we were to apply the steady-flow relationship to determine $\bar{v}$ from the observed time average of the pressure difference (on the assumption that this average pressure is correctly indicated by the manometer), we should obtain a value for $\bar{v}$ which would be too high by the factor $(1 + \lambda^2/2)^{1/2}$. The error is not large for small or moderate values of the amplitude λ – it is only 1 per cent when λ is 20 per cent – but it will depend also on the wave form of the pulsations.

Errors due to the manometer itself are also possible. The preceding paragraphs give the time average of the differential pressure at the pick-up points in the system, e.g. the pressure tappings of an orifice plate. The resistance, inertia, and natural frequency of the manometer and the resistance of its leads will determine whether or not the indicated pressure difference is equal to the time average $\bar{p}$ acting at the pick-up points.

A great deal has been written about the measurement of pulsating flow, mainly in relation to the use of orifice plates, probably the most extensively employed of all methods of flow measurement in industry. A comprehensive summary of the literature was published in 1955[1] and the general conclusions then drawn need little alteration in the light of later publications. Most of the investigations have been concerned with the estimation of the time average $\bar{p}$ of the differential pressure at the pick-up points in orifice installations for different types of pulsation, and not with the separate problem of how the measurement of $\bar{p}$ is affected by the characteristics of the manometer and its connexions.

Measurements with Orifice Plates

On the assumption that the discharge coefficient of an orifice is the same as in steady flow – but there is evidence[2, 3] that there may be differences of the order of 1 per cent or more – it is theoretically possible to calculate $\bar{p}$ if the waveform is known. But this fact is of little practical value since elaborate equipment is usually required to establish the waveform. The most hopeful practical expedient seems to be to reduce the amplitude of the pulsations at the pick-up points to values which several sets of experiments have shown to be associated with small limiting errors. The pulsations can usually be sufficiently reduced by inserting throttling devices into the flow line between the orifice and the source of the pulsations, but only at the expense, very often, of unacceptably high pressure losses. It has been found that large drums or receivers in place of the throttling devices are also effective, and much less costly in pressure loss. In practice this alternative is preferred whenever possible, supplemented sometimes by throttling if the pulsations are so violent that the receiver volume required would be impracticably large (see also pp. 316–18). The necessary volume can be calculated by the use of a non-dimensional parameter known as the Hodgson number after J. L. Hodgson, who was the first to draw attention to its significance in pulsating-flow problems as far back as 1922.[4] It is a tribute to Hodgson that the authors of ref. 1, published over 30 years later, state in their conclusions: "The characteristics of the receiver volume have been firmly established on the basis of the Hodgson number which, according to available evidence, can be accepted as a reliable design criterion."

The Hodgson number H is given by

$$H = \frac{Vf\,\Delta p \varrho}{Mp}, \tag{4}$$

where V is the volume of the receiver, f is the frequency of the pulsations, Δp is the time-average of the pressure difference across the orifice, M is the mass of fluid flowing per second, p is the pressure in the receiver, and ϱ is the mass per unit volume of the fluid in the receiver at pressure p.

The Hodgson-number criterion has been established mainly by investigations undertaken to develop a reliable technique for measuring the air consumption of internal combustion engines or the delivery from air compressors – both cases in which the flow is cut off completely by valves during parts of each cycle. Theoretical and experimental work of this nature by Hodgson,[(4)] Lutz,[(5)] and Herning and Schmid[(6)] indicate that the error in the flowrate will not exceed 1 per cent if $H \nless 1{\cdot}5$, or 0·5 per cent if $H \nless 2{\cdot}3$, and the duration s of the interval during which the flow is not cut off exceeds 10 per cent of the full cycle. This is shown in Fig. 12.1 which is based on the results of the analysis of Herning and Schmid;[(6)] the results of refs. 4† and 5, which covered a smaller range of Hodgson number, agree well with those of Fig. 12.1 where they overlap. Kastner,[(7)] also working with a reciprocating engine, gave the rather more cautious criterion $H \nless 2{\cdot}5$ for a limiting error of 1 per cent.

It should be noted that the dimensional analysis from which the Hodgson-number criterion was deduced took account only of velocity pulsations through the orifice; it ignored any changes that might occur in the discharge coefficient of the orifice in pulsating flow and any errors due to pressure fluctuations in the manometer system. In the various experimental investigations that confirmed the validity of the criterion, however, any effects due to these causes were included in the overall pressure measurements, although in most cases steps were taken to reduce manometer errors as far as possible. Thus, provided that reasonable precautions (see below) are taken in the pressure measurements, the Hodgson number is a reliable practical criterion: it appears that, if the pulsations are sufficiently reduced, the errors due to effects not taken into account in the theory are negligibly small.

Most of the results on which are based the values quoted above for the Hodgson-number criterion were concerned with cases in which the engine or compressor valves reduced the flow to zero for a large portion of each cycle. In many practical cases the flow, although pulsating, possibly with large amplitude, will be zero for much briefer intervals, if at all. In such cases much smaller

† It is pointed out in ref. 1 that the values of H originally given by Hodgson in ref. 4 are too large by a factor of 60.

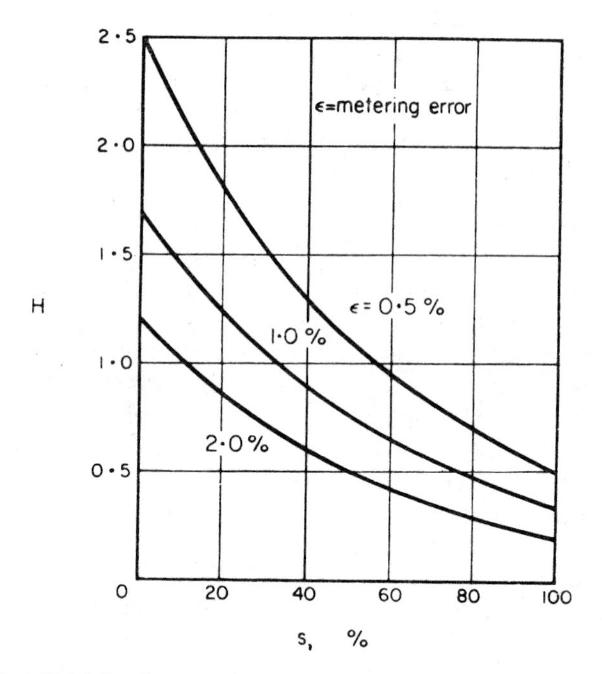

FIG. 12.1. Metering error for various Hodgson numbers and cut-offs.

Hodgson numbers are allowable to limit the errors. This is demonstrated by Fig. 12.2, which shows data given by Schmid[8] for pulsations in which the flow is never zero except instantaneously once per cycle for the limiting case $q_{max/min} = \infty$, where q is the flowrate. It will be seen from this diagram that, for a given metering error, the Hodgson numbers are all considerably less than those shown in Fig. 12.1, except when $s = 100$ per cent, which corresponds to the curve $q_{max}/q_{min} = \infty$ in Fig. 12.2. Another conclusion that can be drawn from all the researches is that the relationship between the metering error and the Hodgson number is much less sensitive to wave shape than to the pulse-duration percentage s. More recent metering-error data have been presented in ref. 9; the conclusions of that paper, however, were partially contradicted by a subsequent investigation,[10] which was itself criticized in certain particulars in ref. 11. In short, no results of general applicability appear to be available for relating the true mean flowrate to that indicated by the meter. These considerations emphasize the wisdom of reducing the amplitude of the fluctuations before the flow reaches the meter rather than attempting to apply a correction to allow for their effects.

If, however, we take into account the more cautious value for the Hodgson criterion recommended by Kastner,[7] it seems reasonable to assume that, if the mean rate of flow is never zero, we can safely take $H \not< 0{\cdot}6$ for the metering

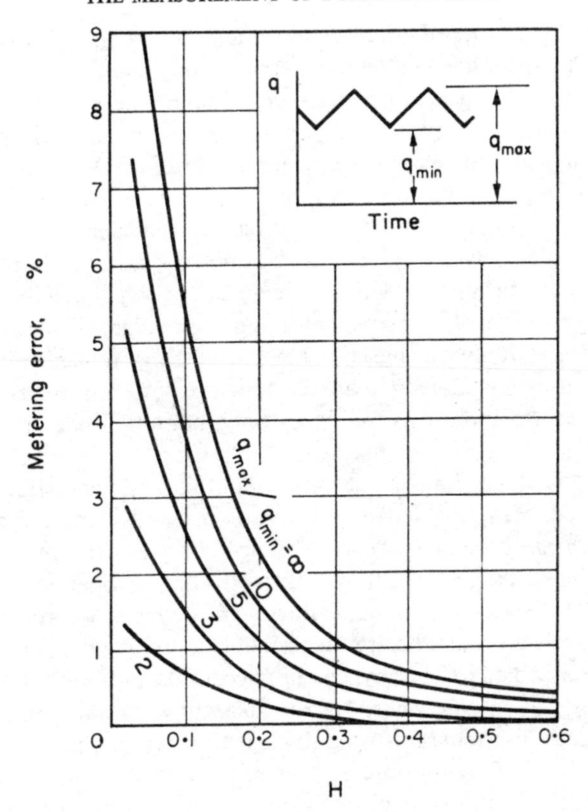

FIG. 12.2. Metering error for various Hodgson numbers and ratios of maximum to minimum flowrate.

error not to exceed 1 per cent, and $H \nless 0{\cdot}8$ for the error not to exceed 0·5 per cent.

But to make use of this criterion in practice presents some difficulties, even apart from the fact that the volume V of the receiver required to make H equal to the appropriate value according to (4) may be impossibly large. We cannot use this equation to calculate V unless we know all the other quantities that occur in it, including, in particular, the frequency f and the mass flow M. The latter cannot be measured accurately until the system includes the receiver of volume V necessary to reduce the pulsation amplitude to the required degree. We start, therefore, by guessing V: we insert in the circuit a receiver of reasonable size between the meter and the source of the pulsations, and then find out whether the pulsations have been substantially reduced. This can be ascertained quite simply by a method of detecting pulsations suggested by Watson and Schofield,[12] used by King,[13] and elaborated into a pulsation-measuring instrument by Kastner.[7] A rubber diaphragm about 0·6 mm thick is stretched, not too tightly, and clamped across one end of a short tube of about 8-cm diameter fixed to the receiver: the other end of this tube is in communication with the inside of the receiver. The diaphragm will be bowed inwards or outwards by the mean suction or pressure in the receiver, and in this position should show no more than a trace of vibration.†

When this condition has been satisfied, the pressures and the mass flow M that occur in (4) will probably be measurable with sufficient accuracy to enable the equation to be solved to give a second approximation to V, provided that the frequency can be measured. Sometimes f can be calculated from characteristics (such as the revolutions per minute of an engine) of the source of the pulsations. If not, it must be measured by one of the various methods available based on the use of an indicating or recording pressure-transducer. The system need not be very elaborate as frequency only, not amplitude or waveform, is required.

Other criteria, depending on pulsation amplitude and not on frequency, have been suggested. Head[17] derived an approximate "pulsation factor" representing the ratio of the mean flowrate indicated by a meter in pulsating flow to the true time average of the flow. The factor itself is open to criticism, as is apparent from the discussion that followed the delivery of Head's paper; but one of the conclusions that he drew is probably a useful guide, namely that the factor can be assumed to be unity, i.e. the effect of the pulsation can be ignored, if the ratio $(q_{max} - q_{min})/\bar{q}$, where $\bar{q}$ is the mean flowrate, does not exceed 0·1. For sinusoidal pulsations, this is equivalent to a limiting value of λ in (3) of 0·05, which, in a differential-pressure meter, will lead to an error of slightly over 0·1 per cent on pressure, or 0·05 per cent on flowrate. It seems probable, therefore, that Head's criterion is on the safe side – too much so, perhaps. Methods of measuring pulsation amplitudes are described in refs. 7, 14, and 15; one of these measures the maximum displacement in either direction of the rubber diaphragm used in the pulsation indicator of Watson and Schofield.[12]

As will be seen from the above, we must accept the fact that, even with the simplification made possible by the use of the Hodgson number or Head's criterion, the accurate measurement of pulsating flow by means of orifice plates is troublesome, largely because a knowledge of pulsation frequency or amplitude is necessary. It becomes easier, however, if possible errors somewhat larger than 1 per cent can be tolerated: in that case it will probably suffice to insert as large a receiver as is necessary to ensure, by reference to a simple pulsation detector such as that of Watson and Schofield, that pulsations have been reduced to a small amount.

Sometimes, however, e.g. when metering high rates of flow of air or gas at high pressure, when M and p in (4) will both be large, the value of V necessary to satisfy the Hodgson criterion may be impracticably large. We then have to

† Other types of pulsation indicator have been suggested, see refs, 14, 15, and 16.

measure the flow by some other device, such as a mass flowmeter (see below), or we can adopt an ingenious type of meter due to Isobe and Hattori.[18] The basic principle of this meter is that the pulsating flow to be measured is caused to generate a secondary air flow which is a definite and known fraction of the main flow; and it is this secondary air flow that is metered. It is subject to exactly the same pressure pulsations as the main flow; but, because it is arranged to be much smaller than the main flow, the receiver volume necessary to achieve the desired Hodgson number for it to be accurately measured is so small that difficulties due to size no longer exist.

Figure 12.3 shows diagrammatically how this meter operates. The upstream and downstream pressures p_1 and p_2 from orifice A in the main flow are transmitted to the upper portions of two compartments B and C which Isobe terms pulsating-flow (PF) transmitters. Each of these is divided into two by means of a slack, impervious membrane, and their lower parts are connected through the pipe D which contains a secondary orifice E. This pipe, together with the lower portions of the PF transmitters, forms part of the secondary-air circuit with atmospheric air entering through the pipe PQ and flowing back to atmosphere through the pipe RS. A pump ensures that the outlet pressure at S is below atmospheric.

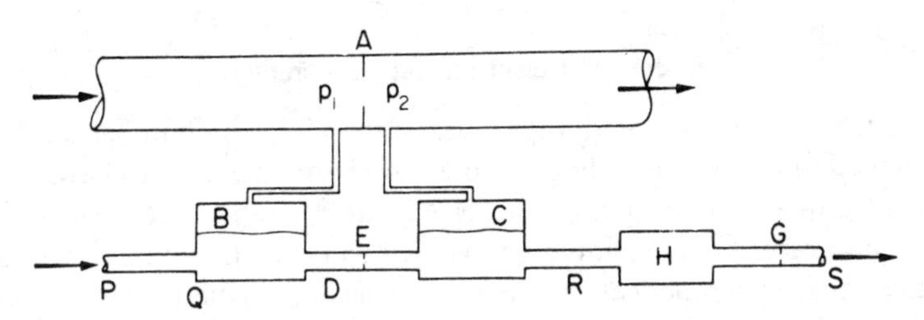

FIG. 12.3. Principle of Isobe flowmeter.

The arrangement of the slack diaphragms and the design of the inlet and outlet ports of the PF transmitters are such that the pressures p_1 and p_2 are accurately transferred to the orifice E, which is therefore subjected to pressure pulsations identical with those of the main orifice. Therefore, on the assumption that pulsations, if they affect orifice discharge coefficients at all, do so equally for both the orifices A and E,† the relation between the main flowrate and the secondary flowrate can be calculated from the geometry of the two orifice installations and measurement of the densities of the fluids in each circuit; or, if thought desirable, established by a calibration under steady conditions. The same factor is then applied to measurement of the secondary flow in pulsating conditions.

† From tests made by Isobe it appears that this assumption was substantially true: differences were between 1 and 2 per cent.

The secondary flow can be measured by a third orifice G in the pipe RS, or by passing the flow from the pipe through any convenient low-rate meter, e.g. a wet gas meter (see Chapter XI). Upstream of whatever metering device is selected is the pulsation-damping receiver H which, in a meter described by Isobe, had a volume of only 4 litres. The capacity of this meter is not stated explicitly, but figures cited in ref. 18 indicate that it was of the order of 3×10^4 m^3/h. The auxiliary orifice E had a diameter of 1 mm, and the ratio of the main flow being metered to the secondary flow through orifice E was about 35,000.

Constructional details of this meter are given in ref. 18. Errors were of the order of 1 per cent.

Another approach to flow measurement when the Hodgson criterion cannot be satisfied is to make use of the choked nozzle (p. 189). Since the reading of this device is independent of downstream conditions, it is evidently well suited to determining the flowrate in the presence of pulsations originating downstream of the meter. In addition, the reading has been found to be affected very little by upstream pulsations of quite large amplitude.[19]

It should be noted that this method can be adopted only if the large pressure loss introduced by a choked nozzle is acceptable, and if the choked condition can be maintained throughout the pulsation cycle.

Effects of Pulsations on Manometers

Although, as already stated, the values of the Hodgson number established for specified limiting errors allow for possible errors due to the characteristics of the manometer and the pressure leads, it is still desirable that sources of error in the manometer system should be avoided as far as possible. A number of investigations, both practical and theoretical, have been made of the behaviour of liquid manometers under the action of pulsating pressure.[9, 20, 21] Effects due to wave action, volumes, and restriction or throttling, all arising in the leads, will cause the pressures transmitted to the surfaces of the manometric liquid to differ from the pressures at the pick-up points; and the final readings will be further affected by damping and inertia of the liquid. Two possible sources of error therefore have to be considered.

(a) *Effects of Pressure Leads*

At one time it was thought that by the insertion into the pressure leads of viscous damping, in which the flow would be laminar, the fluctuations in pressure transmitted to the manometer would be reduced to a negligible amount, and that the mean pressures so transmitted would be the true time averages of the pressures at the pick-up points. This can be deduced from eqn. (17) of Chapter V which shows that the relation between the pressure drop and the flowrate is linear in laminar flow. But that equation was based on the assumption of constant density ϱ; and, as pointed out by Williams,[21] if the amplitude of the pressure pulsations is so great that ϱ varies appreciably throughout the cycle, viscous damping in the leads can give rise to large errors.† Ideally, the leads should be very short; Kastner and Williams[20] describe a sensitive two-liquid manometer in which the air-filled portions of the leads between the pressure pick-up points and the manometer-liquid surfaces were only about 1 cm long.

Usually, however, the leads will, for various reasons, be longer; they should then be of uniform bore and, provided that the pressure fluctations are not too large, viscous damping may be used with success, e.g. in the form of capillary tubing. The viscous element should offer the same resistance to motion in either direction; hence, if capillary tubing is used it should be at least 100 diameters long to reduce possible errors due to dissimilarity between its two ends. If the leads are long, the resistance of the viscous element should be high to prevent wave action in the leads. Other forms of viscous element, such as felt pads[22] or plugs of cotton wool or steel wool[23] in the leads, have been used or suggested, but Williams considers their efficacy doubtful except for small pressure pulsations; the last two of these devices showed linear flow characteristics only for very small rates of flow.[23]

(*b*) *Manometer Errors*

The response of a liquid manometer to an applied pulsating pressure, on the assumption that all connecting-lead errors have been eliminated, has been discussed in a number of papers.[1, 7, 9, 21, 24] Theoretically, this can be calculated if the pulsation characteristics are known; but, since this knowledge is generally not available, the point of practical interest is how errors due to this cause can be avoided. Viscous damping of the manometer liquid is recommended by Williams[21] and Bonnington,[24] and a viscously damped manometer has been successfully used by Kastner[7] in conditions of severe pulsation. The damping should be confined to the manometer liquid and, as in the leads, should offer the same resistance to flow in either direction. In no case should a device such as a tap or valve be used; capillary tubing again is a suitable damping element.

The Viscous Flowmeter

The principle of viscous damping is the basis of the Alcock Viscous Flow Air Meter,[22, 25, 26] made by Ricardo & Co. Engineers (1927) Ltd., originally for

† Actually, Williams related these errors to changes in the kinematic viscosity ν $(= \mu/\varrho)$, but since μ, the viscosity, is unaffected by pressure changes (except at very high or very low pressures), the errors are in effect due to changes in ϱ.

use in their researches on internal-combustion engines. In viscous (laminar) pipe flow the pressure drop varies as the first power of the velocity (see Chapter V); hence, as shown by (2) of the present chapter, the time average of the pressure drop is equal to the pressure drop corresponding to the mean flowrate. The viscous element of the Alcock meter is a honeycomb of long, narrow, triangular passages, about 8 cm long and 0·4 mm in height, within which, throughout the working range of the meter, the flow remains laminar. This viscous element is inserted in the pipeline exactly as if it were an orifice plate or other constriction, and the pressure difference across it is transmitted to a manometer in the usual manner. The pressure tappings are fitted with felt pads to reduce manometer-connexion errors. According to the makers, experiments have shown that with a meter of this type the manometer head is practically proportional to the flow velocity; and that even very irregular flows do not cause significant error.

There is no doubt that, within its sphere, this is a useful meter that deserves wider use than it seems to have had. Unfortunately, its applications are restricted by the high loss of pressure experienced by the air flowing through it; the largest size normally available can deal with a maximum flowrate of about 14 m^3/min, at which the loss of pressure is equivalent to a head of about 33 cm of water.

A simple type of viscous flowmeter consisting basically of a porous plug of glass wool has been described by Fleming and Binder,[(23)] but, as pointed out by Alcock, its limiting rate of flow is considerably less than that of the Alcock meter.

Since the pressure drop across a viscous flowmeter for a specified volume flow is proportional to the viscosity of the air, its readings need correction for temperature changes. From the figures for the variations of viscosity with temperature given in Table 5.1 it follows that a particular meter reading will have to be reduced by about 1 per cent for every 4°C rise in temperature above that at which it was calibrated, and vice versa.

Turbine Meters

For flows in which the density ϱ does not change appreciably, the turbine meter, whose rotational speed is proportional to $v\sqrt{\varrho}$ (see p. 302), can be regarded as a linear-output meter (see p. 311). Therefore if its time constant† is sufficiently low in relation to the periodic time of the pulsations, or, in other words, if its response is sufficiently rapid, its average reading in a pulsating flow

† The time constant is defined as τ in the equation

$$f = f_0(1 - e^{-t/\tau}),$$

where f is the response at time t to a step change at $t = 0$.

will approximate closely to the average flowrate. As a rule, however, it seems unlikely that this condition can be fulfilled, even for slow pulsations, i.e. those with a comparatively high periodic time. In Chapter VIII we saw that a vane anemometer overestimates the average flow by an amount that can be predicted theoretically for rapid fluctuations; and that this overestimation was confirmed experimentally. Ower confirmed it for a frequency of only 2 Hz, and van Mill for 5 Hz. These cannot be regarded as high-frequency pulsations, and it is not easy to imagine a turbine meter with a significantly lower time constant (which depends largely on the inertia), i.e. with a more rapid response, than a vane anemometer. We are therefore forced to the conclusion that turbine meters are unlikely to indicate pulsating flow accurately: they are more likely to be subject to errors of the order of those found for vane anemometers.

However, for water flow Dowdell and Liddle[(27)] found that two turbine meters gave readings within 2 per cent of the average for rapidly pulsating flows.

Mass Flowmeters

In some types of mass flowmeter that produce a linear output in terms of the mass flow, e.g. that described on p. 304, the sensing element does not rotate, so that inertia effects are likely to be very small. Such instruments seem, therefore, to be eminently suited to the measurement of pulsating flow; and in fact the makers of a number of these meters state that they can be used for this purpose. No detailed account of any experimental verification of this statement, however, seems to have been published.

There are certain types of mass flowmeters which are basically volume meters incorporating additional metering devices to measure density, the two measurements being combined in the meter to enable it to indicate mass. Before such a type is adopted for the measurement of pulsating flow, the intending user should satisfy himself that the output of each of the two elements is linear and that any inertia effects are negligible.

Hot-wire anemometers

All the methods so far considered in this chapter are intended to measure the mean rate of flow in a pipe. If measurements of the local average of a pulsating velocity are required, these are probably best carried out by means of a hot-wire anemometer suitably connected to an oscillograph which will record the waveform. A hot wire about 6 mm long and 0·02 mm in diameter will have very small lag and enable the waveform even of rapid pulsations to be accurately recorded.

Reduction of Pulsations by Contraction of the Stream

As remarked on p. 310, the type of flow with which this chapter is concerned can be regarded as having a mean axial velocity v on which is superposed a fluctuating axial component $\pm v_x$; and attention was drawn to the fact that in this respect there is a similarity between pulsating flow and turbulent flow, in which there is a r.m.s. axial turbulent component $\pm\sqrt{(\overline{v_x^2})}$. Now it is known that the ratio $\sqrt{(\overline{v_x^2})}/v$ is reduced by a smooth contraction; there is, therefore, some *prima facie* reason for believing that a contraction will have a similar effect on the amplitude of a pulsating flow.

For turbulent flow the effect can be considerable. Theoretical analyses may be consulted in refs. 28 and 29; experimental results are given in refs. 30 and 31. The reduction in $\sqrt{(\overline{v_x^2})}/v$ arises from a reduction in $\overline{v_x^2}$ as well as the increase in v. For a contraction with area ratio 1/10, for instance, MacPhail[30] found that $\sqrt{(\overline{v_x^2})}/v$ was reduced from about 0·037 at entry to the contraction to about 0·0015 at the outlet.

There appear to be no published experimental data to show whether a contraction reduces the amplitude of a pulsating flow by comparable amounts. Until such evidence is available, all that can be said is that if a smooth contraction such as a nozzle can be fitted in a pipeline carrying a pulsating flow, and if measurements are made downstream of the contraction, conditions are likely to be improved.

References

1. A. K. Oppenheim and E. G. Chilton, Pulsating flow measurement — a literature survey *Trans, A.S.M.E.* **77** (1955) 231.
2. E. Estel, Durchflusszahl von Normdüsen und Druckabfall in Rohren bei pulsierender Strömung, *Physikalische Z.* **38** (1937) 748.
3. J. M. Zarek, Metering pulsating flow, *Engng* **179** (1955) 17.
4. J. L. Hodgson, The metering of steam, *Trans. Instn Naval Architects* **64** (1922) 184; see also *The Orifice as a Basis of Flow Measurement*, Instn Civ. Engrs Selected Engineering Papers No. 31 (1925).
5. O. Lutz, Über Gasmengenmessung bei Kolbenmaschinen mittels Düsen und Blenden, *Ingenieur-Archiv* **3** (1932) 138, 432.
6. F. Herning and C. E. Schmid, Durchflussmessung bei pulsierender Strömung, *Z.V.D.I.* **82** (1938) 1107.
7. L. J. Kastner, An investigation of the airbox method of measuring the air consumption of internal combustion engines, *Proc. Instn Mech. Engrs* **157** (1947) 387.
8. C. E. Schmid, Messfehler bei der Durchflussmessung pulsierender Gasströme, *Z.V.D.I.* **84** (1940) 596.
9. S. W. Earles and J. M. Zarek, Use of sharp-edged orifices for measuring pulsating flow, *Proc. Instn Mech. Engrs* **177** (1963) 997.
10. J. Grimson and N. Hay, Errors due to pulsation in orifice meters, *Ae. J. R. Ae. S.* **75** (1971) 284.
11. B. J. Jeffery, Comments on Grimson and Hay (1971), and Authors' reply, *Ae. J. R. Ae. S.* **76** (1972) 40.
12. W. Watson and H. Schofield, On the measurement of the air supply to internal com-

combustion engines by means of a throttle plate, *Proc. Instn Mech. Engrs* **19** (Parts 1–2) (1912) 517.
13. R. O. King, The measurement of air flow by means of a throttle plate with special reference to the air supply to internal combustion engines, *Engng* **115** (1923) 456, 481.
14. S. R. Beitler, E. J. Lindahl, and H. B. McNichols, Developments in the measuring of pulsating flow with inferential head meters, *Trans. A.S.M.E.* **65** (1943) 353.
15. E. J. Lindahl, Pulsation and its effect on flowmeter, *Trans. A.S.M.E.* **68** (1946) 883.
16. R. Houghton, The detection of pressure fluctuations, *Engng* (1937) 425.
17. V. P. Head, A practical pulsation threshold for flowmeters, *Trans. A.S.M.E.* **78** (1956) 1971.
18. T. Isobe and H. Hattori, A new flowmeter for pulsating gas flow, *I.S.A. Journal* **6** (1959) 38.
19. L. J. Kastner, T. J. Williams, and R. A. Sowden, Critical-flow nozzle meter and its application to the measurement of gas flow rating in steady and pulsating streams of gas, *J. Mech. Engng Science* **6** (1964) 88.
20. L. J. Kastner and T. J. Williams, Pulsating flow measurements by viscous meters, *Proc. Instn Mech. Engrs* **169** (1955) 419.
21. T. J. Williams, Pulsation errors in manometer gauges, *Trans. A.S.M.E.* **78** (1956) 1461.
22. J. F. Alcock, Contribution to discussion of paper by N. P. Bailey, Pulsating air velocity measurements, *Trans. A.S.M.E.* **61** (1939) 301.
23. F. W. Fleming and R. C. Binder, Study of linear resistance flowmeters, *Trans. A.S.M.E.* **73** (1951) 621.
24. S. T. Bonnington, *Investigation of the Measurement of Pulsating Flow*, British Hydromechanics Research Association Publication, R.R. 466 (1953).
25. British Patent 473139 (1937).
26. J. F. Alcock, Contribution to discussion of ref. 23.
27. R. B. Dowdell and A. H. Liddle, Measurement of pulsating flow with propeller and turbine type flowmeters, *Trans. A.S.M.E.* **75** (1953) 961.
28. G. I. Taylor, Turbulence in a contracting stream, *Z. f. angewandte Mathematik u. Mechanik* **15** (1935) 91.
29. G. K. Batchelor, *The Theory of Homogeneous Turbulence*, Cambridge University Press, 1953.
30. D. C. MacPhail, Turbulence changes in contracting and distorted passages, *R. & M.* 2437 (1951).
31. M. S. Uberoi, Effect of wind-tunnel contraction on free-steam turbulence, *J. Ae. S.* **23** (1956) 754.

CHAPTER XIII

EXAMPLES FROM PRACTICE

IN THIS chapter, some typical examples are given from practice of some of the methods of measuring air flow described in the foregoing pages, the methods of calculation being shown in detail. These examples are followed by a few additional practical notes.

Measurement with a Pitot–static Tube of the Quantity of Air Flowing Along a Pipe

In this hypothetical example of the application of the pitot-traverse methods of Chapter III, it is assumed that an estimate is required, subject to an error not exceeding about 2 per cent, of the quantity of air flowing along a circular pipe of 40 cm diameter. The pitot–static combination is of standard type, the differential pressure being equal to the velocity head without correction. The manometer used is of the inclined-tube type, graduated in inches. It is assumed to have been previously calibrated, with the result that a mean factor of 0·351 has been found applicable over the whole scale to convert the readings to centimetres of water.

We shall use the ten-point log-linear method of numerical integration described in Chapter VI, according to which we have to divide the pipe into five zones of equal area and take velocity readings in each zone along two mutually perpendicular diameters. In all, twenty readings will be taken, ten along each diameter; and the distances from the pipe wall of the ten measuring points on each diameter are (see Table 6.1): 0·76, 3·04, 6·12, 8·68, 14·44, 25·56, 31·32, 33·88, 36·96, and 39·24 cm. The velocity-head readings obtained on the manometer are 6·72, 7·75, 8·74, 8·70, 8·40, 7·12, 6·55, 5·66, 4·40, and 2·56 on one of the two diameters and 4·28, 5·85, 6·96, 7·13, 7·42, 7·38, 7·02, 6·88, 5·82, and 4·15 on the other. In order to convert these readings of velocity head into centimetres of water, so that we can use the formulae of Chapter VI, we must multiply them by 0·351, the calibration factor of the manometer. Equation

(13) of Chapter VI gives the velocity head h_m corresponding to the mean velocity in the pipe, and in this case it becomes

$$\sqrt{h_m} = \frac{1}{20}(\sqrt{6{\cdot}72} \times \sqrt{0{\cdot}351} + \sqrt{7{\cdot}75} \times \sqrt{0{\cdot}351} + \ldots$$

$$+ \sqrt{4{\cdot}28} \times \sqrt{0{\cdot}351} + \sqrt{5{\cdot}85} \times \sqrt{0{\cdot}351} + \ldots)$$

$$= \frac{\sqrt{0{\cdot}351}}{20}(\sqrt{6{\cdot}72} + \sqrt{7{\cdot}75} + \ldots + \sqrt{4{\cdot}28} + \sqrt{5{\cdot}85} + \ldots).$$

Thus instead of multiplying each reading by the calibration factor and extracting the square root of the product, we save work by summing the roots of the readings and multiplying the sum by one-twentieth of the root of the calibration factor. Thus

$$\sqrt{h_m} = 0{\cdot}0296 \times 50{\cdot}44 = 1{\cdot}493.$$

The temperature of the air in the pipe is 50°C and the barometer stands at 764 mm. The static pressure in the pipe is less than 5 cm of water. We use eqns. (19) and (20) of Chapter VI to get V, the volume, or Q, the mass, of air flowing, in cubic metres or kilograms per minute respectively. A, the area of the pipe, is $0{\cdot}1257$ m^2.

We have

$$V = 0{\cdot}1257 \times 1234\sqrt{\left(\frac{273+50}{764}\right)} \times 1{\cdot}493 \text{ m}^3/\text{min},$$

$$Q = 0{\cdot}1257 \times 573\sqrt{\left(\frac{764}{273+50}\right)} \times 1{\cdot}493 \text{ kg/mm},$$

i.e.

$$V = 150{\cdot}5 \text{ m}^3/\text{mm},$$

$$Q = 165{\cdot}4 \text{ kg/min}.$$

If the mean static pressure were, say, 30 cm of water below atmospheric pressure, we should have to replace the figure 764 by 764 − 300–13·56, i.e. by 742. Neglect of this correction would lead to an error of about 1·5 per cent in both V and Q. As the static pressure is actually less than 5 cm of water, however, the error is about 0·2 per cent, which is well within the limits desired. If greater accuracy had been required this correction, as well as those for turbulence, wall-proximity, and transverse velocity-gradient effects (pp. 128–32) and for humidity of the air (pp. 337–9) would have had to be applied.

Calibration of a Small Plate Orifice

This example affords a useful illustration of more than one of the points discussed in earlier chapters, and will consequently be treated at some length. The conditions to be met were as follows: for certain purposes, which are not material to this discussion, a stream of air was to be maintained in a pipe of nominally 2·54 cm diameter by means of a small fan; and it was required to provide a ready method of measuring the rate of flow, and of maintaining it at any steady value between limits of 1·5 and 3 m/sec. Further, the scheme adopted was to be such that it would permit the rate of flow to be determined rapidly.

The low rate of flow, coupled with the condition last mentioned, imposed restrictions on the choice of method; and it was decided that the use of some constricting device permanently connected to a sensitive, but not too delicate, differential manometer was the most suitable scheme to employ. Space was too limited to allow sufficient length of pipe upstream and downstream of the constriction for a standard constriction to be used without calibration. It was evident, therefore, that whatever form of constriction was used would have to be calibrated *in situ*. Hence it was decided to use a plate orifice, as being the type of constriction occupying the least space and the cheapest and simplest to install. Moreover, in view of the necessity for calibration, no attempt was made to reproduce any standard form, the orifice consisting merely of a reasonably square-edged circular hole cut in a 1·6-mm brass plate, which was clamped in flanges fixed near the outlet of the pipe. There was a length of straight pipe about 12 diameters long upstream of the plate, and a length of about 3 diameters downstream. Pressure taps for the manometer were provided fairly close to the plate, without any care as to their exact position.

An inclined-tube manometer was chosen on the score of simplicity and cheapness for indicating the orifice head; and it was decided that a convenient length of liquid column for the lowest rate of flow (1·5 m/sec) would be about 5 cm. It seemed reasonable to suppose that the length of the column could easily be read to 1 mm, so that the pressure corresponding to 1·5 m/sec could be read to within 2 per cent. And since the speed is proportional to the square root of the pressure, the accuracy on speed at the lower end of the scale would be within 1 per cent, which was ample for the purpose in view. Alcohol was selected as the manometric liquid for reasons explained in Chapter X; it was coloured with a blue aniline dye in order to make the meniscus more prominent. The slope of the inclined tube was set to about 1 in 10.

Calculations of the necessary size of the orifice were then made as follows: From eqn. (12) of Chapter VII we have

$$a_1 \varrho v_1 = q = \alpha a_2 \sqrt{\left[\frac{2\varrho(p_1 - p_2)}{1 - m^2}\right]}.$$

Remembering that $m = a_2/a_1$, we may reduce this equation to the following form:

$$\frac{a_1}{a_2} = \sqrt{\left[\frac{2\alpha^2(p_1 - p_2)}{\varrho v_1^2} + 1\right]}, \tag{1}$$

which enables us to calculate the area of the orifice a_2 in terms of the area of the pipe a_1 if we know v_1, the velocity in the pipe, and the pressure drop $p_1 - p_2$.

In the case under consideration, $p_1 - p_2$ had been fixed at 5 cm of alcohol in the inclined tube for a speed v_1 of 1·5 m/sec. In (1), $p_1 - p_2$ must be expressed in pascals (N/m^2) if the SI system of units is used. The slope of the monometer tube was 1 in 10 and the relative density of alcohol could be taken as 0·8 approximately.

Hence

$$\begin{aligned} p_1 - p_2 &= 5 \times 1/10 \times 0{\cdot}8 \text{ cm of water} \\ &= 5 \times 1/10 \times 0{\cdot}8 \times 98{\cdot}1 \text{ Pa} \\ &= 39{\cdot}2 \text{ Pa.} \end{aligned}$$

For the approximate calculation of a_1/a_2 it was sufficiently accurate to assume α for the orifice to be 0·6.

Hence (1) became

$$\frac{a_1}{a_2} = \sqrt{\left[\frac{2 \times 0{\cdot}36 \times 39{\cdot}2}{1{\cdot}225 \times 2{\cdot}25} + 1\right]} = 3{\cdot}35,$$

so that d_2, the diameter of the orifice, $= 2{\cdot}54/\sqrt{3{\cdot}35} = 1{\cdot}39$ cm, since the diameter of the pipe was 2·54 cm. Actually, the orifice diameter was made 1·27 cm. Before this dimension could be finally accepted, however, it was necessary to see whether the irrecoverable loss of head that occurs with an orifice was beyond the capacity of the fan available. At the maximum delivery, 3 m/sec, the pressure drop through the orifice will be four times that used in the calculation of a_1/a_2, since the calculation was made for a speed of 1·5 m/sec and $(p_1 - p_2)$ varies as the square of the speed. Thus we obtain at 3 m/sec $(p_1 - p_2) = 4 \times 0{\cdot}5 \times 0{\cdot}8 = 1{\cdot}6$ cm of water. This value requires correction, however, since it involves the assumption that $a_1/a_2 = 3{\cdot}35$, whereas by making the orifice 1·27 cm diameter the ratio actually becomes 4. We see from (1) that $(p_1 - p_2)$ varies as $(a_1/a_2)^2$; hence the true pressure drop at 3 m/sec becomes $1{\cdot}6 \times (4/3{\cdot}35)^2 = 2{\cdot}28$ cm of water. As stated in Chapter VII, the head lost may be taken as $(1 - m)$ times the drop through the orifice, i.e. the head lost will be $\frac{3}{4} \times 2{\cdot}28$ or 1·71 cm of water. The maximum head to be supplied by the fan consists of this lost head through the orifice, the frictional resistance of the total length of the pipe, and the velocity head equivalent to the speed of the air issuing from the open end of the pipe remote from the fan. This head must be included because the energy in the issuing jet is dissipated; it is only recovered when the piping is part of a closed circuit, i.e. when the far end of the pipe is connected back to the fan intake, or if a diffuser is fitted at the outlet.

The maximum velocity head ($\frac{1}{2}\varrho v^2$) to be supplied by the fan is $\frac{1}{2} \times 1{\cdot}225 \times 3^2 \div 98{\cdot}1 = 0{\cdot}056$ cm of water. The head necessary to overcome pipe friction is obtained thus: the value of vd/ν for the conditions of maximum delivery is (3 × 0·0254/0·000015) approximately, or about 5000. From Fig. 5.3, we see that, at this value of vd/ν, γ for smooth pipes is about 0·0095. Also, from eqn. (4) of Chapter V, $(p_A - p_B) = \varrho\gamma p v^2 l/d$, where p_A and p_B are the static pressures at the two ends of the length l of the pipe of diameter d. In the case under consideration, l/d is about 15; thus, since the velocity head $\frac{1}{2}\varrho v^2$ is 0·055 cm of water, the pressure drop due to friction becomes $(p_A - p_B) = 4 \times 15 \times 0{\cdot}0095 \times 0{\cdot}055 = 0{\cdot}031$ cm of water. It should be noted that this result

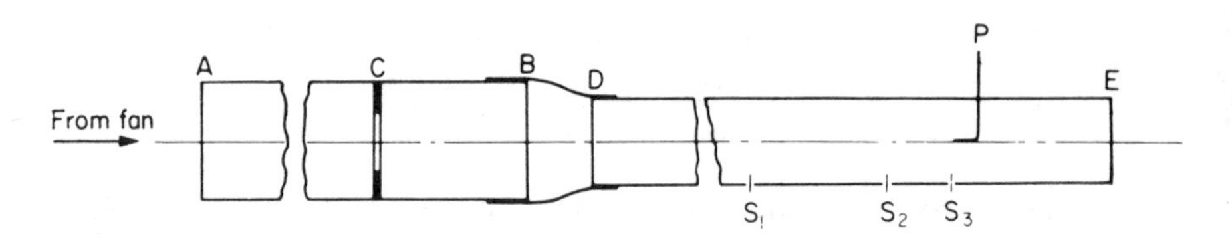

FIG. 13.1. Arrangements for calibrating small plate orifice.

is only an approximation, since the numerical data used relate to long smooth pipes. But even if the pressure loss in the pipe had been many times greater than the loss calculated in this way, say 0·2 cm of water, the total fan head required at the highest speed would have been only 1·71 + 0·06 + 0·2, i.e. about 2 cm of water. This was well within the capacity of the fan available, which was a small fan of the type used in hair-dryers. The proposed orifice size was therefore suitable; and an orifice plate was accordingly made to the specified dimensions.

The next step was to calibrate the orifice over the required speed range. For this purpose, a 1·27-cm (nominal) diameter smooth brass calibrating pipe (see Chapter VI) was used, in which there was provision for traversing a section with a small hypodermic total-pressure tube actuated by a micrometer screw (see Fig. 6.13). A diagrammatic sketch of the arrangement as used for the orifice calibration is given in Fig. 13.1. AB is the 2·54-cm diameter pipe, with the orifice at C, and DE is the 1·27-cm diameter calibrating pipe, connected to the end B of the 2·54-cm pipe by means of a reduction cone and rubber tubing. P is the traversing pitot, and S_1, S_2, and S_3 represent static-pressure taps leading to side holes flush with the inside of the brass tube. The hole S_3 was opposite the mouth of the total-head tube P, and the distances S_1S_2 and S_1S_3 were 53·0 and 5·59 cm respectively. The object of these three static holes has already been explained (Chapter VI).

The calibration resolved itself into two stages. Firstly, a velocity traverse was made at one rate of flow by means of the micrometer pitot P, in order to verify that the ratio of the mean velocity v_m to the velocity v_c at the axis of the pipe at P could be taken from the data for smooth pipes given in Fig. 5.6, in spite of the possible effect of the large pipe attached to the mouth of the calibrating pipe. As will be seen later, this was found to be the case. Hence there was no need to traverse the pipe at P in order to measure the velocity of flow at each speed at which the orifice was subsequently calibrated, as would have been necessary had conditions been such that Fig. 5.6 could not be used. All that was necessary was to set the pitot with its mouth at the axis of the tube, and to observe the velocity head corresponding to v_c at different rates of flow, simultaneous readings being taken of the pressure drop across the orifice C. This constituted the second stage of the calibration. Details of the procedure and calculations for each stage follow.

Determination of v_m/v_c for the 1·27-cm Pipe at One Rate of Flow

We use eqn. (3a) of Chapter VI in one of its non-dimensional forms, namely

$$v_m = 2v_c \int_0^1 \frac{v_r}{v_c}\frac{r}{a}\, d\left(\frac{r}{a}\right) = v_c \int_0^1 \frac{v_r}{v_c}\, d\left(\frac{r}{a}\right)^2, \tag{2}$$

where v_r is the mean velocity at radius r in the pipe of full radius a, and v_c is the velocity on the axis when $r = 0$.
Thus, to obtain the value of v_m, we evaluate

$$\int_0^1 \frac{v_r}{v_c}\frac{r}{a}\, d\left(\frac{r}{a}\right) \quad \text{or} \quad \int_0^1 \frac{v_r}{v_c}\, d\left(\frac{r}{a}\right)^2$$

graphically by measuring v_r at a number of radii and plotting $(v_r/v_c)\,(r/a)$ or (v_r/v_c) against r/a or $(r/a)^2$ respectively. The areas under the resulting curves will give the values of $v_m/2v_c$ or v_m/v_c respectively.

In the work under discussion, the velocity heads at the various values of r were obtained by connecting the pitot P to one side of a Chattock manometer and the static hole S_2 to the other. This gave the total head at P less the static head at S_2, whereas what was required was the total head at P less the static at S_3 in the same plane as the mouth of the total-head tube. False readings would have been obtained, however, had P and S_3 been connected to the manometer; for the pressure at S_3 was affected by the presence of the pitot. On the other hand, S_2 was far enough upstream of the pitot to be out of its zone of influence. The readings of velocity head obtained had therefore to be corrected by the amount of the static pressure drop between S_2 and S_3. This was obtained by measuring the drop between S_1 and S_2 and, since the pressure drop along a smooth straight pipe is proportional to the length, multiplying this drop by the ratio of the length S_2S_3 to S_1S_2, i.e. by 5·59/53·0. Thus the velocity head at P was $(P - S_2) + (S_1 - S_2) \times 5{\cdot}59/53{\cdot}0$. The pressures $P - S_2$ and $S_1 - S_2$ were read on the same manometer, a two-way tap being provided on the one side so that it could be connected either to P or S_1.

The Chattock gauge employed had cup centres 33·22 cm apart, and the dimensions l_2 and d of eqn. (5), p. 256, relating to the Chattock gauge were

25·40 cm and 1·27 mm respectively. Hence one turn of the micrometer wheel of the gauge was equal to a head of (33·22 × 1·27)/25·4 mm of water. And from the equation $p = \frac{1}{2}\varrho v^2$, under standard atmospheric conditions,

$$v = \sqrt{\frac{2p}{\varrho}} = \sqrt{\left(\frac{2 \times 33{\cdot}22 \times 1{\cdot}27 \times 9{\cdot}81}{25{\cdot}4 \times 1{\cdot}225} \times G\right)},$$

where G is the gauge reading in complete turns necessary to balance the head due to the velocity v, i.e.

$$v = 5{\cdot}16\sqrt{G} \text{ m/sec.}$$

Hence, returning to (2) we write this in terms of the gauge reading G and obtain

$$v_m = 2 \times 5{\cdot}16\sqrt{G_0}\, I_1, \tag{3}$$

where

$$I = \int_0^1 \frac{v_r}{v_c}\frac{r}{a}\, d\left(\frac{r}{a}\right)$$

$$= \frac{1}{2}\int_0^1 \frac{v_r}{v_c}\, d\left(\frac{r}{a}\right)^2$$

and G_0 denotes the value of G when $r = 0$ (i.e. 9·51 in Table 13.1).

Table 13.1 gives details of the traverse. The length r is easily obtained by a simple calculation from the pitot micrometer reading when the pitot is touching the far wall of the pipe (see Chapter VI), the radius a of the pipe (in this case 6·27 mm), and the radius of the pitot tube (0·30 mm).

By plotting $(r/a)(\sqrt{G}/\sqrt{G_0})$ against r/a (Fig. 13.2(a)), or $\sqrt{G}/\sqrt{G_0}$ against $(r/a)^2$ (Fig. 13.2(b)), and determining the area beneath the curve, we obtain 0·396† for the value of

$$\int_0^1 (r/a)(\sqrt{G}/\sqrt{G_0})d(r/a) \quad \text{or} \quad \frac{1}{2}\int_0^1 (\sqrt{G}/\sqrt{G_0})d(r/a)^2. \text{ Hence}$$

$$v_m = 2 \times 5{\cdot}16\sqrt{9{\cdot}51} \times 0{\cdot}396 = 12{\cdot}60 \text{ m/sec.}$$

Normally, traverses would be made across several radii, both to check that the radial distribution of velocity is everywhere much the same and to obtain a mean value of the integral. We note also that additional pitot-tube readings near the wall of the pipe would have made it possible to define the shapes of the curves of Fig. 13.2 in that region (r/a near 1) more precisely. The readings near the wall would need correction for wall-proximity effect (p. 139).‡

† See Note on Determination of Flowrate in Peripheral Zone, p. 337.

‡ Corrections should also have been applied, where appreciable, for the effect of the exposed stem of the pitot tube. Their magnitude has only recently been established (p. 131).

TABLE 13.1. TRAVERSE CALCULATIONS

Radius of point of observation r (mm)	$\frac{r}{a}$	$\left(\frac{r}{a}\right)^2$	Pressure readings Turns of Chattock gauge wheel				$\frac{\sqrt{G}}{\sqrt{G_0}}$	$\frac{r\sqrt{G}}{a\sqrt{G_0}}$
			$P - S_2$	$S_1 - S_2$	$\frac{5{\cdot}59}{53{\cdot}0}(S_1 - S_2)$	$(P - S_3) = G$		
0	0	0	8·64			9·51	1·000	0
1·83	0·292	0·085	7·95			8·82	0·963	0·281
3·10	0·494	0·244	6·98			7·85	0·909	0·449
4·37	0·696	0·485	5·70			6·57	0·831	0·578
5·00	0·797	0·635	4·95			5·82	0·782	0·623
5·64	0·899	0·808	3·86	8.21 throughout	0·87 throughout	4·73	0·705	0·634
5·97	0·952	0·906	2·63			3·50	0·607	0·578
5·79	0·923	0·852	3·71			4·58	0·694	0·640
5·16	0·822	0·677	4·79			5·66	0·771	0·634
4·52	0·720	0·519	5·63			6·50	0·827	0·595
3·25	0·518	0·268	6·84			7·71	0·900	0·466
1·98	0·316	0·100	7·82			8·69	0·956	0·302
1·27	0·202	0·041	8·15			9·02	0·974	0·197

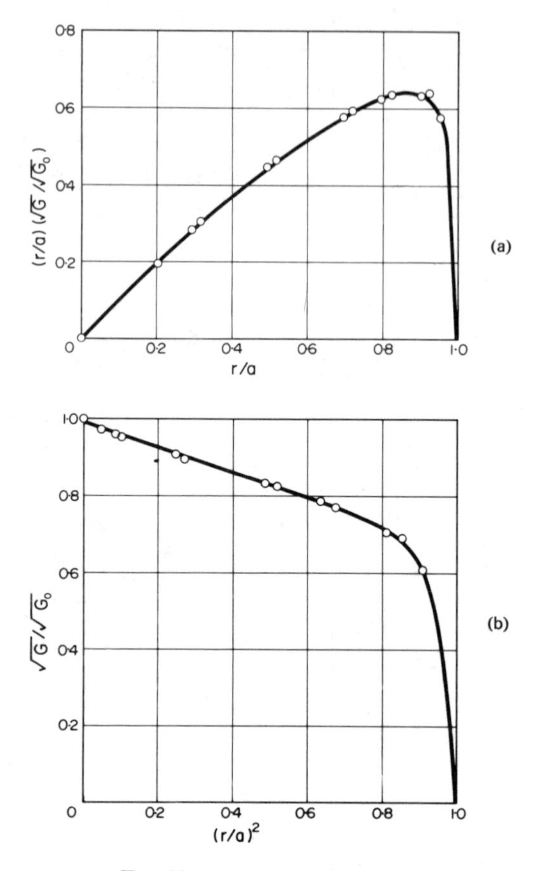

FIG. 13.2 Pipe-traverse data.

Now v_c, the velocity at the axis of the pipe, is the velocity corresponding to the value of G when $r = 0$; hence

$$v_c = 5{\cdot}16\sqrt{G_0}$$

$$= 5{\cdot}16 \times \sqrt{9{\cdot}51}$$

$$= 15{\cdot}91 \text{ m/sec}$$

and $$v_m/v_c = 0{\cdot}792.$$

The value of $v_c d/\nu$ for this flow in the calibration pipe is $(15{\cdot}91 \times 0{\cdot}01254)/(1{\cdot}46 \times 10^{-5})$, i.e. $1{\cdot}37 \times 10^4$, so that $\log v_c d/\nu = 4{\cdot}14$. At this value of the abscissa, we find from Fig. 5.6 that the value of v_m/v_c would be expected to be 0·792 from Nikuradse's curve, which agrees exactly with the value determined experimentally. Hence, for the subsequent orifice calibration it was considered necessary to take only one reading of the velocity pressure at each rate of flow with the pitot tube at the axis of the calibration pipe, values of v_m/v_c from Nikuradse's curve in Fig. 5.6 being used to convert the measured velocity here ($r = 0$) to the mean value v_m. Measurement of the diameter of the pipe AB (nominally 25·4 mm) showed it to be 25·35 mm; hence the mean velocity in that pipe was given by $(12{\cdot}55/25{\cdot}35)^2 \times v_m$, i.e. by 0·245 v_m. The corrections applied for the effects of turbulence on the readings of the pitot tube and for the displacement of its effective centre (because of the velocity gradient) are indicated below.

Orifice Calibration

The orifice pressure taps were connected to the inclined-tube manometer, which had previously been filled with alcohol and set to the desired slope. A series of simultaneous measurements was then made of the axial velocity in the 1·27-cm pipe and the differential pressure across the orifice in terms of the manometer reading. The results are shown in Table 13.2.

The values of the mean velocity given in the last column of Table 13.2 now need correction in the light of Fage's work on the effect of turbulence and of size of total-pressure tube. The values of $v_m d/\nu$ in the 1·27-cm pipe ranged from about 4400 to 12,700, for which the corresponding corrections to allow for turbulence for measurements with total-pressure tube and side hole (see Table 6.2(b), p. 129) are 0·9 and −0·75 per cent respectively. A correction of −0·8 per cent to all readings therefore appears reasonable. The ratio of the diameter of the pitot to that of the measuring pipe in the present example was 0·05, and, by extrapolation of the data given on p. 131, we find that, with a value of one-sixth for m (determined from the observed velocity profile in the 1·27-cm pipe), the appropriate correction is about −1·7 per cent. This is reduced by the wall proximity effect (see p. 139) to −1·3 per cent. It should be noted that the work described was done before Fage's results had been published; otherwise a smaller pitot tube would have been used to reduce this correction due to size. However, taking the results as they stand—they lose nothing of their value as an illustrative example because of the large value of the correction—we see that the figures in the last column of Table 13.2 must be reduced by 0·8 + 1·3 per cent, i.e. by 2·1 per cent.† We thus obtain finally the results given in Table 13·3, from which we can plot a calibration curve of v_1 against manometer reading, which can be used to determine the mean rate of flow in the pipe AB for any reading (within the calibration range) of the inclined-tube manometer.‡

† A further correction should be applied for stem blockage (see p. 133).

‡ The reader should not be misled by the number of figures quoted in the above calculations and in Table 13·2; the final results are probably not more accurate than to within about 98 per cent, if only because of the lack of definition of the curves of Fig. 13.2 at the outer radii.

TABLE 13.2. ORIFICE CALIBRATION

Inclined tube manometer reading (mm)	Chattock gauge readings				Axial velocity v_c in 1.27-cm pipe = 5·15 $\sqrt{G}$ m/sec	Log $v_c d/\nu$	v_m/v_c from Fig. 5.6	Mean velocity v_m in 1·27-cm pipe (m/sec)	Mean velocity in pipe AB = 0·245 v_m (m/sec)
	$P - S_2$	$S_1 - S_2$	$\frac{5·59}{53·0}(S_1 - S_2)$	$P - S_3 = G$					
3·10	1·50	1·64	0·17	1·67	6·66	3·76	0·771	5·13	1·26
5·41	2·51	2·66	0·28	2·79	8·60	3·87	0·778	6·69	1·64
8·59	3·84	3·91	0·41	4·25	10·62	3·96	0·784	8·33	2·04
11·91	5·12	5·13	0·54	5·66	12·25	4·02	0·786	9·63	2·36
13·18	5·66	5·61	0·59	6·25	12·88	4·04	0·787	10·13	2·48
15·29	6·53	6·44	0·68	7·21	13·83	4·07	0·788	10·90	2·67
16·99	7·28	7·05	0·74	8·02	14·58	4·10	0·790	11·52	2·82
22·00	9·26	8·81	0·93	10·19	16·44	4·15	0·792	13·02	3·19
24·38	10·25	9·65	1·02	11·27	17·29	4·17	0·793	13·71	3·36
28·04	11·86	10·78	1·14	13·00	18·57	4·20	0·794	14·74	3·61

TABLE 13.3. CORRECTED ORIFICE CALIBRATION

Inclined-tube manometer reading (cm)	Corrected mean velocity in pipe AB (m/sec)
3·10	1·24
5·41	1·61
8·59	2·01
11·91	2·32
13·18	2·44
15·29	2·63
16·99	2·77
22·00	3·14
24·38	3·31
28·04	3·55

One further word of explanation should be added. It will be seen that all the calculations have assumed that the air density ϱ was 1·225 kg/m^3. The calibration curve just mentioned, therefore, is valid only for this value of ϱ. We know, however, that the pressure drop through the orifice is proportional to ϱv^2. Hence, if the barometric pressure and the temperature of the air in the pipe differ appreciably from standard conditions, so that ϱ is not 1·225, we have to multiply the velocity obtained from the calibration curve for any manometer reading by $\sqrt{(1{\cdot}225/\varrho)}$. Equation (7) of Chapter VI enables the value of ϱ to be calculated for any observed temperature and barometric pressure.

Determination of the Orifice Coefficients

For the purpose for which this apparatus was required it was not necessary to determine the discharge coefficients of the orifice. But, since the results given provide all the data for the calculation, it will be useful to show how the coefficients are deduced.

The inclined-tube manometer was calibrated against the Chattock gauge by the method indicated in Chapter X. In this way, it was found that 2·5 cm observed head on the manometer was equivalent to 1·700 turns of the Chattock-gauge wheel, i.e. to

$$\frac{33{\cdot}22 \times 1{\cdot}27 \times 9{\cdot}81}{25{\cdot}4} \times 1{\cdot}700 = 27{\cdot}70 \text{ Pa}$$

Hence 1 cm on the manometer was equivalent to 11·08 Pa.

From (1) we have

$$\alpha = v_1 \sqrt{\left\{\frac{\varrho[(a_1/a_2)^2 - 1]}{2(p_1 - p_2)}\right\}}. \tag{4}$$

The orifice diameter was measured and found to be 12·7 mm, so that $a_1/a_2 = (25{\cdot}35/12{\cdot}7)^2 = 3{\cdot}984$, and $(a_1/a_2)^2 - 1 = 14{\cdot}87$.

Hence, (assuming that $p = 1{\cdot}225\ \text{kg/m}^3$),

$$\alpha = 3{\cdot}02v_1 \sqrt{\left[\frac{1}{(p_1 - p_2)}\right]}, \tag{5}$$

where v_1 is the mean velocity in the pipe AB. Values of v_1 and the pressure drop (the latter in terms of inclined-tube gauge readings) across the orifice are given in Table 13.3. In order to obtain the pressure $(p_1 - p_2)$ in Pa for use in (5), we have to multiply the observations by the calibration factor 11·08 of the manometer (see above). Table 13·4 gives details of the calculation of the orifice coefficients for the values of v_1 taken from Table 13.3.

TABLE 13.4. CALCULATION OF ORIFICE COEFFICIENTS

v (m/sec)	Manometer reading M (cm)	$(p_1 - p_2)$ Pa $= 11{\cdot}08\ M$	$\sqrt{\frac{1}{p_1 - p_2}}$	$\alpha\left(= 3{\cdot}018v_1\sqrt{\frac{1}{p_1 - p_2}}\right)$
1·24	3·10	34·3	0·1706	0·638
1·61	5·41	59·9	0·1292	0·628
2·01	8·59	95·1	0·1025	0·622
2·32	11·91	132·0	0·0870	0·609
2·44	13·18	146·1	0·0827	0·609
2·63	15·29	169·4	0·0768	0·610
2·77	16·99	188·3	0·0729	0·609
3·14	22·00	243·7	0·0640	0·606
3·31	24·38	270·2	0·0608	0·607
3·55	28·04	310·7	0·0567	0·606

The values of the discharge coefficient in Table 13.4 afford an instance of the feature noted in Chapter VII, namely a rise at low Reynolds numbers. At a mean speed of 1·2 m/sec in the pipe AB, the value of $v_1 d/v$ is about 2100, and we saw in Chapter V that this is just about the value of the critical Reynolds number at which the flow changes from the streamline to the turbulent variety. Figure 5.6 shows that the ratio of the mean velocity in the pipe to the maximum rises steeply in the transition range of $v_1 d/v$. Thus, at the lower rates of flow, the velocity at the centre is higher in relation to the mean velocity than at the higher rates. In other words, at the lower rates of flow relatively more air passes through the orifice at the centre of the pipe than at the higher rates, and so we find the discharge coefficients higher.

It will be noted that no measurements of barometric pressure or of air temperature were necessary in order to determine the values of α by this method of calibration. A standard value of α of 1·225 kg/m³ was assumed throughout. The reason for this will be clear from consideration of (4) from which α was calculated. Here ϱ occurs in the numerator under the radical sign, but it must be remembered that v_1 in (4) was determined from pitot–static pressure observations, and hence is of the form $KP/\sqrt{\varrho}$, where K is some constant and P is a pressure. Hence $\sqrt{\varrho}$ cancels out of the equation. The evaluation of v_1 in metres per second would, in fact, have been an unnecessary step if the sole object of the work had been the determination of α. Since, however, v_1 had been determined (for another purpose) on the assumption that ϱ was equal to its standard value, the same value had to be inserted in (4).

Note on Determination of Flowrate in Peripheral Zone

For the graphical determination of the flowrate in the circular pipe described on pp. 329 ff the area of the part of the curve (Fig. 13.2a or b) from $r = 0$ to the outermost measurement station (radius r_1 say) is readily determined by any of the usual methods. But because the shape of the curve to be integrated is difficult to define with precision close to the wall, it is recommended in ref. 1 that the part of the area corresponding to the annulus between $r = r_1$ and $r = a$ should be evaluated on the assumption that the velocity near the wall varies according to the power law $v_r/v_c = (1 - (r/a)]^m$ (see (10), p. 87), with a value of m deduced from the velocities recorded at the measurement stations nearest to the wall. If $\log v$ is plotted on a base of $\log (r/a)$, m will be the slope of the line as $r/a \to 1$. The contribution ΔV to the flowrate corresponding to the annulus between r_1 and a is then obtained by substituting this value of m in the equation

$$\Delta V = 2\pi a^2 v_c \int_{r_1/a}^{1} \frac{r}{a}\left(1 - \frac{r}{a}\right)^m d\frac{r}{a},$$

from which we obtain, on integration and simplification,

$$\Delta V = 2\pi a^2 v_1 \left(1 - \frac{r_1}{a}\right) \frac{1 + (m+1)\, r_1/a}{(m+1)\,(m+2)},$$

where v_1 is the velocity measured at r_1.

Reference 1 gives the equation

$$\Delta V = \pi(a^2 - r_1^2) v_1 m/(m+1),$$

an approximation that is permissible close to the wall. In our example discussed on p. 329, $m \doteqdot 1/6$ and ΔV amounts to about 5 per cent of the total flowrate.

Effect of Humidity on Air-speed Measurement

As a rule, the air will contain a significant amount of moisture and, for high accuracy, this must be allowed for in calculating the air density. The established method of doing this is based on the law of partial pressures, which states that for the total pressure of a mixture of gases (in this case air and water vapour) is the sum of the partial pressures of its constituents. Making use of this law, and of the fact that the density of water vapour is 0·622 times that of dry air at the same pressure and temperature, we can obtain the following formula for the density of moist air:

$$\varrho_w = \varrho_d \frac{P - 0{\cdot}378p}{P}, \tag{6}$$

where ϱ_d and ϱ_w are the densities of dry and moist air respectively, P is the absolute pressure, and p is the partial pressure of the water vapour.

At 15°C and 1013·25 millibars (760 mmHg), $\varrho_d = 1{\cdot}225$ kg/m³. Therefore at any other temperature t°C and pressure P millibars,

$$\varrho_d = \frac{1{\cdot}225 \times 288}{1013{\cdot}25} \frac{P}{(t + 273)}$$

$$= 0{\cdot}3482 \frac{P}{(t + 273)}. \tag{7}$$

Hence, from (6),

$$\varrho_w = \frac{0{\cdot}3482(P - 0{\cdot}378p)}{(t + 273)}. \tag{8}$$

The vapour pressure p can be obtained by observing the wet-bulb and dry-bulb temperatures t_w and t°C and referring to tables (covering a wide range of t_w and t) published by the British Standards Institution[(2)] or the Institution of Heating and Ventilating Engineers.[(3)] The density of the humid air can then be calculated from (8).

Alternatively, ϱ_w can be obtained directly from ref. 2, in which are tabulated (*inter alia*) the specific volume v_s of the humid air and its water content g for a range of t and t_w. The specific volume is defined as the volume of air in cubic metres that contains 1 kg of dry air plus g kg of water vapour, so that the density ϱ_w is given by

$$\varrho_w = \frac{1 + g}{v_s}. \tag{9}$$

These values of v_s and g are all given for a pressure P of 1013·25 mbar, and they are stated in ref. 2 to be sufficiently accurate for all practical purposes within the range of P from 950 to 1050 mbar. Within this range, therefore, ϱ_w when $P \neq 1013{\cdot}25$ will be $P/1013{\cdot}25$ times that given by (9).

Reference 2 gives a formula yielding better accuracy outside this range, but recourse to this will seldom be necessary for incompressible-flow conditions.

It should be noted that in all these calculations for finding ϱ_w, P is the barometric pressure plus or minus as the case may be the static pressure in the pipe or duct referred to P as datum.

The following numerical example demonstrates the procedure:

Assumed Conditions

Barometer 739 mm Hg = 985·2 mbar.

$$\left.\begin{aligned} t &= 18°\text{C} = 291\text{ K} \\ t_w &= 13{\cdot}5°\text{C} \end{aligned}\right\} \quad t - t_w = 4{\cdot}5°\text{C}.$$

Static pressure in pipe is −600 mmH_2O, i.e. −58·8 mbar.

Hence P in (7) and (8) = 985·2 − 58·8 = 926·4 mbar.

From Ref. 1, section 4.1.4, p. 10:

$$p = p_s - 6{\cdot}66 \times 10^{-4} \times 926{\cdot}4\,(t - t_w), \qquad (10)$$

where p_s is the saturated vapour pressure at temperature t_w, and is obtained from Table 5 of ref. 1 as 15·47 mbar. Hence, from (10),

$$\begin{aligned} p &= 15{\cdot}47 - 2{\cdot}78 \\ &= 12{\cdot}69; \end{aligned}$$

so that, from (8),

$$\begin{aligned} \varrho_w &= \frac{0{\cdot}3486(926{\cdot}4 - 0{\cdot}378 \times 12{\cdot}69)}{291} \\ &= 1{\cdot}103\ \text{kg/m}^3. \end{aligned}$$

From Ref. 2:

At t = 18°C, t_w = 13·5°C, and 1013·25 mbar, the tables for sling wet bulb give

$$v_s = 0{\cdot}8347 \text{ and } g = 0{\cdot}00779.$$

Hence, from (9),

$$\varrho_w \text{ (at 1013·25 mbar)} = \frac{1{\cdot}00779}{0{\cdot}8347} = 1{\cdot}207,$$

and therefore

$$\varrho_w \text{ at 926·4 mbar} = 1{\cdot}207 \times \frac{926{\cdot}4}{1013{\cdot}25} = 1{\cdot}104\ \text{kg/m}^3.$$

Although the value of P in this example is rather outside the range (950–1050 mbar) for which the tables of ref. 2 are stated to be "accurate for practical purposes", the value of ϱ_w obtained agrees to better than 0·1 per cent with that obtained above using the method and data of ref. 1.

Note on the Changes of Static and Total Pressure Along a Pipe System and on the Measurement of Resistance

A thorough understanding of the changes of static and total pressures that occur along a system of piping is of fundamental importance to mining and ventilating engineers, and others faced with similar problems; the following amplification of the discussion given on pp. 282–4 may therefore be found useful. Consider again the case illustrated in Fig. 11.3 (p. 283) in which air, starting from rest at O, is caused to flow along a length AX of pipe by a fan at B. It was shown that when the fan is sucking from A to X the total pressure is zero at A,† whereas when the fan is blowing from X to A it is the static pressure at A that is zero. The way in which this happens can be explained as follows. When the fan is sucking there would be nothing to cause air from the atmosphere to enter the pipe unless the mouth formed a low-pressure region. In fact we may regard the flow as being started by the formation by the fan of a region of low pressure at A, into which air is forced by the higher atmospheric pressure outside. On the other hand, when A is on the outlet side the air emerges into the surrounding atmosphere by its own momentum which the fan has generated. Immediately outside the boundary of the jet, as it issues from A, the pressure is atmospheric, and as the jet is parallel just as it leaves the pipe, the static pressure across it must be equal to the pressure just outside, i.e. atmospheric pressure. When A is on the inlet side, the air does not enter as a parallel jet but is sucked in at all angles, and the pressure in the entering stream is not the same as the pressure immediately outside the pipe.

The analysis shows also the origin of the misleading statement already referred to in Chapter V, namely "the resistance of a system of piping on the inlet side of a fan is measured by a total-head tube, and on the outlet side by a static tube". In a sense, this statement is true, for eqns. (2) and (3) of Chapter XI show that on the inlet side the frictional resistance f_2 of the length of pipe AX is equal to the total pressure at X, while on the outlet side f_2 is equal to the static pressure at X. This, however, is only because certain pressures are zero at the open end (and so need not be measured), according to whether the fan is sucking or blowing. It must be remembered also that the analysis has been based on the assumption that the pipe is parallel throughout, and that the velocity is uniform across the section. In practice it is unusual for the latter condition to be even approached, and there will also often be various changes of pipe diameter in the complete system. The true resistance (i.e. a measure of the energy expended in maintaining the flow) can then be obtained only by taking the difference between the mean total pressures across the sections A and X in the manner described in Chapters V and VI. If X is on the inlet side of the fan, since the total pressure is zero, this means that we need only make a total

† Neglecting the small losses that may arise in starting the air from rest outside the system.

pressure traverse across the section X, but when X is on the outlet side, total pressure traverses must be made at both X and A.

It is true that in some cases, where there is no change in pipe diameter and the velocity profiles across different sections do not differ much, and there is not much swirl, we can obtain an approximate measure of the resistance by averaging a few static-pressure or total pressure observations taken at different points in the section X, according to whether X is on the inlet or outlet side of the fan. But to accept such procedures unquestioningly as generally true will cause confusion and errors. Obviously, the resistance of a system of piping for given conditions of flow cannot depend on whether it is on one side of a fan or the other; fundamentally, the resistance must be measured in the same way whether the fan is sucking or blowing.

Another source of confusion arises from the fact that, when the system of piping is on the pressure side of the fan, some engineers include in the resistance of the system the velocity head at the outlet, where discharge takes place into the atmosphere. This leads to the anomaly that a given piping system, through which a given rate of flow is taking place, has a different resistance according to whether it is on the pressure or the suction side of the fan. It is true that the energy represented by the velocity head is lost at efflux, but this quantity should be included in calculations of the fan duty and not in the resistance of the piping system.

References

1. B.S. 1042: Part 2A: 1973, *Methods for the Measurement of Fluid Flow in Pipes*, Part 2, *Pitot Tubes, 2A Class A accuracy*, British Standards Institution, London, 1973.
2. B.S. 1339: 1965, *Definitions, Formulae and Constants Relating to the Humidity of the Air*, British Standards Institution, London, 1965.
3. *I.H.V.E. Guide*, Book C, 1970. Institution of Heating and Ventilating Engineers, London, 1970.

APPENDIX I

METRIC UNITS[1]

THE units adopted in this book are everywhere metric and, with a very few exceptions, are those of the *Système International d'Unités* (*SI*): this is based on the metre (m), the kilogram (kg), the second (s, although "sec" is the abbreviation used in this book), the ampere (A), and the kelvin (K) (unit of temperature), together with two other base units not used in this book (namely the candela and the mole). The Celsius scale of temperature (formerly called the Centigrade scale, and still written °C) can be used consistently with the SI: a temperature *interval* of 1 K is identical with that of 1°C, but the zero of the Celsius scale is 273·15 K.

Derived units in the SI which have been given special names include the following:

Electrical potential, potential difference, electromotive force	volt (V)	(W/A)
Electrical resistance	ohm (Ω)	$\frac{V}{A}$
Energy, work, quantity of heat†	joule (J)	(N m)
Frequency	hertz (Hz)	(cycle per second)
Force	newton (N)	(kg m/sec^2)
Power	watt (W)	(J/sec)
Pressure, stress	pascal (Pa)	(N/m^2)

The relationships between pressures and heads of liquid column are as follows:‡

$$1 \text{ mm } H_2O = \mathbf{9{\cdot}80665} \text{ Pa}$$

$$1 \text{ mm Hg} = 133{\cdot}322 \text{ Pa}$$

Some measuring instruments graduated in British units are likely to remain in use for some time to come. For this reason, some useful conversion factors

† One consequence of identifying the unit of heat quantity with that of energy and work is that equations involving both thermal and mechanical quantities are no longer encumbered with the mechanical equivalent of heat.

‡ The use of bold type indicates exact values; others are accurate to the number of figures quoted.

are set out below, and are followed by skeleton tables for the conversion of lengths in feet and inches to equivalent metric values.

Area

1 ft^2 = 929·030 cm^2
1 in^2 = **6·4516** cm^2

Density

1 lb ft^{-3} = 16·0185 kg m^{-3}
1 slug† ft^{-3} = 515·379 kg m^{-3}

Energy

1 ft lbf = 1·35582 J
1 ft pdl‡ = 0·0421401 J
1 kWh = **3·6** MJ§

Force

1 lb f = 4·44822 N
1 pdl = 0·138255 N

Heat Quantity

As Energy; also the international table calorie (**4·1868** J), the 15°C calorie (4·1855 J), and the thermochemical calorie (**4·184** J).

Length

1 ft = **30·48** cm
1 in = **2·54** cm
(See also Table, p. 346).

Mass

1 lb = **0·45359237** kg
1 slug = 14·5939 kg

Moment of Force

1 lbf ft = 1·35582 N m
1 pdl ft = 0·0421401 N m

† 1 slug = **9·80665/0·3048** lb.
‡ 1 pdl (poundal) = 1 lb ft sec^{-1} = 0·138255 N.
§ M = 10^6.

Pressure

1 lbf ft^{-2} = 47·8803 Pa
1 pdl ft^{-2} = 1·48816 Pa
1 lbf in^{-2} = 6·89476 kPa
1 mm H_2O = **9·80665** Pa
1 mm Hg = 133·322 Pa
1 in H_2O = 249·089 Pa
1 ft H_2O = 2·98907 kPa
1 in Hg = 3·38639 kPa

Speed

1 ft sec^{-1} = **0·3048** m sec^{-1}
1 ft min^{-1} = **0·508** cm sec^{-1}

Torque

As Moment of Force.

Velocity

See Speed.

Viscosity, Dynamic

1 slug ft^{-1} sec^{-1} = 47·8803 kg m^{-1} sec^{-1}

Viscosity, Kinetic

1 ft^2 sec^{-1} = 0·0929030 m^2 sec^{-1}

Volume

1 ft^3 = 0·0283168 m^3

Work

As Energy.

Reference

1. Pamela Anderton and P. H. Bigg, *Changing to the Metric System: Conversion Factors, Symbols and Definitions*, H.M.S.O., London, 1972.

Skeleton Tables for the Conversion of Lengths in Feet and Inches to Equivalent Metric Values (Exact)

ft	m
1	0·3048
2	0·6096
3	0·9144
4	1·2192
5	1·5240
6	1·8288
7	2·1336
8	2·4384
9	2·7432
10	3·048
20	6·096
30	9·144
40	12·192
50	15·240
60	18·288
70	21·336
80	24·384
90	27·432
100	30·480

in.	cm	m
1	2·54	
2	5·08	
3	7·62	
4	10·16	
5	12·70	
6	15·24	
7	17·78	
8	20·32	
9	22·86	
10	25·40	
20	50·80	
30	76·20	
40	101·60	1·016
50	127·0	1·270
60	152·4	1·524
70	177·8	1·778
80	203·2	2·032
90	228·6	2·286
100	254·0	2·540

in.	mm	cm
0·1	2·54	
0·2	5·08	
0·3	7·62	
0·4	10·16	1·016
0·5	12·70	1·270
0·6	15·24	1·524
0·7	17·78	1·778
0·8	20·32	2·032
0·9	22·86	2·286
1·0	25·40	2·540

in.	mm
1	25·4
0·1	2·54
0·01	0·254
0·001	0·0254
0·0001	0·00254

in.	mm
1/16	1·5875
1/8	3·1750
3/16	4·7625
1/4	6·3500
5/16	7·9375
3/8	9·5250
7/16	11·1125
1/2	12·7000
9/16	14·2875
5/8	15·8750
11/16	17·4625
3/4	19·0500
13/16	20·6375
7/8	22·2250
15/16	23·8125
1	25·4000

APPENDIX II

NUMERICAL EVALUATION OF NON-DIMENSIONAL QUANTITIES

WHEN non-dimensional products are evaluated numerically, all the quantities concerned should be measured in units that are mutually consistent. Thus if ν in an expression for Reynolds number (vl/ν) is measured in m^2/sec, v should be measured in m/sec and l in metres; if ν is in cm^2/sec, v should be in cm/sec and l in cm. Similarly, if ϱ in the pressure coefficient $p/\varrho v^2$ is in kg/m^3, p should be in pascals (N/m^2) and v in m/sec; if ϱ is in g/cm^3, p should be in $dynes/cm^2$ and v in cm/sec.†

Provided that mutually consistent units are used (and provided also that the units systems have been similarly based),‡ the numerical values of non-dimensional products are independent of the units of measurement.

When evaluating vl/ν, for example, measuring quantities in the c.g.s. system instead of in the m.k.s. system results in a 10^2-fold increase in the measures (numerical values) of v and l, accompanied by a 10^4-fold increase in that of ν (now in cm^2/sec instead of m^2/sec), leaving the value of vl/ν unchanged. Similarly, the measure of p (in $dynes/cm^2$) is 10 times that in N/m^2; at the same time, the measure of ϱ in g/cm^{-3} is 10^3 times that in kg/m^3 and that of v in cm/sec is 10^2 times that in m/sec, so that the value of vl/v likewise remains unchanged. (Similar remarks apply to comparisons of results expressed in

† Corresponding mutually consistent units in the British systems were ft^2/sec for ν, ft/sec for v, and feet for l, together with either $lb/mass)/ft^3$ for ϱ and $poundals/ft^2$ for p or $slug/ft^3$ for ϱ and $lb\text{-}force/ft^2$ for p.

‡ The units systems must be similarly based in the sense that they use the same physical laws for establishing derived units in terms of the base units and assign the same values to the proportionality factors in these laws. In mechanics, for instance, the units of force (F) and mass (M) are in practice always related through the second law of motion (law of acceleration (a): $F \propto Ma$); the proportionality factor is assigned the value unity in the Système International, as in the c.g.s. system (and also in the ft/lb (mass)-sec-poundal and ft-slug-sec-lb (force) systems). Values obtained for vl/ν are identical in all these systems, but would be different if in the k.m.s. system the newton (unit of force) were replaced by the kilopond (kg-force, equal to 9·80665 N): the proportionality factor would then have been set equal to 1/9·80665. Measures of force, stress, and viscosity would be correspondingly less than in the *Système International*; so also, therefore, would be values of vl/ν (although still non-dimensional). Values of pressure coefficient would be affected in the same way. (Another example of alternative values of proportionality factor in a law used for establishing units occurs in the so-called rationalized and unrationalized systems of electromagnetism.)

metric units with those in British units, although the various conversion factors† are no longer integral powers of 10; just as the newton is identified with k.m.s. and the dyne with g.c.s., the poundal is identified with lb(mass)-ft-sec^2 or the slug with lb(force) ft^1 sec^2, and numerical values of the non-dimensional products again remain unchanged.)

Conversely (again subject to the use of mutually consistent units and similarly based systems) all results in non-dimensional form can be applied to obtain the values of the various physical quantities in any such units systems, and not only in the particular units system in which the non-dimensional results were obtained.

† Conversion factors for the units of quantities that occur frequently in the measurement of air flow are to be found in Appendix 1.

APPENDIX III

STANDARD ATMOSPHERIC PROPERTIES AT SEA-LEVEL

PROPERTIES of the internationally agreed standard atmosphere are set out in ref. 1. Values at sea-level are as follows.

Pressure: 101·325 kPa, corresponding to 760 mm Hg.

Temperature: 288·15 K (15°C).

Density: 1·2250 kg m^{-3}

Speed of sound: 340·294 m sec^{-1} (corresponding to a value of 1·4 for the ratio of the specific heat capacities).

Viscosity, dynamic: $1{\cdot}7894 \times 10^{-5}$ kg m^{-1} sec^{-1}.

Viscosity, kinetic: $1{\cdot}4607 \times 10^{-5}$ m^2 sec^{-1}.

Acceleration due to gravity: 9·80665 m sec^{-2}.

The value of the gas constant R in the ideal-gas equation ($\varrho = pM/RT$, where M is the molar mass) was taken to be 8·31432 J K^{-1} mol^{-1}; the value of M was $28{\cdot}9644 \times 10^{-3}$ kg mol^{-1}.

Reference

1. International Civil Aviation Organization: *Manual of the ICAO Standard Atmosphere extended to 32 km (105,000 ft)*, ICAO Document 7488/2 (1964).

NAME INDEX

SUBJECT INDEX